Strategies for Media Reform

Des Freedman, Jonathan A. Obar,
Cheryl Martens, *and* Robert W. McChesney
Editors

Strategies for Media Reform

INTERNATIONAL PERSPECTIVES

FORDHAM UNIVERSITY PRESS · NEW YORK · 2016

Fordham University Press has no responsibility for the persistence or
accuracy of URLs for external or third-party Internet websites referred
to in this publication and does not guarantee that any content on such
websites is, or will remain, accurate or appropriate.

Fordham University Press also publishes its books in a variety of
electronic formats. Some content that appears in print may not be
available in electronic books.

Visit us online at www.fordhampress.com.

Library of Congress Control Number: 2016942153

Printed and bound in Great Britain by
Marston Book Services Ltd, Oxfordshire

18 17 16   5 4 3 2 1

First edition

CONTENTS

PREFACE

ROBERT W. MCCHESNEY *(University of Illinois at Urbana-Champaign)*

This volume is a testament to the emergence of media reform as a concrete area of political activity and as an emerging and important field of intellectual inquiry and scholarly research. Inside these covers you will find a heterogeneous collection of essays by some of the leading media reform activists and scholars of our times. In this preface I offer a few observations on the emergence of media reform and how to consider it.

Communication and media systems have come to play a central role in contemporary societies. Media reform is premised on a simple notion: "the problem of the media" (McChesney 2004). This phrase refers to the fact that communication and media systems are *always* the result of government policies, rules, regulations, and subsidies, both direct and indirect. There is no natural media system; it is always created. It is a problem to be solved, like an algebraic equation, with the difference being that there is no "right" answer, only a range of answers that reflect different values and priorities.

That does not mean the existing political economy does not greatly influence what a media system will look like. A capitalist society will have pressure to adopt a commercial system. But, as the stark differences in broadcasting systems adopted by various capitalist nations demonstrates, there remains a certain range in how the media system can be structured. Most important, there is no "default" position. Even if one wishes to have a profit-driven communication system in a capitalist society, it requires extensive government policymaking and involvement to make it practical.

Put another way, media and communication have significant power and influence in society, and the systems are the result of government policies. Then the question is where do these policies come from? Political deliberation and debate. That is the nucleus of the media atom. If one wishes to change the media, one needs to change the system and that means engaging in the political work to change the policies that create the system. That is the stuff of media reform. One could conclude then that the more democratic a nation, the more democratic the media policymaking process would be, and the more likely a resulting media system would reflect popular concerns.

It sounds pretty sexy at this point, but, in fact, media policymaking has generally been a boring political backwater for much of its history. The reason is that once media systems get established with successful enterprises and practices they tend to be immune to any sort of fundamental change. Media *reform*, in such a context, generally becomes the more generic field of *media policy*; it involves marginal changes or changes that suit the interests of the dominant players and that have little or no appeal to the general public. Policy "debates" deal with technocratic and administrative issues, or seemingly minor squabbles between self-interested commercial parties. To the extent the public knows anything about media policy issues—and the public usually is in the dark—the prospective reforms likely appear inconsequential. It makes more sense to put one's energies elsewhere.

Even in ostensibly democratic societies, the media policymaking process tends to be the domain of wealthy and self-interested elites. In a word, it tends to be corrupt. The difference between media reform and mainstream media policy studies has little to do with the subject but everything to do with attitude. Mainstream media policy research—the kind often found in business schools, law schools, and communication programs that want to lure the funding that goes to business schools and law schools—makes its peace with the status quo and happily operates inside the parameters that are given. Media reform, on the other hand, works to break down the barriers that limit public participation and the range of alternatives. It questions the integrity of the process as a key part of its work. Only in an insane world is the former termed *scientific* and the latter *ideological*.

Media reform becomes imperative when there is a *critical juncture*. This term has been appropriated from historical sociology and refers to calamitous periods when the status quo is in crisis and fundamental change is likely to happen (McChesney 2007). The only question is what sort of change is likely to happen? The notion of critical junctures was meant for societies as a whole; they referred to periods of great social upheavals, like the 1930s or the 1960s. Applied to media systems, critical junctures refer to those periods when media systems are in crisis and on the verge of being replaced. This most often occurs with the development of a radical and sweeping new communication technology, like radio and television broadcasting or the Internet. Old industries are threatened and new industries have yet to emerge. What the state does will largely determine who will win and what values and practices will be privileged.

At moments like these, the options before society tend to be many, as there are fewer and weaker entrenched interests to dominate deliberations. If a society is also in broader social and political turmoil, the options grow that much larger, as the ruling elites stand on weaker ground. The outcome can go a long way toward influencing how society will develop.

By this framework, for example, the United States had media-related critical junctures with the nation's founding and constitution, the emergence of professional journalism in the early twentieth century, and the establishment of radio broadcasting in the late 1920s and early 1930s. Most nations had their own critical junctures with the emergence of broadcasting. Many newly liberated "third-world" nations had them in the 1960s and 1970s, where new technologies met traditional patterns of colonial domination. Indeed, the movement for a New World Information and Communication Order (NWICO) was the result of this critical juncture.

It is during critical junctures that media reform moves from the shadows into the full light of day. It was during the critical juncture of the 1960s that Raymond Williams, with the aim of influencing Labour Party policies, wrote a series of pioneering books, pamphlets, and articles that remain breathtaking for their prescience. Williams made a powerful case that the "means of communication" were central to advanced societies and how they were structured was a fundamental political issue. It would go a long way toward determining how democratic a nation would be (Foster and McChesney 2013).

Williams highlighted the deep flaws within a commercial media system, and he advocated removing culture and communication to a significant extent from the capital accumulation process. He regarded the battle to democratize the communication system, to create well-funded, independent, uncensored, participatory media as the foundation of a democratic society. But Williams was more than a democrat; he was a socialist, and this drove his treatment of media and communication. Williams argued that a vibrant independent participatory media was not merely a requirement for democracy, but for socialism. His was a direct critique of the Soviet model of one-party dictatorship with a state-run monopoly media.

Williams was first among a coterie of Western Left intellectuals who pondered the striking importance of media, communication, and culture by beginning of the second half of the twentieth century. These writers included Ralph Miliband, C. Wright Mills, E. P. Thompson, Paul Baran, Paul Sweezy and Herbert Marcuse. It defined the importance of media to the New Left. When, however, the New Left collapsed and was replaced by the triumph of neoliberalism, the notion of radically transforming media, not to mention society, became academic. And in the academy, research that went against the neoliberal tide was hardly encouraged. In this context media policymaking, which enjoyed a burst of popular political activity in the 1960s and 1970s, returned to its status of being a by-invitation-only playground for political elites, commercial interests and their allies.

This all began to change quickly by the turn of the century. Indeed, we are now in the midst of the mother of critical junctures, at least for media and possibly for all of society, and that explains the explosion in media reform. I was involved in forming the US media reform group Free Press in 2003, which surprised me and everyone else with how much support it attracted. The US experience demonstrates that there is a wellspring of popular interest in a number of media issues. It also demonstrates how utterly powerful and determined commercial interests have become.

The primary reason for the media critical juncture is the Internet and the digital revolution. It has simultaneously turned nearly all traditional media industries upside down and created both crises and extraordinary opportunities. It has alleviated and tremendously aggravated the traditional problems of commercial media systems, the kinds that proliferate.

The current critical juncture includes two particular recent developments that deserve mention, as they pose daunting challenges to scholars and proponents of media reform. First, with the Internet, communication has been insinuated in the bone marrow of modern capitalism, and the market into the sinews of everyday life, far beyond anything that ever existed previously. It is in part that the business logic leads to a corporate (as well as governmental) surveillance that makes every possible moment pregnant with the possibility of commercial exploitation. Even more important, the Internet's vast commercial bounty has been captured by a small number of gigantic firms that are monopolistic by economic standards (McChesney 2013). By June 2014 there were 32 publicly traded US corporations with a market value of at least $125 billion. Fourteen of these 32 firms were primarily Internet firms, most of them monopolies in the sense that John D. Rockefeller had a monopoly with Standard Oil in the late nineteenth century. Just 15 years ago observers like myself bemoaned the implications of having a half-dozen media conglomerates ranking among the 200 largest firms in the United States. That looks like a golden age of small-scale competition today.

What this means is that these firms have almost unimaginable political power along with mind-boggling economic power. On any issue that affects them—and few issues do not—when they are in agreement they will be an unbeatable force, at least in the United States. It is a direct challenge to almost any conventional notion of the type of economy that is compatible with a credibly democratic society.

Media reformers ignore these developments at their peril. The global capitalist economy is stagnant and there appears to be little hope for much of a life for large sectors of the population (Foster and McChesney 2012). Inequality is growing to dangerous and unacceptable levels, by nearly any standard besides Ayn Rand's. Democratic practices are under pressure to conform to the needs of investors. I believe that there needs to be a recognition that the creation of credible democracy increasingly demands we consider moving past capitalism as we know it (McChesney 2015). Where that leads exactly, I do not know. But I do know if we do not broach the subject for fear of antagonizing powerful interests, we are not being honest intellectuals.

Second, commercial journalism is in freefall collapse and disintegration. American-style professional journalism always had its flaws—most significantly, a reliance upon political and economic elites to set the agenda and the range of legitimate debate—but at its peak, it also provided for significant resources to cover communities and politics. It began to crumble with commercial pressure to lowball editorial budgets to increase profits in the merger wave of the 1980s and 1990s. But the Internet, by removing advertising as a credible source of revenues, has permanently ended the commercial model of popular journalism aimed at the entire population. I have examined this crisis and its implications (McChesney and Nichols 2010). The bottom line is this: if there is going to be credible widespread democratic journalism, it will require significant public investment and very wise policymaking; it is a priority for media reform.

With regard to journalism, media reform is best understood as being concerned with providing a core element of the infrastructure of democracy. This refers to those institutions and processes that empower citizens to be effective members of a democratic polity, where power emanates from the decisions made by an informed citizenry. This means not only guaranteeing the right of all adults to vote, but also guaranteeing that the vote matters. This means taking money out of politics such that "one person, one vote" replaces "one dollar, one vote." It means restructuring the electoral governing system to enhance public participation. This means eliminating corruption; that is, lessening the inordinate power of the wealthy and privileged in the budgetary and decision-making processes. It also means having a credible system of universal public education (Nichols and McChesney 2013).

And, most important, it means devoting the public resources required to provide a truly great independent news media. An informed citizenry is the foundation of democracy and there is no route to such an outcome that does not require a strong news media system, or communication system writ large. This is the framing mechanism for the work of media reform.

Understood this way, media reform is not a partisan movement. The movement is all about establishing the institutions and rules for effective self-government whatever the country and whatever the specific and most pressing challenge to a democratic media system. Whether the results of effective self-government go left or right, socialist or free market conservative, or some combination thereof will be the result of the system working, as it should be.

At the same time, the democratic infrastructure of the United States and elsewhere is not merely atrophying, nor is it merely being neglected—it is under sustained attack. Those making the attack know full well what they are doing and its significance. There are powerful forces—generally comprised of corporations and the wealthy—that benefit from the status quo and prefer the system as it is. A world where the wealthy guide the ship of state, where the government budget is a private feeding trough, where there is precious little journalism, and where most people rationally tune out politics is A-OK with them. It is not, however, OK with anyone else, nor should it be. When it comes to participatory democracy, media reform is anything but neutral or agnostic.

REFERENCES

Foster, J. B., and R.W. McChesney, 2012. *The Endless Crisis: How Monopoly-Finance Capital Produces Stagnation and Upheaval from the USA to China.* New York: Monthly Review Press.

———. 2013. "The Cultural Apparatus of Monopoly Capital: An Introduction." *Monthly Review* 65, no. 3 (July–August): 1–33.

McChesney, R.W. 2004. *The Problem of the Media: U.S. Communication Politics in the 21st Century.* New York: Monthly Review Press.

———. 2007. *Communication Revolution: Critical Junctures and the Future of Media.* New York: New Press.

———. 2013. *Digital Disconnect: How Capitalism is Turning the Internet Against Democracy.* New York: New Press.

———. 2015. *Blowing the Roof off the Twenty-First Century: Media, Politics, and the Struggle for Post-Capitalist Democracy.* New York: Monthly Review Press.

McChesney, R. W., and Nichols, J. 2010. *The Death and Life of American Journalism: The Media Revolution that Will Begin the World Again.* New York: Nation Books.

Nichols, J., and R. W. McChesney. 2013. *Dollarocracy: How the Money-and-Media-Election Complex is Destroying America.* New York: Nation Books.

Strategies for Media Reform

Introduction

CHAPTER ONE

Media Reform

An Overview

DES FREEDMAN, *Goldsmiths, University of London*
JONATHAN A. OBAR, *York University, Canada*

Media reform is a great and formidable challenge. Across international contexts, reformers are inspired by what the late C. Edwin Baker (2007, 7) referred to as the democratic distribution principle for communicative power: "a claim that democracy implies as wide as practical a dispersal of power within public discourse." The challenge is made manifest in battles over the future of investigative journalism, media ownership, spectrum management, speech rights, broadband access, network neutrality, the surveillance apparatus, digital literacy, and many others waged in pursuit of the normative ideals at the heart of Baker's vision. At the same time, those committed to media reform confront formidable challenges: entrenched commercial interests and media conglomerates; sometimes powerful, sometimes disorganized, and sometimes neoliberal governments; a general public often disenfranchised, digitally illiterate and not focused on issues of media reform; and always, the uphill battle of organization, mobilization, and influence that is the work of any activist.

In light of these significant challenges, the central question addressed by this volume is: what strategies might be utilized to overcome these obstacles in the pursuit of media reform?

Sharing a list of strategies, however, will not ensure their effectiveness. We need to close the gap between a strategy's supposed potential and the realization of its effective implementation. Recent high-profile examples of social media activism demonstrate the presence of this gap. For instance, the uses of Twitter during the Arab Spring, Facebook by the Occupy movement, YouTube

by the KONY 2012 campaign, and the blogosphere by the SOPA/PIPA uprising—all generated excitement about social media's potential to strengthen
digital activism efforts. What remains, however, is a widespread lack of understanding of how best to design and implement an effective social media strategy, and how to ensure the most productive return on investment. A recent
survey of sixty-three activist organizations operating in Canada clearly articulated this concern (Obar 2014). A number of the groups noted that they are
relatively new to social media, and that the process of integrating the technologies and accompanying strategies into established methods and routines poses
a variety of considerable challenges. A member of one group said: "There
appears to be little information or instruction available on how to use social
media, particularly Twitter and Facebook. . . . We have goals and a strategy for
the use of social media, and understand the technical side, but the actual
operational side . . . is largely guesswork" (ibid., 225). Activist organizations
also said that social media's constant updates amplify these challenges. As noted
by a representative from a second group: "Right now I am experiencing issues
drawing members from the face book [*sic*] group to the new page. Also if the
technology changes (like the new face book [*sic*] timeline and you are not keen
on the changes) it is difficult" (ibid.). Indeed, the process of integrating a new
social media strategy into established routines poses considerable challenges.
According to a representative from a third organization: "As with any new
development, it has taken care and attention to work out the kinks in our system. A fair amount of time that could have been spent doing other things is
now directed at social media use, and it is difficult to measure that impact.
Inevitably, this medium will continue to change and necessitate further analysis and labour hours" (ibid.).

The lack of information and training available to activists, the constant
changes in the landscape and the tools, and the best combination of strategies
for moving forward, while especially relevant to the social media question, are
challenges to overcome when considering any number of media reform strategies. Indeed, although the successful integration of digital media into the work
of media reform activists is one of the more prominent organizational challenges
currently faced by groups worldwide, the same difficulty of bridging the gap
between excitement and effectiveness is common to a wide variety of potential
media reform strategies (both offline and online). The chapters included in this
volume aim to contribute to the closing of these gaps through the presentation
and analysis of successful strategies that have helped to advance media reform
efforts in a variety of countries and contexts throughout the world.

A DEFINITION OF MEDIA REFORM

As public communication systems continue to face underfunding and political
interference while corporate media giants increasingly cement their rule; as state

elites concurrently construct more sophisticated means of surveilling their populations through digital technologies originally anticipated to help liberate them; and as burgeoning media systems struggle against established power structures for independence of voice, there is an urgent need to reflect on and to share effective strategies for media reform. Despite the rhetoric about the threat to established power posed by social media innovation, in many countries throughout the world, communication choices are dominated by the likes of Google, Facebook, 21st Century Fox, CCTV, Brian Roberts, Carlos Slim rather than the voices of individual consumers; our media environments continue to be shaped by the interactions of political and corporate elites rather than the collaborative spirit of user communities.

Media reform movements are, therefore, a response to expressions of concentrated media power and develop in the context of ongoing struggles over the distribution of communicative resources. We have seen vibrant campaigns *for* net neutrality, press freedom, affordable broadband, community radio, publicly funded broadcasting, ownership transparency, and media diversity. We have also seen determined battles *against* cyber-surveillance, unethical and inaccurate journalism, ownership deregulation, Internet censorship, and state intimidation. These struggles for communication rights are part of a wider challenge to social and economic inequalities and an essential component of a vision for a just and democratic society.

Given the diverse array of challenges facing media reformers worldwide, media reform can sound a little tentative, a bit polite, and, most importantly, rather ambiguous. After all, contemporary neoliberal projects to restructure, for example, health, education, and welfare in order to introduce market disciplines into these areas, are usually described as "essential reforms." Furthermore, democratic attempts to "reform" industries and institutions that are so intertwined with established commercial and political interests is often seen as, at best, severely limited and, at worst, utterly futile—partly because of the ability of the media to marginalize and undermine campaigns that call for their very reform.

Yet media reform remains both an effective mobilizing paradigm and a strategic course of action. It is this latter approach—an emphasis on strategies for media reform—on which this collection focuses: a set of international prescriptions that allow individuals and organizations to work together and "to engage with and transform the dominant machinery of representation, in both the media and political fields" (Hackett and Carroll 2006, 16). Media reform cannot be reduced to a single mode of operation, to a narrow set of predetermined objectives, or to a limited number of channels and spaces. It is rather a field in which multiple actors use a range of techniques and capacities to restructure communication systems in the interests not of privileged elites (whether state or market) but of ordinary users, viewers, readers, and listeners so that they might participate better in and make informed choices about the world in which they live.

Media reform is, therefore, likely to be very messy. It contains, in the words of Joe Karaganis (2009, 4), two orientations: "one originating in a consumer-rights-based model of policy advocacy; the other emerging from predominantly civil-rights-informed concerns with accountability, representation, and voice in the media." From our own perspective, these two "geographies of activism" (ibid.)—one more formally connected to official processes of lobbying and campaigning, the other to concepts of media justice and social movements—are copresent, although very often in a tense relationship, in the way in which we understand media reform. So without wanting to tidy up the field of media reform and to reduce it to a neat and predictable area of activity, let us consider three dimensions (or rather *demands*) of media reform, taken from the chapter by Cross and Skinner in this collection and realized in the annual *Media Democracy Day* events in Canada. There are, of course, other typologies of media reform that may be equally useful—for example, Hackett and Carroll's (2006, 52) definition of democratic media activism as based on internal reform of the media, the creation of new spaces, and the transformation of contextual factors; or Napoli's (2009, 385) conception of public interest media advocacy based around "framing processes, political opportunities and mobilizing structures"—but we feel that the following categories provide a framework that speaks to the emphasis on strategic action that is at the heart of this book.

KNOW THE MEDIA

The first definition of media reform relates to attempts to critique the content and structures of the mainstream media and therefore to delegitimize them as ideologically committed to supporting the power relations to which they are tied. In many countries, for example, we have broadcast systems dominated by commercial interests, overwhelmingly formulaic in their efforts to minimize risk and desperate to chase ratings at any cost, or public broadcast outlets that are institutionally tied to elite structures and that increasingly mirror the outputs and genres of commercial rivals or act as a mouthpiece for privileged political interests. We have newsroom cultures that, again, stick very closely to consensual agendas and established political discourse while, at times, as was uncovered following the phone hacking crisis in the United Kingdom, demonstrates "a recklessness in prioritising sensational stories, almost irrespective of the harm that the stories may cause and the rights of those who would be affected . . . all the while heedless of the public interest" (Leveson 2012, 10). Furthermore, we have online environments that all too often are replicating the monopoly structure and commodity basis of the analogue systems to which they are connected (Curran, Fenton, and Freedman 2016).

There are different schools and approaches from political economy critics who focus on the unequal distribution of resources and structural constraints that shape the dynamics of media systems—including the levels of ownership concentration, meaning that media are more likely to reflect the corporate and ideological interests of their owners—to more culturally focused accounts that talk about the distortions of symbolic power and the exclusion of voices and audiences from creative decision making.

The most influential account of mainstream media critique remains the propaganda model, developed by Ed Herman and Noam Chomsky in their powerful condemnation of US elite media coverage of foreign affairs in *Manufacturing Consent* (1989). Herman and Chomsky argue that the media produce propaganda: sets of ideas that are necessary to secure, in the words of Walter Lippmann, "the manufacture of consent." Propaganda is used to naturalize the ideas of the most powerful groups in society and to marginalize dissent. Their propaganda model depends on five "filters" working on the media that ensure a structural bias in favor of dominant frames: concentrated private ownership, the power of advertising, the domination of elite sources, the use of "flak" (sustained attacks on oppositional voices), and the construction of an enemy, whether Communism during the Cold War or fundamentalist Islam today. Mainstream media perform an ideological role—none more so than the so-called liberal media, which foster the greatest illusions precisely because their liberalism produces a deceptive picture of a pluralistic media system when, in reality, there is none. All media, whether liberal or conservative, are tied to current relations of power and involved in distorting, suppressing, and silencing alternative narratives to capitalist power.

Despite criticisms that it is too focused on the United States and that it does now allow sufficiently for challenges to media power, the propaganda model has been taken up by an increasing number of academics and activists, from the prolific watchdog site Media Lens in the United Kingdom to academics working on Hollywood (for example, Alford 2010) to the excellent annual attempt to map "the news that didn't make the news" by *Project Censored*.

Robert McChesney (2008) describes another particularly valuable form of critical knowledge for media reform activists: that the system is not "natural" but was created to reflect particular capitalist interests. Just as it was structured to benefit private interests, one of the main jobs for the media reform movement is "to make media policy a political issue" (ibid., 57) so that publics can demand that it should be reconstructed instead to benefit the public interest.

BE THE MEDIA

The second dimension of media reform acknowledges that we cannot rely on the mainstream media to adequately represent our lives as they are lived, to hold power to account and to reflect honestly on the media's own interconnections with established power; we are forced to make our own media. This relates to the theory and practice of alternative media (Atton 2002; Downing 2000) which draws on participatory accounts of democracy to produce media that better engage *and* reflect the diversity of the population. Alternative media outlets aim to produce content that forgoes the false objectivity of mainstream news and the sensationalist formats that dominate schedules through methods that are more democratic and where institutions are focused not on profit or control but empowerment.

Alternative media is not new and indeed has always provided a historic counterpoint to mainstream media. For example, the Chartist press in the United Kingdom in the 1840s provided early trade union activists with an essential organizing tool given the hostility toward their campaign from the more official titles (Harrison 1974). Buoyed by developments in electronic technology in the early 1970s, the German activist and theorist Hans Magnus Enzensberger (1970) wrote of the possibility of public mobilization through emerging media: "For the first time in history, the media are making possible mass participation in a social and socialized production process, the practical means of which are in the hands of the masses themselves" (ibid., 13). Contrasting the "repressive" and depoliticized uses of traditional media to the "emancipatory" possibilities of what he saw as new, decentralized media like pirate radio and community video, he urged activists to build new channels of communication on the basis that "every receiver is a potential transmitter" (ibid., 16).

In the age of digital media, we are better able to realize Enzensberger's vision of a horizontal and interactive communications system that allows for the mobilization of audiences as producers and for the possibilities of content that defies an artificially narrow consensus. Social movement theory has a particular role to play here in considering the communicative competences, performances, and structures that are necessary to publicize, organize, and galvanize movements for social justice (Atton 2002; Castells 2012; Downing 2000). We have a whole host of platforms, technologies, and practices in place—from hacktivism to citizen journalism, from protest masks to protest music, and from culture jamming to community media—that both challenge the agendas and narratives of mainstream media and that allow ordinary media users to take control of the technologies.

CHANGE THE MEDIA

Building on a critique of the limitations of the mainstream media and buoyed by efforts to communicate in our own terms, there is a third strand of media reform which is perhaps the most contentious: efforts to democratize actually existing media through initiatives like diversifying ownership, campaigning for new forms of funding for marginalized content, challenging existing copyright regimes, pressing for more ethical forms of journalism, and more recently, opposing forms of surveillance. This requires an engagement with official structures—with formal legislative processes, with parliaments and policy makers, with lobbyists and lawyers—in other words, with the very constituents of the system that are responsible for a diminished and degraded media culture. Perhaps not surprisingly then, it is this dimension of media reform—the one that occupies the attention of most of the contributors to this collection— that is most noticeably absent from much social movement theory that considers the role of the media.

For example, the introduction to *Mediation and Protest Movements* (Cammaerts, Mattoni, and McCurdy 2013) identifies four key themes for activists in relation to the media: questions of visibility, the nature of symbolic power, the possibilities afforded to protest by networked technologies, and the role of audiences and publics. The authors note that capturing the attention of mainstream media is often just as crucial as producing our own images and performing our own communicative practices but there is no attention paid to the fact that we might want or be able to change the practices and priorities of a capitalist media.

Donatella della Porta (2013), in a very interesting chapter in the same book that explores the relationship between social movements and the media, makes the important point that both media studies and social movement theory "consider both political institutions and mass media as given structures" (ibid., 28), but then herself fails to consider what the consequences of this might be in terms of struggles over the shape of the media. She concludes her chapter by arguing that we need to get to grips with "the agency of social movements in the construction of democracy and communications" (ibid., 33); of course this is a very tricky task if, by and large, social movement activists and social movement theory have little interest in trying to modify the structures and institutions as they are currently organized. Does that mean that we have to reconstruct communications exclusively from the bottom up and that every existing media institution needs to be abolished? Even if that is the case, which many reform activists may believe (though, in private, some may come to miss selected HBO programs), what does this mean in terms of building effective and relevant reform movements?

For many social movement activists, this type of media reform—of trying to democratize the media—is seen as potentially counterproductive in that activists are likely either to be incorporated into official channels or to tailor their demands to meet the values and demands of vested interests. Thinking in particular of the US media reform group Free Press, Mickey Huff (2011) of *Project Censored* warns of the dangers of "working through the system" and of attempting merely to fix, rather than to replace, a social system that has been found to be demonstrably unfair and unequal. This lends itself to reformist illusions that the system can indeed be repaired and that media institutions, even if we do fix them, will ever deliver social justice within the existing frame of capitalism. As Huff argues, we need to "be the Media in word and deed . . . not lobby those in power to reform their own current establishment megaphones for their own power elite agendas, as that will not happen, and indeed, has not, for the most part, in the past."

Yet media reform efforts do not need to take the form simply of a polite request for minor changes to media behavior or structure. Instead, we need to broaden the debate and to deepen the crisis by campaigning for specific remedies to, for example, online discrimination, media concentration, press scapegoating, and the decline of local news while recognizing that these failures are indeed systemic and not incidental or peripheral to the core operations of the media. We need, in other words, to bring more radical politics to questions of reform.

Of course, reform movements are always fraught with tensions and contradictions: they combine people who are happy to stick to immediate and winnable demands with those who want to go much further; they consist of fragile coalitions between people who think that the system as it exists can deliver reforms that will satisfy enough people and those who think that there are structural inequalities that cannot be ironed out given the priorities of capitalism.

In these circumstances, the best tactic for those who want to see radical and durable change is not to withdraw from reform-minded movements but to demonstrate that reforms can only be won and protected through systemic critique and radical action such as the boycotts, marches, occupations, and direction action that have won the greatest victories in struggles for social justice.

Media reform, as with many other campaigns for social reform and justice, is therefore a way of reaching out to those people who have a healthy and often instinctive critique of the status quo but who maintain some illusions in the status quo. It is a way of working with those who want to see meaningful change but are not yet prepared to junk existing institutions and traditional forms of political campaigning as in parliamentary methods.

According to Robert Hackett and William Carroll (2006, 13), democratic media activism is both defensive and proactive; in other words, it is both reform-oriented in practice but also revolutionary and autonomist in spirit. Media reform for them involves a redefinition of the very idea of democracy to include new rights such as the right to share meaning as well as an increased emphasis on participation and equality through acts of media-making. Indeed, it is harder and harder to insulate media reform from political reform in particular because of the lack of autonomy of the media field from the actions of the state and the market despite the fact that the media still retain the power to affect the operations of other social actors.

McChesney (2008) in his work on media reform movements echoes this link between media and political reform as well as the need to connect both the insider and media justice elements of media reform. He argues that the contemporary US media reform movement was triggered by the antiglobalization struggles that took place from the late 1990s and which raised serious questions about the incorporation of the right to communicate within neoliberal frames and policies. The movement had to "bed in" before taking to the streets but was also inspired by radical critique of mainstream media performance (54). This is why media reform activists need to employ an inside/outside perspective—producing research, engaging in official lobbying, attempting to influence politicians and regulators—as well as applying external pressure and participating in direct action to transform the wider political climate. That is why all three dimensions of media reform outlined in this chapter—to know the media, to be the media, and to change the media—are so crucial if we are to build effective coalitions to transform media institutions and processes.

For media reform activists, this means organizing in two distinct but complementary ways. First, we need to engage with the process as it is and not simply as we would like it to be—or rather we use our vision of what the media *might* look like in order to deal with how they are currently constituted. We have to use all available channels to spread our messages including more formal political channels inside Parliament or Congress. We need to understand, if not actually to speak, the language of our opponents; to grasp the nature of the political cycles and opportunities that exist; to provide facts and data to back up our case and develop objectives that are not just a series of ultimatums.

Second, effective media reform also requires a radical perspective as no meaningful campaign for media reform is likely to be supported by the media itself or, indeed, by hardly any people in positions of power. Of course there are exceptions such as the recent movement against NSA surveillance where some corporate entities have an interest in being part of a coalition, not least in order to win back some credibility. The point is to create the conditions not simply in which we frame modest demands in the hope of them being accepted but to campaign hard for a shift in the public's attitude to these issues precisely in order to apply pressure on the politicians and regulators who have the formal power to act. Confining the movement to pretty modest proposals is unlikely to stave off opposition by the media or politicians. Indeed, the primary audience for media reform activists is not necessarily politicians, and certainly not the media, but publics, other activists, ordinary citizens whose needs are not being met and whose rights are being undermined.

What this means is that to secure a fundamental shift in media power, we need to engage in media reform but not from a reformist perspective. We can learn a lot from the German revolutionary Rosa Luxemburg who distinguished between revisionist strategies for reform, which attempt to administer palliative care to the capitalist system, and more radical strategies that seek to win reforms as a fundamental part of a revolutionary strategy to transform the status quo. While the former wants "to lessen, to attenuate, the capitalist contradictions" in order to stabilize society and produce consensus (Luxemburg 1989 [1899], 51), the latter seeks to struggle for reforms as part of a more widespread challenge to capitalist hegemony. The crucial point for Luxemburg however was that movements for reforms were central to a more profound social struggle: "Between social reforms and revolution there exists for the revolutionary an indissoluble tie. The struggle for reforms is its means; the social revolution, its aim" (ibid., 21).

Media reform, like all other forms of social reform, is a contradictory and uneven process in which different groups are involved and different strategies are involved. There is a world of difference between a reform campaign which calls on a handful of the "great and the good" to plead its case and one which seeks to mobilize greater numbers of people using all the tactics that are available—a difference perhaps between "reform from above" and "reform from below." You do not have to believe exclusively in parliamentary change to fight for reforms though it would be equally short-sighted to refuse to engage with parliamentary processes as part of a reform campaign. Just as there is little point in aiming only at the band-aid, there is also little point in refusing at least to treat the wound. We need to delegitimize and pose alternatives to the power structures that created these problems in the first place. Media reform allows us to do this *if* we build the necessary coalitions and *if* we pursue the right strategies.

AN OVERVIEW OF THE BOOK

This edited collection brings together strategies for advancing media reform objectives, prepared by thirty-three activists and scholars from more than twenty-five countries, including Canada, Mexico, the United States, Argentina, Bolivia, Brazil, Chile, Colombia, Ecuador, Guatemala, Uruguay, Venezuela, Iceland, Germany, Switzerland, the United Kingdom, Myanmar, Taiwan, Thailand, the Philippines, Egypt, Ghana, Israel, and Qatar. Contributors first presented their ideas in the summer of 2013 at a preconference of the International Communication Association, hosted by Goldsmiths, University of London in the United Kingdom. The goal then, as it is now, was to bring together successful and promising strategies for media reform to be shared across international borders and media reform contexts. There are, of course, gaps—campaigns that are not mentioned, countries that are passed over, strategies that are not fully developed, and themes that are not sufficiently explored (for example, the relationship between media reform and gender and between media reform and LGBT issues). We hope that this volume will serve as a useful resource for scholars and activists alike, looking to better understand the concept of media reform, and how it is being advanced around the world.

PART I: INTRODUCTION

The collection is organized into four parts. Part I provides a general introduction followed by an essay by Becky Lentz that presents the first strategy for media reform by asking us to consider the role of media policy literacy in the ongoing battle for media policy advocacy. Lentz suggests that media reform movements should be viewed as an "ongoing and cumulative knowledge-building process" where reformers are continuously evaluating and building upon their reform strategies. In so doing, reformers strengthen their efforts as they traverse the long and dangerous terrain of media policy-making. While battles lost should not be seen as failures, battles won should not be seen as stand-alone victories. No matter the outcome, the labor from past campaigns should serve future policy efforts pedagogically by forcing reformers to learn from past mistakes and successes in order to better understand how reforms can be won in the future.

Parts II, III, and IV focus on Internet activism, media reform movements, and media reform in relation to broader democratic efforts, respectively. Each part contains essays from scholars studying media reform within a specific international context, followed by a collection of commentaries by media reform activists from around the world. Activist voices need to be circulated inside scholarly communities if we are to realize the pedagogical benefits of sharing strategies for media reform. To help readers focus on the strategies shared, the chapters begin with a brief summary of the media reform strategies discussed at length by the authors.

PART II: INTERNET ACTIVISM FOR MEDIA REFORM

The Internet is relevant to current battles for media reform in two ways: first, as a tool for advancing activist objectives thus enhancing the capabilities of media reform organizations; and second, as a site of struggle over the future of the media. As the media landscape shifts online and as users increasingly access the Internet as a primary source for news and information, media reform efforts are addressing how the Internet affords the normative ideals central to earlier campaigns, as well as newer aspirations that extend from more traditional efforts. As a result, we are seeing media reform battles waged over issues including Internet access, equality of online opportunity, the protection of online privacy and speech, and a host of other civil liberties relevant to the digital citizen.

The scholarly essays in this part address the previously mentioned roles of the Internet in relation to contemporary media reform. Jonathan A. Obar and Leslie Regan Shade discuss how the digitally mediated, networked society referred to as the Fifth Estate presents users with an opportunity to reinvigorate the public watchdog function. Presented in the context of a victory over attempts by the Canadian government to extend its digital surveillance capabilities, the authors describe specific strategies that emphasize how Internet technologies can be harnessed to hold those in power to account. Christian Christensen describes how media reform debates are emerging as a by-product of the work of WikiLeaks—one of an evolving group of "digital defenders"— that highlights both the capacity of the Internet to hold corporate and state power to account, but also how contemporary and future media reform efforts must do more to protect whistleblowers and promote transparency. M. I. Franklin focuses on the development of the *Charter of Human Rights and Principles for the Internet* and *Ten Internet Rights and Principles*, framing the Internet as a battleground for contemporary media reform efforts that exemplify the tensions between rights-based agendas and those championed by a techno-commercial elite.

The activist commentaries include four essays from media reform organizations. Rainey Reitman from the Electronic Frontier Foundation writes about the lessons to be drawn from the battle over the infamous Stop Online Piracy Act (SOPA) in the United States, emphasizing five strategies drawn from the effort: decentralize, speak up quickly and often, be visible, use Internet communities as a political force, and cross political lines. Craig Aaron and Timothy Karr from Free Press, the most visible media reform organization in the United States, discuss the role of outside-in grassroots strategies in the crucial battle over network neutrality in the United States. Joshua Breitbart, formerly of the New America Foundation, outlines how the technical and policy expertise of the Open Technology Institute contributed to the US Broadband Technology Opportunities Program. And David Christopher discusses how Canadian media reform group Openmedia.ca fights for the future of the Internet by placing citizens at the heart of the debate.

PART III: THE POWER OF THE MEDIA REFORM MOVEMENT

Our starting point in this part is that media reform is not simply a noun but a verb—not simply a normative attachment to change but an active engagement with the circumstances in which we find ourselves. This means that we need to focus on the *agents* of change and to consider the most relevant organizational models as well as the most effective strategies in the pursuit of media reform. As with any progressive social movement, the capacity for change is related both to objective conditions (the scale of the problem, the political and economic contexts, the salience of the issues involved) and to more subjective factors concerning how best to mobilize in support of reform and what kind of campaign structures are necessary to organize opposition. Media reform is not a spontaneous outcome but the result of individuals working together to identify the problem, suggest alternatives and then to put into action the agreed strategies.

The chapters in this part emphasize strategies from different international contexts associated with specific media reform movements. Alejandro Abraham-Hamanoiel writes about AMEDI, the most prominent media reform group in Mexico, and identifies the multiple strategies it uses—from litigation and lobbying to publishing and media monitoring. Benedetta Brevini and Justin Schlosberg focus on the role of the Media Reform Coalition in the United Kingdom in the aftermath of the phone hacking scandal, arguing that successful movements need to both engage with mainstream media and official policy debates and resist the terms on which these debates and agendas are premised. Hsin-yi Sandy Tsai and Shih-Hong Lo write about media reform in Taiwan and describe a number of strategies that groups like the Media Watch Foundation and the Campaign for Media Reform have utilized to defend public broadcasting, oppose monopolies, and press for free speech in an emergent liberal media landscape. Kathleen Cross and David Skinner outline the activities of three different Canadian campaigns. They emphasize four insights relevant to organizing media reform initiatives: the need to understand regional and national contexts, to seek collaborations and coalitions, to pursue links with academic institutions, and to use multiple approaches and modes of engagement.

In the commentary section, the contributions from media reform organizations help to flesh out precisely how the strategies work in practice in the context of the struggle for low-power FM in the United States and in attempts to open up the airwaves and to protect speech rights in West Africa. Hannah Sassaman and Pete Tridish from the Prometheus Radio Project describe strategies that yielded one of the biggest media reform victories in recent history in the United States, the battle for low-power FM radio. Sanjay Jolly, also of the Prometheus Radio Project, writes about more recent developments in the continued struggle for low-power FM in the United States. Kwame Karikari describes the work of the Media Foundation for West Africa in building the necessary coalitions to open up the airwaves in the Republic of Guinea and to secure communication rights in the conflict-ridden context of Liberia.

PART IV: MEDIA REFORM AS DEMOCRATIC REFORM

The chapters in this part demonstrate the close interrelation and historical contingency of media and democratic reform. The authors draw on case studies from the United States, Israel, South America, Egypt, and South East Asia to illustrate how media reform activism is intimately linked to social movements' demands for democratic and pluralistic forms of representation and broader communication rights.

Among the scholars, Victor Pickard argues that at least three lessons can be drawn from the post–World War II media reform movement, including the importance of inside-outside strategies, the importance of structural reform efforts and the connections between media and other political and social reform battles. Arne Hintz describes how civil society actors can advance reform objectives by engaging in do-it-yourself policy-making through a process he calls "policy-hacking." Presenting examples from a variety of international contexts—including Argentina, Germany and Iceland—Hintz explains how reformers can repackage and upgrade policy utilizing a DIY framework. Manuel Puppis and Matthias Künzler argue that academic input into the policycymaking process often coincides with the interests of political and media elites but suggest that scholars should use all opportunities to break out of this impasse and to play a role both in identifying policy alternatives and helping to produce more democratic media systems. Noam Tirosh and Amit Schejter consider the roles of unions, professional associations, alternative media, and civil society organizations in media reform in Israel. The authors emphasize that the strategy that has had the most success in the Israeli context has been the utilization of alternative media in more extreme forms—by breaking the law and communicating via unlicensed media services. Cheryl Martens, Oliver Reina, and Ernesto Vivares consider the important role of social movements and focus on the cases of Argentina and Venezuela to discuss how media reform in South America is challenging neoliberal models of development and communication and the redistribution media power. Rasha Abdulla describes the steps necessary to establish a public broadcaster in Egypt, underscoring that media reform is contingent on state processes, which may lag behind public demands. And Lisa Brooten examines media reform efforts in Thailand in relation to those of Myanmar and the Philippines, illustrating the interrelationship between media reform and broader social efforts, including those to end impunity in cases of violence and corruption, and the strengthening legal and regulatory frameworks.

In the commentary section, Peter Townson of the Doha Centre for Media Freedom describes media literacy initiatives and campaigns to defend media freedom in the Arab world and to equip journalists with the resources necessary to provide effective coverage of the conflicts in which they are involved. Mark Camp of Cultural Survival discusses the use of community radio by Indigenous people in Guatemala to secure their rights to freedom of expression and culture and identifies the legislative struggles to secure meaningful media reform. And Marius Dragomir of the Open Society Foundations lists the steps taken to democratize the Mexican media environment, focusing on key strategies including research, coalition-building, information dissemination, and lobbying.

Media reform is a constituent part of contemporary struggles for democracy and social justice. Sometimes, it is carried out in its own right—for example, to campaign for independent voices and grass roots outlets in contexts where media spaces are largely dominated by vested interest. At other times, it grows organically out of prodemocracy campaigns that are stifled by the lack of communicative capacity and where demands for media reform arise more spontaneously in order to carry the struggle forward. In both cases, it is vital that researchers, scholars, activists, and ordinary users come together to reflect on and to develop the most effective strategies to safeguard our communication systems for the public good. We hope that this volume makes at least a small contribution to an objective that is likely to prove crucial in establishing democracy and equality in the twenty-first century.

REFERENCES

Alford, M. 2010. *Reel Power: Hollywood Cinema and American Supremacy*. London: Pluto.

Atton, C. 2002. *Alternative Media*. London: Sage.

Baker, C. E. 2007. *Media Concentration and Democracy: Why Ownership Matters*. New York: Cambridge University Press.

Cammaerts, B., A. Mattoni, and P. McCurdy, eds. 2013. *Mediation and Protest Movements*. Bristol, UK: Intellect.

Castells, M. 2012. *Networks of Outrage and Hope: Social Movement in the Digital Age*. Cambridge, UK: Polity.

Curran, J., N. Fenton, and D. Freedman, eds. 2016. *Misunderstanding the Internet*, Second edition. London: Routledge.

della Porta, D. 2013. "Bridging Research on Democracy, Social Movements and Communication." In *Mediation and Protest Movements*, edited by B. Cammaerts, A. Mattoni, and P. McCurdy, 23–37. Bristol, UK: Intellect.

Downing, J. 2000. *Radical Media: Rebellious Communication and Social Movements*. London: Sage.

Enzensberger, H. M. 1970. "Constituents of a Theory of the Media." *New Left Review*, no. 64: 13–36.

Hackett, R. A., and W. Carroll. 2006. *Remaking Media: The Struggle to Democratize Public Communication*. London: Routledge.

Harrison, S. 1974. *Poor Men's Guardians*. London: Lawrence and Wishart.

Herman, E., and N. Chomsky. 1989. *Manufacturing Consent: The Political Economy of the Mass Media*. New York: Pantheon.

Huff, M. 2011. "Beyond Boston and Media Reform for 2012: Supposed 'End of Times' Should Marshal a New Beginning for Media Democracy in Action." *Project Censored*, April 12. www.projectcensored.org/new-beginning-for-media-democracy-in-action/.

Karaganis, J. 2009. "Cultures of collaboration in media research. Social Science Research Council." Retrieved from http://papers.ssrn.com/sol3/papers.cfm?abstract_id=148518.

Leveson, B. 2012. *An Inquiry into the Culture, Practices and Ethics of the Press, Executive Summary*. London: Stationery Office.

Luxemberg, R. 1989 [1899]. *Reform or Revolution*. London: Bookmarks.

McChesney, R. W. 2008. "The U.S. Media Reform Movement: Going Forward." *Monthly Review* 60, no. 4: 51–59.

———. 2013. *Digital Disconnect: How Capitalism Is Turning the Internet Against Democracy*. New York: New Press.

Napoli, P. M. 2009. "Public Interest in Media Advocacy and Activism as a Social Movement." In *Communication Yearbook 33*, edited by C. Beck, 385–429. New York: Routledge.

Obar, J. A. 2014. "Canadian advocacy 2.0: An Analysis of Social Media Adoption and Perceived Affordances by Advocacy Groups Looking to Advance Activism in Canada." *Canadian Journal of Communication* 39, no. 2: 211–233.

CHAPTER TWO

Media Policy Literacy

A Foundation for Media Reform

BECKY LENTZ, *McGill University, Canada*

MEDIA REFORM STRATEGY

To enact an advocacy campaign aimed at changing the shape of media systems requires engaging in myriad forms of communicative work, which in turn demands knowledge of policy-making processes, the regulatory and legal history of the issues at stake, activism and advocacy practices, and the contemporary legal and political environment. My intent here is to re-conceptualize media policy advocacy by foregrounding not only the work that it involves but the multiple forms of knowledge on which it is based, and to argue that media reform is, therefore, centrally connected to media policy literacy. Media policy literacy can be engendered through a combination of critical media policy pedagogy and opportunities for situated learning.

As Philip Napoli (2009) has illustrated in his capacious overview of media reform scholarship, there are meaningful differences in how scholars have understood the political stakes, moments of opportunity, and mobilizing structures and strategies of media activism efforts. These works have made visible the alternate imaginings of the socio-political functions of media and have exposed the ideological and structural impediments to citizen group influence in the policy-making process. In so doing, these studies have denaturalized both media systems and media policies and, through their analysis, have flagged important lessons for future media policy advocacy efforts (McChesney 1993; Fones-Wolf 2006; Pickard 2010a, 2010b). This chapter discusses the challenges associated with promoting similar forms of media policy literacy amongst media reformers. A case study about the Consortium on Media Policy Studies (COMPASS) program in the United States provides one example of a situated learning program contributing at least some of the grounded understanding essential to promoting successful media policy literacy efforts.[1]

WHAT IS MEDIA POLICY LITERACY?

For Lunt and Livingstone (2011, 118), media literacy refers to the "policies and initiatives that have been designed to bridge the gap between what people know about and what they need to know about media in an increasingly liberalised and globalised environment." Media *policy* literacy, in the context of media reform, refers to what people need to know about the governance of media institutions, legal principles undergirding media regulation, and the processes by which media policies and laws are formed, debated, enacted, and changed. An emphasis on media policy literacy seeks to make visible the practices by which media policy advocates and would-be advocates (i.e., members of the general public) acquire the knowledge necessary to devise and participate in media reform and media justice efforts, the work that constitutes media policy advocacy, and the multiple spheres and strategies that comprise the scope of these practices.

While administrative law in the United States requires that administrative agencies like the Federal Communications Commission (FCC) solicit public input as they consider new rules, numerous scholars have noted that structural obstacles often impede the opportunity of members of the public to have a comparable impact to that of other stakeholders. These obstacles include not only discrepancies in access to government deliberations but also knowledge-based resources like those associated with media policy literacy (Gangadharan 2009; Gangadharan 2013; Brown and Blevins 2008; Schejter and Obar 2009; Obar and Schejter 2010).

Beyond an understanding of policy-making processes, media policy literacy is strengthened through competency in the workings of media systems at the macro, mid-range, and micro levels. That is, one needs understanding not only of the macro, such as the political economy of media industries and their regulation, but also their institutional practices (including hiring, promotion, and labor policies), and the impact of their work on communities and individuals. An appreciation of all three levels is required, as well as recognition of their interconnectedness. Media policy literacy is also strengthened through developing understandings of a range of issues, for example, contemporaneous interpretations of freedom of expression in specific legal environments; the limitations of access to media outlets, services and content; the parameters of media diversity; and the impact of electronic media on privacy. Media policy literacy thus refers to a capacity to recognize and act in the public interest on relevant legislative, legal, and regulatory precedents and rulings, as well as the premises and logics upon which these are based, for given policy issues.

The acquisition of these forms of knowledge is necessary but not sufficient for acquiring media policy literacy, however (Lentz 2014b). The other crucial component is gained by on-the-ground experience through participation in media reform efforts—experience that contributes to a grounded understanding of what is possible to achieve given the actual operations of the policy-making process at a particular moment in time. The strategic moments of opportunity to enact change, the most effective course of action for a given policy concern at a particular historical juncture, the potential (but not promised) long-term benefits of immediate participation in a reform effort, and which organizations or communities one should align with to achieve desired outcomes are all examples of the ways in which intellectual and experiential knowledge co-produce media policy literacy..

CHALLENGES TO ADVANCING MEDIA POLICY LITERACY

One challenge to those trying to create such educational experiences is the interdisciplinary breadth and depth of this policy field, which encompasses broadcasting, telecommunications, the Internet, and information policy, not to mention a good bit of social science, political science, economics, computer engineering, and legal expertise. The diversity of issues traveling under the umbrella term *media policy* contributes to an incoherence that challenges peda-gogic efforts. For instance, although students in my media governance courses take many other media-related classes, each year they will remark, several weeks into the term, "This is so important! Why haven't we heard about any of this before in any of our *other* classes?"

Braman's essays about the field's definitional problem are a case in point, as are the variety of encyclopedic entries and books on the topic (see Napoli 2008). In the US context, for instance, Braman notes that after covering media policy's links to the Bill of Rights and the First Amendment, things get compli-cated quickly, especially after adding technology to the mix (Braman 2004). In twenty-first-century media policy scholarship, she argues, our focus is now more on the constitutive role in social and political life of content and distribu-tion infrastructures. I have argued that in the digital media environment, content and conduit are co-constitutive, thus equal attention must be given to issues of infrastructure alongside the political economy of content production, distribution, reception, and exchange (Lentz 2010).

Braman (2004, 161–176) illustrates this classificatory dilemma by describing how media policy embraces several related problems: technology-based conver-gence problems; practice-based problems; policy-based problems; and issue-based problems. She also points out a multitude of definitional perspectives used to map this field of inquiry, for instance, by topical area, legacy legal categories (e.g., statutory versus regulatory), media industry type, stage of an information production chain (e.g., creation, processing, or storage), or intended societal impact (e.g., protecting children from harmful content). She settles on a broad vision that accepts "a multiplicity of definitional faces, each of utility at a different stage of the policymaking process," recasting media policy as information policy: "all law and regulation that deals with information creation, processing, flows and use."

These and other well-intended definitional efforts contribute more, not less, to what I call the media policy tower of babble (Lentz 2014a). Confusion and debate over terminology, entry points, and theoretical foci only obscure rather than clarify what one must know to intervene in policy areas like those mentioned earlier. This is not to suggest by analogy that arriving at an overarching definition of media policy is hubristic, but the fact remains that the sheer complexity of issues, institutions, aspirational goals, threats, concerns, legal doctrines, and technologies requiring mastery confounds efforts to design an education that equips graduates—and others—with the capacity to *produce* policy change.

As with media and communications studies more generally, media *policy* studies are defined less by a canon of core texts than by their objects of study. Permeable intellectual boundaries demand rigorous interdisciplinary training, and some measure of practical experience. Media policy scholarship matters to democracy, freedom of information and expression, access to knowledge, economic and community development, cultural identity, and personal privacy—all offer a suitable vantage point for study. Communities of practice are an equally good entry, be they media producers, media audiences, cultural workers, consumer activists, communication rights advocates, journalists, open source developers, children's and women's right activists, or social justice activists. Or, it can be approached according to key concepts such as neoliberalism, democratic theory, public sphere theory, economic development, globalization, international development, or human rights. Finally, courts, legislatures, regulatory agencies, or international bodies are another entry point for study.

Media policy education interrogates history, legal doctrine, and frameworks, institutions, and aspirational goals; but without direct experience, media policy remains relatively abstract, its impact being less visible than, say, pollution or the explicit violation of human rights. Exceptions include infrequent but vitally important spikes in public awareness like the millions of signatures gathered during the media consolidation debate of 2003 (Moyers 2003), the online protests during the SOPA/PIPA debates, or the extensive media coverage of the NSA/PRISM affair that has made massive government surveillance an open secret.

Events like spectrum auctions have considerable implications for national treasuries, but few laypeople even know such a marketplace exists. Newcomers have to connect media policies to lived experience to understand what is at stake. Yet media policy scholars, practitioners, and public interest advocates speak the languages of many different disciplines. The overall project enjoys no coherent moniker like climate change or human rights. It is referred to as information and communications technologies (ICT) policy, communication and information policy (CIP), communication law and policy, or Internet governance; all encompass multiple, intersecting objects of study.

Media policy literacy also requires changing traditional notions of media literacy, which, critics observe, are often limited to "the ability to access, analyse, evaluate and communicate messages in a variety of forms" (Lunt and Livingstone 2011, 122). Whilst media literacy agendas are now being expanded by governments to include *digital* media literacy, the overall focus remains, with a few exceptions, protectionist (Wallis and Buckingham 2013). "Media literacy here seems individualised, prioritising consumers and consumer choice over citizens and citizens' rights, and prioritising protection over participation" (Lunt and Livingstone 2011, 125).

A robust media policy literacy agenda offers a potential counterweight to such a neoliberal media education agenda. Whereas digital media literacy may distract attention from deregulation, critical media policy literacy would develop capacity to intervene on deregulation or liberalization agendas (see Shade and Shepherd 2013). Yet to do so, critically minded media policy scholars and advocates need to cultivate viable working partnerships with public interest policy practitioners and advocates to create, and help to fund, opportunities for students to participate, at least peripherally, in communities of practice engaged in media reform work. According to Lunt and Livingstone (2011, 135), "the critical analysis of media literacy not as an inert skill or the property of an individual but rather as a social, contextualised capability."

Traditional public sphere theory, that is, the Habermasian normative framework, argues that the best defense against threats to democratic media is an informed public adequately equipped for rational public debate. To be sure, there are many critics of the Habermasian framework, but donor institutions and philanthropists determined to salvage journalism or promote social media's promise do not appear to be among them based on funding patterns noted in annual reports.[2] An exception is the Media Democracy Fund (MDF) with its support for the Media Literacy Project based in New Mexico, whose mission "through education, programs, and grassroots campaigns . . . [is to] transform people into critical media consumers and engaged media justice advocates who deconstruct media, inform media policy, and create media that reflect their lived experience."[3]

Shoring up journalism builds better content enterprises, be they public, private, or some combination of the two; however, it only builds part of the necessary movement infrastructures needed for sustainable media reforms in the public interest. If university faculty hiring and course offerings are another indicator of priorities, there is much respect for all things digital: digital journalism, digital culture, digital humanities, or digital media education. Yet despite its importance to the future of democracy, we see scant attention paid not only to questions of media reform but also to digital media policy or even policy literacy. Being called upon to participate in episodic media policy struggles, sign online petitions, contribute to issue campaigns, re-Tweet arguments, watch mesmerizing TED Talks, or take part in protests at opportune flashpoints does not an educated digital citizen make.

We need to cultivate in university students a sense of citizenship (analog and digital) beyond voting or conscientious consumption, and inculcate media policy literacy in the same way we do, for example, "environmental education" as an established topic area in secondary and postsecondary education as well as for the general public. The World Environmental Education Congress is just one example of what is needed in the field of media policy studies,[4] enhancing deliberate attention to policy issues in the way that environmental education teaches skills to mitigate environmental problems alongside raising awareness and sensitivity about environmental challenges. Environmental education includes attention to governance issues involving international treaties, agreements, and norms. So should media policy literacy educational agendas.

The core argument here is that media policy literacy is a precondition for informed engagement in media reform struggles. While media education's goal may be to produce more effective or responsible ways of *using* media and technology, media *policy* pedagogy engenders the capacity to recognize and mobilize what Bitzer (1992) argues is "the rhetorical situation" or *kairos*, those propitious moments for decision or action, in this case, in media policy struggles. Media policy apprentices need to be able to navigate the temporal and discursive complexity of policy advocacy practice while doing policy advocacy work alongside experts who are already part of a community of practice (Postill 2013; Streeter 2013). It follows then that media policy literacy emerges not just from the classroom or from responding critically to scholarly writing, but directly from experience, something the media policy "tower of babble" with its multiplicity of approaches, entry points, and objects of study cannot facilitate unless it explicitly embraces practices. Media policy education, therefore, must draw on a *situated* theory of learning that offers forms of "legitimate peripheral participation" in the sense advanced by Lave and Wenger (1991). Substantively, this type of participation has to include attention to institutions, legal doctrines, aspirational goals, the history and practice of social movements, constituencies of interest, and specific policy issues like communication rights, freedom of expression, and broadband inclusion. In these circumstances, internships, fellowships, exchange, or residency programs may help to bring academic study to life.

Yet despite media policy's clear political significance, there are too many academic programs featuring critique and too few programs preparing students to become participants in media policy or to represent the public interest in government, corporations, or media policy advocacy organizations. Trends related to media studies in higher education raise the question: why aren't more public affairs and public policy programs teaching media policy? By the same token, why aren't more academic programs looking beyond the study of journalism or new media to address policies related to information, communication, and technology? Finally, why are there still so few handbooks, encyclopedias, readers, and journals dedicated to media policy studies and media reform itself?

Perhaps part of the answer rests in the rather long shadow cast by journalism studies and media studies programs, which have traditionally emphasized content over infrastructure concerns where content is linked to freedom of expression, namely, freedom of the press. More recently, attention to issues of code, infrastructure, and digital forms of activism, have been on the rise since Lessig published *Code* (1999).

Even still, the tendency is to feature primarily instrumental concerns—how media and technologies enable social and political change—such as activists' use of Twitter in Arab Spring uprisings or social movements' uptake of technology. Less common is news coverage of media policy as an object of activism itself (Mueller, Pagé, and Kuerbis 2004).

There are also too few civil society organizations, sustainable academic centers, institutes, or think tanks in media policy-making. Why the media policy and media reform fields remain so under-resourced confounds many working in this area. Donors that include the Ford Foundation, the Open Society Institute, and the Media Democracy Fund have already invested vast amounts of money. So why is there still a mere handful of media policy courses in information schools, media and communication studies programs, and political science departments? Why not degrees or certificates in media policy and media reform? Why aren't more public affairs and public policy programs teaching media policy? Why aren't more programs looking beyond the study of journalism or faddish new media phenomena to address information, communication, and technology policy? Why are there still so few handbooks, encyclopedias, readers, and journals dedicated to media reform?

This points to what I have previously called a "blind spot in higher education" related to media policy studies (Lentz 2014a). Despite this study area's relevance to the issues and concerns mentioned earlier, the field of media policy studies remains on the margins of institutions of higher education. If measured by degrees or certificates granted in media policy studies, it has no significant uptake in undergraduate or graduate curricula that mirrors other public policy areas like environmental protection, public safety, or public health.

Deficiencies in attention to media policy pedagogy are particularly noticeable in professional schools of business, international development, law, and public affairs, policy, administration, and service. There are, of course, exceptions. Those related to law programs in the United States include the Berkman

Center at Harvard University, the Yale Information Society Project, George-town University's Institute for Public Representation and American University, which now features Internet governance as does McGill University. Schools of Communication and Information Studies, on the other hand, have offered academic homes to many media policy studies scholars and their students.[5] Still, there are relatively few centers or institutes where the histories of media policy studies and media reform movements are being taught, or where gradu-ating students with an interest in this area of public policy formation and issue advocacy might intern, take up a fellowship, or eventually work.[6]

Rather than privileging the gendered design, uneven deployment, or undemocratic governance of media resources that occurs, many media stud-ies and communications departments instead offer courses and programs of study featuring the uses or abuses of media and technology resources. Com-pelling titles of courses, books, and articles in media studies feature the FUD factor—fear, uncertainty, and doubt—rather than how public interest advocates are working to address these concerns and reform institutions and practices (Hindman 2009; Landau 2010; McChesney 2013). Courses also feature aspects of media and cultural industries, media literacy, and more recently, digital activism. Such resources are essential in pointing out to stu-dents the current and potential harms to freedom of expression accompany-ing media and technology innovations. Yet students also need to be aware of scholarship that chronicles how public interest advocates work to address media misrepresentations, violations of privacy, and other media-related policy concerns (e.g., Bennett 2010; Franklin 2013; Hackett and Carroll 2006; Jorgensen 2006; Klinenberg 2007; McChesney, Newman, and Scott 2005; Mueller, Pagé, and Kuerbis 2004). I would argue that these endeavors need supplements to advance the civic dimension of media reform efforts.

Additionally, given that this work is situated primarily in information and communications schools means that the policy studies sector does not neces-sarily engage with media policy scholars or students. Students trained in such schools tend not to get the broader public policy and legal grounding that a law or public policy school offers, both of which would be helpful to them in media policy work.

In summary, a media future that inspires and cultivates innovation, open-ness, and participation depends on the capacity of people to consider them-selves as more than merely digital consumers. Critically minded media policy scholars have an obligation to team up with open source, media democracy, media justice, and freedom of expression communities of practice to render classrooms as collaborators of theory and practice. Universities can also open their doors to advocates as "professors of practice," creating "sabbaticals for radicals" that allow scholars and students to learn from practitioners, and for practitioners to enjoy opportunities for reflection. University professors teaching media policy courses can also create educational consortia to share syllabi, with oral history projects as possible points of praxis that bring together teaching, research, and practice. Indeed, there are many opportunities to work through the media policy tower of babble that undermines advancing public engagement in one of the most important area of policy studies of our time.

The following section explores one of several examples where efforts are being made to situate MA and PhD students in experiential learning settings where they can practice policy advocacy alongside seasoned citizen and consumer policy advocates.

MEDIA POLICY PEDAGOGY IN PRACTICE

The Consortium on Media Policy Studies (COMPASS) program provides MA and PhD students in communication studies with opportunities to work in the summer months (June through August) in Washington, DC, with governmental agencies, companies, and public interest organizations that match the students' topical research interests. Since the program's inception around 2002 when the first students were placed as a pilot effort, close to forty fellows have benefited from the opportunity to spend a period of eight weeks between June and August working alongside experienced policy practitioners and advocates. Partnering organizations so far have included the Alliance for Community Media, the Aspen Institute, the Center for Democracy and Technology, the Center for Digital Democracy, Common Cause, Consumers Union, the Federal Communications Commission, Free Press, Internews, the Media and Democracy Coalition, the New America Foundation, the offices of Representatives Diane Watson (D-CA) and Maurice Hinchey (D-NY) and Senators Bernie Sanders (I-VT) and Dick Durbin (D-IL), the US State Department, and the World Bank.

According to a 2009 assessment (Newman 2009), the program evolved out of conversations among senior scholars in the field of media and communication studies at a time of renewed interest in the field of media policy, due to the groundswell of public engagement in the US media consolidation debate of 2003. This renewed interest was prompted in large part by the 1996 Telecommunications Act, which resulted in the largest overhaul of communications policy legislation in years. Since that time, the media policy advocacy sector has witnessed the emergence of several new advocacy organizations and coalitions made possible by a considerable injection of new funding from institutions that include the Markle Foundation, the Ford Foundation, and the Open Society Institute, and family foundations like the Haas Charitable Trust and the Benton Foundation. A key theme among one donor institution in particular, the Ford Foundation, was a renewed role for academic institutions in media policy practice (Pooley 2011).

At present, the Consortium is led by the two Annenberg Schools of Communication (USC and University of Pennsylvania) and by the University of
Michigan. According to the program's website:

> The purpose of the consortium is to build bridges between the academic
> study of the mass media and policymakers. COMPASS seeks to train more
> graduate students in the areas of media policy, law and regulation; we are
> dedicated to making the academic study of the mass media and communica-
> tion systems more relevant to and informing of national and international
> policy planning and regulatory proposals.[7]

For the purposes of this project, past COMPASS fellows were contacted,
reaching out to fellows going as far back as 2002 and as recently as 2013 to
supplement interviews conducted by Russell Newman in 2009. Past COM-
PASS fellows were asked to either complete an e-mail survey or to participate
in a brief Skype interview about their fellowship experience. At the time of
writing, thirteen of the thirty-three past fellows have responded to our request
for information, for a total of eight completed e-mail questionnaires and seven
Skype interviews.

According to Newman (2009) and to accounts by various early fellows, the
COMPASS program began rather informally, with communication studies
professors using their institutional resources and contacts to connect students
to Washington, DC. According to one fellow who continued to work in the
advocacy sector after his own fellowship experience, and who later helped to
facilitate COMPASS placements, congressional fellowship arrangements were
made "on an ad hoc basis with particular members of congress." According to
another fellow, the program's "lack of formalization is what enabled it to exist
in the first place" with placements being established "behind the scenes." However, since 2008, and perhaps due to the fact that the Democrats are now in
power, there have been essentially no congressional COMPASS fellowship
placements; all placements have been at NGOs based in Washington. According to Newman's report, "there is widespread agreement that some form of
administration of the program and formalization of the fellowships in Washington is necessary for it to continue."

All COMPASS fellows interviewed and surveyed were pursuing graduate
degrees at the time of their COMPASS fellowships. While the majority of fellows heard about the COMPASS opportunity through their educational institutions—either via mailing lists or being advised of the opportunity by faculty
members—a couple of fellows learned about the COMPASS fellowship program through contacts from past internships in the nongovernmental sector.
One fellow expressed that he specifically chose to pursue his graduate degree
because the institution allowed him access to the COMPASS fellowship
program. Before beginning their fellowships, respondents' knowledge about
policy-making varied widely, from "very little" to "moderate" to "quite knowl-

edgeable" and "very advanced." Those who were quite advanced in their knowledge of policy-making had past fellowships at nongovernmental organizations.

According to the COMPASS website, fellows receive support from the core institutions to locate an appropriate placement, a stipend of $5,000, and travel expenses for attending a follow-up retreat. Students' home graduate institutions are expected to provide $2,000 for housing expenses. Of those interviewed/surveyed, three fellows said that they received $6,000 to cover the costs of their fellowship experience, which they said was sufficient for their needs at the time. Two fellows did not indicate the amount of money they received for participating in the fellowship program, but indicated that the amount was "sufficient." One fellow received $2,000 from the COMPASS program and $2,500 from the student's home institution—this fellow said that the amount received was insufficient and "did not even cover my rent." Lastly, one fellow received no funding as she lived at home and commuted on a daily basis.

Survey responses so far indicate that fellows' day-to-day work responsibilities varied significantly, depending on the fellowship placement. Fellows in the NGO sector reported that their daily tasks included attending media policy events at other organizations, conducting research and compiling data, organizing interviews, designing a survey, making contributions to larger projects, reading policies, writing draft reports, attending meetings, working with community groups, and doing educational work. Fellows placed in congressional offices were responsible for a wide range of daily tasks, including writing "dear colleague" letters, writing resolutions, drafting statements and speeches, reading and sorting constituent mail, and attending events such as hearings, briefings, and events held by public interest and lobbying groups. One congressional fellow reported being given "the full responsibilities of a legislative aid," and was responsible for working on broadcast regulation, saying "I was able only to work on that issue, researching that issue, reading all the news that was coming out about that issue, studying the bills that were being drafted about that issue."

As fellows' tasks differed, so too do their assessments of the importance of their fellowship work. Some found their work to be essential and important to the organization they were working for while others found their work to be not very essential or had trouble assessing the impact of their work. For the most part, writing and research skills, as well as a prior knowledge of policy-making, were found to be the most essential skills needed for these fellowships—capabilities that fellows often use in doctoral programs. Fellows also noted organizational skills, academic review, survey design, editing, critical reading, "big picture thinking," and statistical analysis as other essential skills for such a fellowship.

Nearly all fellows indicated an interest in getting to experience and observe how power operates in Washington, DC; they all took advantage of their time there by attending events, talks, briefings, and hearings in their off-work hours, all of which contributed to their learning and understanding of how the Hill works. For some fellows, being able to glean how power operates in the US capitol was the most substantial take-away from their fellowship experience. For example, one fellow had this to say about his congressional office fellowship:

> For me the more interesting part was learning the kind of behind-the-scenes politics; for example, how various congressional offices would strategize, or rather the various congressional staffers, how they would coordinate and try to message particular things. We also interfaced with the press fairly often so it was interesting to see that side of things. But there was just so much that you would never hear about, even just seeing things like how most of this language was being put together by twenty-five-year-old congressional staffers, seeing the relationships that these congressional offices had with outside groups and other institutions ranging from public interests, nonprofit groups to corporate lobbyists.
>
> I personally was very interested in seeing how corporate lobbyists were taking part in these daily activities, just seeing them in the halls of congress; so just basically seeing how power operates on a daily basis.

While the majority of fellows agree that they did all their learning about policy-making from working alongside and observing congressional staffers, one fellow said that he learned the most by attending meetings, while others said they learned by having conversations and asking questions, or as they went along by completing tasks independently and then seeking feedback from colleagues.

Given that the COMPASS fellowship lasts such a short time, i.e., only during the summer months, the majority of fellows did not feel they became an authentic member of their community of practice, the term mentioned earlier in this paper by those working from a situated learning perspective following Lave and Wenger. By the end of the program, most said they gained useful knowledge about both policy-making and how the Hill works. Fellows also said that their COMPASS experience gave them new confidence in their academic studies, and gave them opportunities to attend events and meet new people, future contacts, and supportive faculty—all listed as strengths of the COMPASS program.

However, several felt that they had too little contact with other fellows, that they did not make the connections they were hoping to make, and that they had a sense of doing work that was not useful to the organization where they were placed. That said, nearly all fellows described their COMPASS experience as positive.

With their fellowship experience behind them, some past fellows say their work now is not specifically related to anything they learned during their fellowship. However, other fellows report that their fellowship experience played a large role in the work that they are doing now. One fellow said that the fellowship was formative: "[It has] really shaped the way that I think about policy and about activism as well and the goals of activism and what they can be and what they should be." Other fellows said that their work would be much less policy oriented had they not completed a COMPASS fellowship.

A deeper examination of the actual work that fellows perform during their fellowships alongside seasoned experts could also offer insights into how to structure approaches to media policy pedagogy both within and outside higher education. For example, fellows report in interviews and surveys that they participated in many routine activities that at first seemed strange but later made sense to them. Such tasks included writing reports, preparing public testimony, developing white papers and policy briefs, and meeting with public officials and their staff. Fellowship experiences that offer opportunities to participate in advocacy practice fosters a sense of belonging and legitimacy that is not possible otherwise. While this point may seem obvious to practitioners (including law faculty in clinical legal programs), it may not be clear to those involved with media and communications studies teaching and scholarship that is de-linked from practice contexts and professional training programs. For this reason, an examination of syllabi from COMPASS and similar programs could be a valuable tool for the media reform movement.

CONCLUSION

What is gained by looking at media reform through this lens of media policy literacy? First, the approach broadens how we understand the labors of media policy advocacy communities. Rather than honing in on particular campaigns, and narrating and explicating their successes or failures, examining media reform as imbricated in media policy literacy requires that media reform movements view advocacy work as an ongoing and cumulative knowledge-building process, one in which advocates continually learn, interpret, and revise how to intervene in the policy-making process. The campaigns media reformers mount hinge on and are informed by previous experiences with advocacy efforts. Accordingly, media reform should be seen as an active pedagogic process, one in which advocacy groups both acquire the skill sets and resources necessary to intervene in policy-making while, at the same time, adjusting or modifying their expectations of what can be accomplished at particular historical junctures.

Second, looking at media reform through this lens of media policy literacy allows us to reimagine, and perhaps to decenter, questions of success or failure in media reform efforts that elide those episodes of media advocacy practice that serve as teaching moments for practitioners themselves. Deliberate attention to reflecting on practice serves to advance the work of policy advocacy in the public interest. Media policy advocacy groups may participate in a reform effort with the expectation that their immediate actions may not achieve their stated goals. However, to see these campaigns as failures often is to misunderstand their purpose or the way they function for advocacy groups themselves.

Finally, to highlight the centrality of media policy literacy to media policy advocacy is to recognize the long-term commitment and tenacity of media reform and media justice activists. It is to be wary of notions of advocacy constituted by quick fix actions to address what seem like exigent policy questions. It is to see that media reform efforts are borne not only of impassioned beliefs and commitments to a more democratic or participatory media, but of the tremendous amount of work required to gain the many literacies required to make such efforts effective. A commitment to media policy literacy, therefore, is essential for the potential success of media reform campaigns; that is, if by success we mean contributing to building ongoing capacity for seasoned advocates to nurture newcomers, for scholars to learn from practitioners by chronicling their various policy wins as well as losses, for practitioners to obtain time to reflect on these experiences, and for students of media studies to be able to pursue opportunities to engage directly as coparticipants in media reform efforts.

NOTES

1. The case study is licensed under the Creative Commons Attribution Noncommercial No Derivatives (by-nc-nd); see http://ijoc.org.

2. See Freedom of Expression, Media and Justice, www.fordfoundation.org/issues/freedom-of-expression/exploring-issues-of-justice-through-media; "Nonprofit Journalism and the Need for Policy Solutions," www.freepress.net/blog/11/12/05/nonprofit-journalism-and-need-policy-solutions; and the Knight Foundation, Media Innovation, www.knightfoundation.org/what-we-fund/innovating-media.

3. See the Media Literacy Project, http://medialiteracyproject.org; Journalismfund.eu, www.journalismfund.eu/; and "Philanthropic Foundations: Growing Funders of the News," http://communicationleadership.usc.edu/pubs/PhilanthropicFoundations.pdf.

4. See World Environmental Education Congress, www.environmental-education.org/.

5. In the United States, examples include the Center for Information Policy Research (CIPR) at the University of Wisconsin-Milwaukee, the Telecommunications and Information Policy Institute (TIPI) in the Moody School of Communication at the University of Texas at Austin, the University of Michigan, the Annenberg School for Communication at the University of Pennsylvania, Syracuse University School of Information Studies, the Institute for Information Policy at Pennsylvania State University, the Quello Center at Michigan State University, American University, the University of Virginia, and Duke University's DeWitt Wallace Center. Also noteworthy are the Oxford Internet Institute's grad-

uate program in Internet Studies where a policy focus is included as part of the curriculum.

6. A study of university-based degree- or certificate-granting programs related to media policy literacy (broadly conceived to include Internet governance, ICT, and other related subspecialties) would be a worthy undertaking for future media reform infrastructure-building work.

7. Consortion of Media Policy Studies, COMPASS, http://compass consortium.org/.

REFERENCES

Bennett, C. J. 2010. *The Privacy Advocates: Resisting the Spread of Surveillance.* Cambridge, MA: MIT Press.

Bitzer, L. F. 1992. "The Rhetorical Situation." *Philosophy and Rhetoric* 25: 1–14.

Braman, S. 2004. "Where Has Media Policy Gone? Defining the Field in the Twenty-First Century." *Communication Law and Policy* 9, no. 2: 153–182.

Brown, D., and J. Blevins. 2008. "Can the FCC Still Ignore the Public?" *Television and New Media* 9, no. 6: 447–470.

Fones-Wolfe, E. A. 2006. *Waves of Opposition: Labor and the Struggle for Democratic Radio.* Urbana: University of Illinois Press.

Franklin, M. I. 2013. *Digital Dilemmas: Power, Resistance, and the Internet.* New York: Oxford.

Gangadharan, S. P. 2009. "Public Participation and Agency Discretion in Rulemaking at the Federal Communications Commission." *Journal of Communication Inquiry* 33, no. 4: 337–353.

———. 2013. "Toward a Deliberative Standard: Rethinking Participation in Policymaking." *Communication, Culture and Critique* 6, no. 1: 1–19.

Hackett, R. A., and W. K. Carroll. 2006. *Remaking Media: The Struggle to Democratize Public Communication.* New York: Routledge.

Hindman, M. 2009. *The Myth of Digital Democracy.* Princeton, NJ: Princeton University Press.

Jørgensen, R. F., ed. 2006. *Human Rights in the Global Information Society.* Cambridge, MA: MIT Press.

Klinenberg, E. 2007. *Fighting for Air: The Battle to Control America's Media.* New York: Metropolitan Books.

Landau, S. 2010. *Surveillance or Security? The Risks Posed by New Wiretapping Technologies.* Cambridge, MA: MIT Press.

Lave, J., and E. Wenger. 1991. *Situated Learning: Legitimate Peripheral Participation.* New York: Cambridge University Press.

Lentz, B. 2010. "Media Infrastructure Policy and Media Activism." In *Sage Encyclopedia of Social Movement Media*, edited by J. Downing, 323–326. Thousand Oaks, CA: Sage.

———. 2014a. "Building the Pipeline of Media and Technology Policy Advocates: The Role of 'Situated Learning.'" *Journal of Information Policy* 4: 176–204.

———. 2014b. "The Media Policy Tower of Babble: A Case for 'Policy Literacy Pedagogy.'" *Critical Studies in Media Communication* 31, no. 2: 134–140.

Lessig, L. 1999. *Code and Other Laws of Cyberspace.* New York: Basic Books.

Lunt, P., and S. Livingstone. 2011. Media Regulation: Governance and the Interests of Citizens and Consumers. London: Sage.

McChesney, R. W. 1993. *Telecommunications, Mass Media and Democracy: The Battle for the Control of U.S. Broadcasting, 1928–1935.* New York: Oxford University Press.

———. 2013. *Digital Disconnect: How Capitalism Is Turning the Internet Against Democracy.* New York: New Press.

McChesney, R. W., R. Newman, and B. Scott, eds. 2005. *The Future of Media: Resistance and Reform in the 21st Century.* New York: Seven Stories Press.

Moyers, B. 2003. "Bill Moyers on Big Media." *NOW.* October 10. www.pbs.org/now/commentary/moyers27.html.

Mueller, M., C. Pagé, and B. Kuerbis. 2004. "Civil Society and the Shaping of Communication-Information Policy: Four Decades of Advocacy." *Information Society* 20, no. 3: 169–185.

Napoli, P. M. 2008. "Media Policy." In *The International Encyclopedia of Communication,* edited by W. Donsbach. Oxford: Blackwell Publishing.

———. 2009. "Public Interest Media Advocacy and Activism as a Social Movement." In *Communication Yearbook 33,* edited by C. Beck, 394–401. New York: Routledge.

Newman, R. A. 2009. "COMPASS: An Assessment." White paper. Ford Foundation. December 1.

Obar, J. A., and A. M. Schejter. 2010. "Inclusion or Illusion? An Analysis of the FCC's Public Hearings on Media Ownership 2006–2007." *Journal of Broadcasting and Electronic Media* 54, no. 2: 212–227.

Pickard, V. 2010a. "Reopening the Postwar Settlement for U.S. Media: The Origins and Implications of the Social Contract between Media, the State, and the Polity." *Communication, Culture and Critique* 3, no. 2: 170–189.

———. 2010b. "'Whether the Giants Should Be Slain or Persuaded to Be Good': Revisiting the Hutchins Commission and the Role of Media in a Democratic Society." *Critical Studies in Media Communication* 27, no. 4: 291–411.

Pooley, J. 2011. "From Psychological Warfare to Social Justice: Shifts in Foundation Support for Communication Research." In *Media and Social Justice,* edited by J. Pooley, L. Taub-Pervizpour, and S. Curry Jensen, 211–240. New York: Palgrave Macmillan.

Postill, J. 2013. *The Multilinearity of Protest: Understanding New Social Movements Through Their Events, Trends, and Routines.* Melbourne: RMIT University.

Schejter, A. M., and J. A. Obar. 2009. "Tell It Not in Harrisburg, Publish It Not in the Streets of Tampa: Media Ownership, the Public Interest and Local Television News." *Journalism Studies* 10, no. 5: 577–593.

Shade, L. R., and T. Shepherd. 2013. "Viewing Youth and Mobile Privacy Through a Digital Policy Literacy Framework." *First Monday* 18, no. 12 (December 2). http://firstmonday.org/ojs/index.php/fm/article/view/4807/3798.

Streeter, T. 2013. "Policy, Politics, and Discourse." *Communication, Culture and Critique* 6, no. 4: 488–501.

Wallis, R., and D. Buckingham. 2013. "Arming the Citizen-Consumer: The Invention of 'Media Literacy' within UK Communications Policy." *European Journal of Communication* 28, no. 5: 527–540.

PART TWO

Internet Activism for Media Reform

CHAPTER THREE

Activating the Fifth Estate

Bill C-30 and the Digitally Mediated Public Watchdog

JONATHAN A. OBAR, *York University, Canada*
LESLIE REGAN SHADE, *University of Toronto, Canada*

MEDIA REFORM STRATEGY

Operating outside the framework of traditional systems of governance and civic engagement, the digitally mediated, networked society referred to as the Fifth Estate presents the general public with a unique opportunity to reinvigorate the public watchdog function. While previous discussions of the Fifth Estate have emphasized that the communicative power it enables can help to hold government accountable, specific strategies have yet to be clearly identified. This chapter presents three strategies for activating a digitally mediated Fifth Estate: (1) building an online community of networked individuals, (2) shaping preexisting digital platforms to enable members of the public to contribute focused and pointed user-generated content, and (3) developing targeted content to be shared and distributed. Examples of these three strategies, respectively, include (1) the Stop Online Spying coalition (building community); (2) online petitions, digital form letters, and the #TellVicEverything Twitter attack (shaping platforms); and (3) OpenMedia's Stop Online Spying web materials, various online videos, and the Vikileaks Twitter attack (targeted content). This discussion is presented in the context of the successful media reform battle to defeat Canada's Bill C-30, an attempt by the Canadian government to expand upon its cyber-surveillance capabilities.

In February 2013, when the Canadian government canceled its plans to pass Bill C-30, it appeared to be a victory for the digitally mediated, networked society referred to as the Fifth Estate. Commonly referred to as the Lawful

Access Bill, C-30 was an attempt to expand upon the cyber-surveillance capabilities of the Canadian government. After introducing the bill a year earlier, there were no public hearings, no formal calls for public comment, and no reports of Members of Parliament being flooded with phone calls;[1] yet on February 11, 2013 when Justice Minister Rob Nicholson announced the government's decision to cancel the bill, he explained, "we've listened to the concerns of Canadians who have been very clear on this" (Payton 2013). This raises the question, if no traditional methods of public consultation were pursued, how was such a clear and convincing message of public dissent advanced?

This chapter discusses the public outcry generated in opposition to Bill C-30, led by the media reform group OpenMedia, and suggests that the uprising provides an example of a digitally mediated Fifth Estate influencing government policy. It is also suggested that this influence may have been advanced by a watchdog function exercised by the Fifth Estate in which members of the general public, independent of a watchdog press, contributed to holding the government to public account. While previous discussions of the Fifth Estate have emphasized that the communicative power it enables has the ability to hold government to account (e.g., Dutton 2008, 2009), specific strategies for activating the Fifth Estate have yet to be clearly identified. As a result, while media reformers may draw a sense of excitement from success stories like the battle over the Stop Online Piracy Act/Protect IP Act (SOPA/PIPA), the methods and strategies for replication need further clarification.[2] In what follows, we present a brief discussion of the Fifth Estate, followed by the context of the Bill C-30 debate. Next, we present three strategies for activating a digitally mediated Fifth Estate, drawn from the C-30 case study. These three strategies include: (1) building an online community of networked individuals, (2) shaping preexisting digital platforms to enable members of the public to contribute focused and pointed user-generated content, and (3) developing targeted content to be shared and distributed.

THE FIFTH ESTATE

The historical conceptualization of the three "estates of the realm" described the hierarchically conceived power structure of feudal society. At the top of the hierarchy was the First Estate, referring to the clergy, the Second Estate referred to the nobility, and the Third Estate referred to the commoners. The extent to which this hierarchy has evolved into more contemporary forms of societal governance has been a point of contention (e.g., Pollard 1920). For instance, it has been suggested that in the United Kingdom, the three estates evolved into the leaders of the Church of England (clergy), the House of Lords (nobility), and the House of Commons (commons); whereas in the United States, the estate model contributed to the separation of powers in the executive, legislative, and judicial branches of government. Whether or not the concept of the estates of the realm is still relevant, their depiction of a hierarchical societal power structure has contributed to relevant conceptualizations of a Fourth and now Fifth Estate that highlight newer collaborative opportunities for challenging the established societal order, as highlighted by the case study presented here.

THE FOURTH ESTATE: THE PRESS AS WATCHDOG

Drawing from the "estates of the realm" conceptualization more than two hundred years ago, Edmund Burke used the term *Fourth Estate* to describe an independent press operating outside of the established societal order (Carlyle 1905). In years following, autonomy, independence, and the ability to directly and antagonistically challenge dominant groups operationalized a watchdog function central to the Fourth Estate (Donohue, Tichenor, and Olien 1995). This watchdog function describes an intermediary role played by the press that helps to facilitate representative democracy. In this role, the press surveils and reports on entities that comprise the societal power structure to hold them to public account. As described by Donohue et al. (1995, 112), "Such an idea of press surveillance has been basic to the ideology of popular and representative government, because it springs from the idea of the populace as sovereign and entering into a social contract with a governing establishment that will serve popular interests."

THE FIFTH ESTATE: THE NETWORKED PUBLIC AS WATCHDOG

Whether it is a response to the inability of the Fourth Estate to maintain the watchdog function in light of commercial or political conflicts of interest, or an extension of a longstanding democratic ideal requiring public involvement in societal governance, the Fifth Estate refers to the notion that members of the public can exercise a similar watchdog function independent of the Fourth Estate. While some scholars attribute the Fifth Estate designation to a variety of influential groups including scientific experts (Little 1924), nongovernmental organizations (Eizenstat 2004), news media critics (Hayes 2008), and even satirical news show like Jon Stewart's *The Daily Show* (Reilly 2010) others see the Internet as central to contemporary conceptualizations, suggesting that networked individuals operating in the digital space represent the latest iteration of the Fifth Estate (Cooper 2006; Dutton 2008; Dutton 2009). These networked individuals build relationships, share information, and create, collaborate, and mobilize online, and in the process, reconfigure access to alternative sources of information, individuals, and resources. As Dubois and Dutton (2014, 239) note, "in doing so (this digitally-mediated Fifth Estate) can fulfill many of the same functions of holding up the activities of government, business, and other institutions to the light of a networked public" (ibid., 239). Indeed, it appears that among the political benefits of the Internet is the notion that this remarkable increase in our ability to create, collaborate, share, and mobilize, grows increasingly possible outside the limited framework of traditional institutions and organizations (Shirky 2008).

This leads us to ask, how can media reformers activate the Fifth Estate? How can they engender a focused and pointed public watchdog? The Bill C-30 case study presents an example of how media reformers might attempt to activate the power of such a networked public.

THE CONTEXT: CANADA'S BILL C-30

On Valentines Day (February 14) in 2012, an effort led by Public Safety Minister Vic Toews saw the Canadian Federal Government introduce Bill C-30 into the House of Commons of Canada (HCC 2012). Commonly referred to as the Lawful Access Bill, C-30 was an attempt to expand upon the search and seizure, interception, surveillance, collection, and decryption capabilities of Canadian law enforcement, with the primary goal of removing the legal and technical barriers inhibiting seamless access to information held in private Internet and mobile accounts. The latest incarnation of a string of attempts to pass lawful access legislation dating back to 1999,[3] the bill proposed the enactment of the Investigating and Preventing Criminal Electronic Communications Act as well as amendments to the Criminal Code of Canada and two other complementary acts.[4] Specific provisions included:

The Canadian government would be authorized to force telecommunication companies to install surveillance equipment (hardware and software) to enable a constant flow of data between the providers and law enforcement agencies. The companies would be required to cover the cost of these installations (which would likely be passed to the consumer).

The mandating of warrantless disclosures of subscriber information and data.

An enhanced ability to obtain transmission data warrants granting real-time access to all information generated during the creation, transmission, or reception of a communication, and preservation orders requiring providers to preserve that information.

The granting of these capabilities to law enforcement agencies, the Canadian Security Intelligence Service (CSIS), the Competition Bureau of Canada as well as any public officer "appointed or designated to administer or enforce a federal or provincial law."

A gag order that would have limited privacy protections provided in PIPEDA (Canadian privacy legislation) prohibiting telecommunication providers from informing users that they are the subject of a lawful access investigation. (HCC, 2012)

In case the Canadian government found themselves limited by the changes made by this bill at some point in the future, a catch-all was also included in Section 64, providing additional regulatory power "generally, for carrying out the purposes and provisions of this Act" (Geist 2012).

During the C-30 debate, the Canadian government cited the need to ratify the Council of Europe Convention on Cybercrime and increase cooperation on international investigations as one impetus for the bill. The other was to protect children, even though the actual content of the bill contained no mention of children or predators except in its eventual title. The latter point became central to the debate for two reasons. The first was that the title was apparently changed after the bill was sent to the printer,[5] and coupled with the lack of language in the bill about protecting children, suggested that the Conservative government was acting disingenuously. The second reason resulted from a related comment made by Public Safety Minister Vic Toews. While introducing the bill in the House of Commons, Toews shot back at a Liberal opponent who critiqued it, quipping that "he can either stand with us or with the child pornographers" (CBC 2012). It is possible that Toews's comment, and the disingenuous nature of the name change contributed to the considerable backlash (against Toews in particular) that eventually led to the bill's demise.

One year later, without having formally consulted with the Canadian public, Canada's Conservative government canceled its plans to pass C-30. On February 11, 2013, Justice Minister Rob Nicholson said: "We will not be proceeding with Bill C-30, and any attempts . . . to modernize the Criminal Code will not contain the measures contained in C-30, including the warrantless mandatory disclosure of basic subscriber information or the requirement for telecommunications service providers to build intercept capability within their systems" (Payton 2013). The minister offered a clear justification for this decision, "We've listened to the concerns of Canadians who have been very clear on this" (ibid.).

What contributed to this about-face? The remainder of this chapter suggests that the clear and convincing message of public dissent may have been advanced by a digitally mediated Fifth Estate. Three strategies media reformers should consider when trying to activate the Fifth Estate, drawn from the C-30 case study will be discussed: (1) building an online community of networked individuals, (2) shaping preexisting digital platforms to enable members of the public to contribute focused and pointed user-generated content, and (3) developing targeted content to be shared and distributed.

STRATEGY 1: BUILDING AN ONLINE COMMUNITY
OF NETWORKED INDIVIDUALS

Activist groups are using the Internet to advance organizational goals, and have been doing so for more than fifteen years. Since the earliest instances of online activism, one of the central strategies deployed has been the building of online communities. For example, in 1997, the Preamble Collaborative organized a network of groups via its website, in opposition to the OECD's Multilateral Agreement on Investment (MAI), an agreement that dealt with trade between the twenty-nine OECD countries (Kobrin 1998). When an early draft of the MAI was leaked, the preexisting network allowed the draft to be shared quickly and widely, contributing to an outpouring of opposition to the OECD's plans. These efforts contributed to the halting of the MAI negotiations, and then the closing of the negotiations (Neumayer 1999).

Since the MAI, a wide variety of similar efforts have demonstrated how the development of online communities allows networks of individuals and groups to quickly and effectively mobilize around a particular issue. Another early example was the "Battle of Seattle" where a community of advocacy organizations combined online and offline activism efforts to successfully halt a WTO event (Carty 2010). One other early example of online community building took place when some of the groups involved created the Independent Media Center website (now referred to as Indymedia). An early version of an interactive social media site, the Indymedia website was designed to allow anyone to post information, whether it was text or audiovisual material in real-time, "and set the model for a many-to-many use of the media whereby activists can subvert the traditional one-to-many approach under mainstream and corporate media" (ibid., 2).

The Stop Online Spying campaign (see Table 1) that developed in opposition to Canada's Bill C-30 serves as a more recent example of how building an online community of networked individuals can be an effective strategy for activating the Fifth Estate.

THE STOP ONLINE SPYING CAMPAIGN

For months leading up to the anticipated introduction of Bill C-30 into Parliament (February 2012), a protracted campaign involving a range of civil libertarian and public interest groups, alongside an array of citizens and academics spoke out against the bill. The effort coalesced under the Stop Online Spying campaign[6] organized by Canadian media reform group OpenMedia—a small, Vancouver-based civil society organization that has been in operation since 2008. They joined legal experts who raised serious concerns about the bill's violation of the Canadian Charter of Rights and Freedoms, and privacy commissioners from across the country, concerned about the significant impact on privacy rights. OpenMedia, which developed and now maintains an online community of more than six hundred thousand members, has quickly established itself as one of the major players in the battle to reform the Canadian media and telecommunications systems. In June 2011, months before the introduction of Bill C-30, OpenMedia launched the Stop Online Spying campaign by bringing together a coalition of forty-two organizations to serve as official members of the campaign (see Table 1). To join the coalition, members had to agree to a position statement articulating that the government should reconsider their lawful access legislation.

TABLE I. STOP ONLINE SPYING COALITION, OFFICIAL MEMBERS

Charter Members

Association for Progressive Communications

British Columbia Civil Liberties Association

BC Freedom of Information and Privacy Association

Canadian Civil Liberties Association

Canadian Council on American-Islamic Relations

Canadian Federation of Students

Canadian Internet Policy and Public Interest Clinic

Electronic Frontier Foundation

Greenpeace

International Civil Liberties Monitoring Group

Ligue Des Droits Et Libertés

OpenMedia.ca

Rocky Mountain Civil Liberties Association

The Tyee

Union Des Consommateurs

Associate Members

Agentic

BC Association for Media Education

Canadian Association of Campus and Community Television User Groups and Stations

Canadian Association of University Teachers

Canadian Dimension

Canadian Union of Public Employees

Civil Liberties Association, National Capital Region

Council of Canadians

Canadian Media Guild

Community Media Education Society

Democracy Watch

The Edmonton Small Press Association

Hospital Employees' Union

The Infoscape Research Lab

Koumbit

Media Action

Megaphone Magazine

National Campus-Community Radio Association

National Union of Public and General Employees

NewsWatch Canada

Obercor Technologies

Sorrell Financial

The Vancouver Observer

W2 Community Media Arts

White Rose Coffeehouse

Wilderness Committee

World Association of Christian Communication North America

The Stop Online Spying campaign was much larger than these forty-two groups however. Academics, activists, journalists, lawyers, media makers, media reformers and a wide variety of individuals and organizations connected to the campaign through the various digital platforms that were shaped and widely shared to enable participation in the debate. As discussed in the next section, by providing platforms and outlets for individuals to express themselves and organize, a wide variety of individuals and organizations (officially and unofficially) joined the online community of networked individuals that added their voices to the Bill C-30 opposition.

STRATEGY 2: SHAPING PREEXISTING
DIGITAL PLATFORMS

Two recent surveys of advocacy organizations operating in Canada and the United States assessed the extent to which social media afford opportunities for advancing activism objectives (Obar, Zube, and Lampe 2012; Obar 2014). Of the various beneficial affordances identified, the most common affordance expressed by the 116 groups surveyed suggests that social media can amplify an organization's message by strengthening outreach efforts and facilitating communication with a large number of individuals. OpenMedia was one of the groups surveyed and noted: "As an organization that facilitates social change through participatory decision-making, it is hugely important to have these low-cost, low-barrier tools to allow us to reach thousands, to receive feedback, and to build communities online" (Obar 2014, 222).

At the same time, among the concerns expressed was the notion that while social media can make it easy to participate and provide opportunities for creating all types of user-generated content, from a multitude of sources, this ease of use and access can make it difficult for organizations to speak with a unified, coherent voice. One large advocacy organization operating in the United States said, "in large organizations use of social media can be diffuse, with not all users being qualified 'spokespeople,' meaning that there may be a well-meaning person who says the wrong thing" (Obar, Zube, and Lampe 2012, 17).

While the Fifth Estate introduces the exciting possibility of an expanded marketplace of ideas operating outside the purview of the traditional outlets for news and information, message focus is introduced as a relative drawback to this expanded access and ease of use. When Justice Minister Rob Nicholson stated, "We've listened to the concerns of Canadians who have been very clear on this," he suggested that the dissent in opposition to Bill C-30 had been relatively clear. The shaping of preexisting digital platforms to enable focused and pointed user-generated content presents one strategy that may have contributed to this outcome. The following sections summarize a number of examples of this strategy that appeared during the C-30 debate.

ONLINE PETITIONS

One of the more straightforward methods was to create online petitions to be shared and signed by the online community. Petitions allow for the shaping of a clear position statement as well as a limited and focused way for supporters to connect their voice. The benefits of focus resulting from these limitations could also be considered drawbacks, as limited participation can be perceived as "slacktivism" or an inaccurate reflection of the public's opinion (Obar 2014; Obar and Argast 2014). This emphasizes one reason why petitions serve an important purpose, but are strengthened by complementary strategies.

Both the Stop Online Spying coalition and the Liberal government released online petitions that were shared by their online networks, with the coalition's petition being signed by more than 150,000 individuals, and the Liberal government's by more than 35,000.[7] It is worth noting that this considerable difference in the number of signatures hints again at the role of the Fifth Estate in the C-30 debate, with more members of the general public seeming to engage with the coalition of activists, rather than with government representatives.

TWITTER ATTACKS—#TELLVICEVERYTHING

A few days after C-30 was introduced, Robert Jensen, a member of the Canadian public living on the eastern coast of Canada (Prince Edward Island), initiated a focused and pointed Twitter attack against Vic Toews, the public face of

C-30. Jensen encouraged his Twitter followers to flood Toews' account with tweets, organized around the #TellVicEverything hashtag. Jensen described his strategy in an interview, "The logic was: Vic, if you want our personal information we will give it to you, here are the mundane details of our lives" (Dubois and Dutton 2014, 88). Jensen went on to say that the goal was to "laugh Vic out" (ibid., 89). #TellVicEverything quickly went viral and became a top-trending topic in Canada, with celebrities and politicians also getting involved. Thousands of messages were posted providing inane and mundane commentary on everyday experiences:

LAURA: I'm so sleepy today but I have to wake up and get on with my
schoolwork so I can save the world in future.
LUCY: . . . drinking some coffee. Might go hiking if it's not too slippery
out. I have a blister on my heel :(
JASON: Morning . . . I've showered, shaved, and made toast. The peanut
butter was chunky . . .
MICHAEL: You're so vain You probably think my email's about you . . .

Some of the tweets more directly combined the use of irony with a pointed critique of the lawful access legislation:

KEVIN: 'Hey @ToewsVic, I lost an email from my work account yester-
day. Can I get your copy?'
STEVE: 'Hey @ToewsVic I have 2 confess, I was a over speed-limit a bit
on 404 today. U won't tell the @OPP_GTATraffic will you?'

While Jensen's strategy appeared to involve humorously giving Toews a taste of his own medicine, the broader results of the attacks included a focused and pointed expression of public opposition, as well as evidence of a groundswell of dissent. Similar to the petition strategy, by creating the #TellVicEverything hashtag and the humorous and ironic theme of revealing personal information about yourself, Jensen shaped the overall message, while providing a platform for members of the general public to convey the strength behind that message.

SEND VIC TOEWS A VALENTINE

Anticipating the introduction of the bill into Parliament in early February, OpenMedia set up a website that offered individuals the ability to print out a Valentines Day card to be sent to Vic Toews. OpenMedia's site noted: "This month, send Hon. Vic Toews a Valentine informing him that you oppose the Online Spying legislation. Follow the link, print out the valentine, and send off your most intimate feelings regarding the legislation. No postage required!"[8] The card itself included the following statement: "Dear Minister Toews: I oppose mandatory internet surveillance. This scheme is poorly thought out, costly, and will leave my personal information less secure. Unchecked mass surveillance is a breach of my fundamental right to privacy."

OpenMedia's Valentines Day card was another example of people using the Internet to provide a platform for members of the Fifth Estate to express a focused and pointed opinion about C-30. Both the form text and the irony of the card's theme focused the pointed message.

E-MAIL YOUR MEMBER OF PARLIAMENT

OpenMedia added an online tool to their website that allowed members of the public to write their Member of Parliament (MP). The site stated:

> The government has just tabled an online spying plan that will **allow authorities to access the private information of any** Canadian at any time, **without a warrant.** If they are successful your personal information could be caught up in a digital dragnet and entered into a giant unsecure registry of private data. . . . If **we care about privacy, the open Internet, our economy and our basic democratic rights,** it's time to tell Ottawa to stop this irresponsible plan NOW.[9]

The site then included a form letter (that could not be modified) to be sent to an MP. Users were asked to input their names, addresses, and postal codes, and after clicking "Send it now," it is assumed that the message was sent by OpenMedia to various MPs associated with the postal code.

WRITE A LETTER TO THE EDITOR

OpenMedia created an online tool similar to the e-mail your MP tool after the Bill C-30 battle that remains on their website as a part of the ongoing campaign to combat lawful access initiatives.[10] This online tool enables members of the public to easily write letters to newspaper editors. Once on the site, users type in their postal code and are then provided with a list of newspapers in the user's area. Checkboxes are included next to each newspaper which, when selected, direct the tool to automatically send the letter to the corresponding editors via e-mail. Once the checkboxes are selected, users are prompted to include an e-mail subject line with the suggestion "Letter to the Editor: Protect Our Privacy." Below the subject box is a statement which reads: "Please personalize your letter by editing the talking points below so they are in your own words. This will give your letter more impact and make it much more likely to get published." As you scroll down the page there is a large text box allowing for three hundred words to be typed. Below the box is a list of key talking points and tips that users are encouraged to consider while writing their letters. When finished, users then fill in some personal information (name, e-mail, and phone number) and submit the letter. It is unclear how many individuals have used this online tool to speak out against government surveillance; however, the tool serves as an interesting example of a platform for encouraging focused and pointed user-generated content.

STRATEGY 3: DEVELOPING TARGETED CONTENT TO BE SHARED AND DISTRIBUTED

A related strategy for activating the Fifth Estate and also for ensuring a focused and pointed message is to create targeted content to be shared and distributed. This strategy predates Web 2.0, and is very similar to many offline strategies that have involved the sharing of predetermined text, images, audio, video, and so on. The digitally mediated examples drawn from the C-30 debate do emphasize an expanded set of opportunities for creating digital content as well as sharing that content via social networks.

ONLINE VIDEOS

University of Toronto researchers Andrew Clement and Kate Milberry produced an early video as part of the New Transparency: Surveillance and Social Sorting research project. Titled *(Un)Lawful Access: Experts Line Up Against Online Spying*, the video was meant for viral and educational distribution and was also part of screenings held across the country.[11] OpenMedia also produced professional online videos that were designed to go viral. In OpenMedia's videos, citizens were seen doing ordinary things, when members of Canadian law enforcement surprised them with a form of surveillance. One video showed a woman walking alone down the street when her cell phone rings. She picks it up and starts talking. As she holds the phone to her ear an officer's hand comes in from the left side and pulls the phone over to his ear (see Figure 3-1). A caption on the screen then reads: "You wouldn't let a police officer do this without a warrant . . . So why should your online communication be any dif-

FIGURE 3-1.   Still from an OpenMedia video (YouTube).

ferent. . . . Find out more at stopspying.ca." OpenMedia's videos have been viewed more than one hundred thousand times.

Hacker group Anonymous also released a related video under the project name "Operation White North." Their videos were clustered on the YouTube channel, Operation VicTory. Their first video release featured a narration rendered in their distinctive (male) robotic voice:[12]

> Hello Vic Toews: We are Anonymous. Recently you have tried to reintroduce the Internet spy bill known as Bill C-30. You claim this bill is to protect children from Internet predators. All this legislation does is give your corrupted government more power to control its citizens. Anonymous will not stand for this.
>
> Now after a severe Internet backlash you have decided to backtrack and bring this before a committee. This is not good enough. We demand that you scrap the bill in its entirely and step down as safety minister. We know all about you Mr. Toews, and during Operation White North we will release what we have, unless you scrap this bill.
>
> We told you to expect us.
>
> We are Legion.
>
> We do not forgive.
>
> We are here.

WEBSITES AND MEMES

Websites, graphics and other imagery peppered the Internet. One website, isvictoewswatchingme.com, displayed the line: "yes. but he didn't know" (linking to a CBC news article where Toews denied knowing the content of the bill), and "you can still stop him" (linking to the stop online spying petition). Another site featured the popular "ceiling cat" meme, with Toews image in place of the cat's (see Figure 3-2).

FIGURE 3-2. The Vic Toews "ceiling cat" meme.

Vikileaks

A few days after the announcement of C-30 in early February 2012, an anonymous individual created the Twitter account @vikileaks30 (see Figure 3-3). The account name was a direct reference to Wikileaks, a highly publicized effort to bring information kept secret by various powerful institutions to public light. The purpose of the account was made clear on the Twitter page: "Vic wants to know about you. Let's get to know Vic." Vikileaks tweeted over ninety personal and professional details about Minister Toews, airing his dirty domestic laundry: his divorce from his former wife after an extramarital affair with a younger woman who was perhaps his child's babysitter, resulting in the birth of a child, his restaurant expenditures, and so on. The Vikileaks content quickly became a nationwide story, receiving a considerable amount of popular press coverage. Parliamentary investigation into the origins of the Twitter account was to be initiated by the Speaker of Parliament, but it was soon revealed that a Liberal Party staff member in the party's research bureau was the perpetrator—and he quickly resigned his position (Fitzpatrick 2012).

FIGURE 3-3.   Vikileaks Twitter account.

OTHER FORMS OF CONTENT

OpenMedia's Stop Online Spying website hosted a variety of different types of content that could be shared. This included targeted social media notes (prewritten Facebook posts and tweets), code for embedding a petition on a website, code for displaying badges and ads of the Stop Online Spying images (see Figure 4), audio PSAs about lawful access, digital posters and handbills, FAQs,

FIGURE 3-4.   Stop Online Spying campaign.

and talking points. OpenMedia also shared a report by the BC Civil Liberties Association called "Moving Toward a Surveillance Society,"[13] and included a variety of other links to articles and websites of interest.

CONCLUSION

Public involvement in government decision-making has been lauded historically for its ability to act as a government watchdog, contributing to protections against economic, social, and moral stratifications, narrow-mindedness, bias, impropriety, and even tyranny (Obar 2010; OECD 2001; Walker 1966). Articulated in many forms over many years going back at least to ancient Athens, the watchdog function of the public has been seen as central to the democratic concept, and necessary for a free and just society. As Thomas Jefferson wrote in the US Declaration of Independence:

> Governments are instituted among Men, deriving their just powers from the consent of the governed—That whenever any Form of Government becomes destructive of these ends, it is the Right of the People to alter or to abolish it, and to institute new Government, laying its foundation on such principles and organizing its powers in such form, as to them shall seem most likely to effect their Safety and Happiness.[14]

A digitally mediated Fifth Estate presents a unique opportunity for members of the public to exercise this watchdog function. This renewed opportunity for public involvement in societal governance is especially welcome in the twenty-first century, at a time when there continue to be many barriers to public participation that limit democratic possibilities (Schejter and Obar 2009; Obar and Schejter 2010). The SOPA/PIPA victory in the United States generated a sense of excitement over the democratic possibilities of the Fifth Estate. What remains however, is the need for a set of strategies that can enable replication. This chapter suggested three strategies that media reformers can consider when attempting to activate the Fifth Estate. The first strategy, building an online community of networked individuals, has, since the earliest days of online activism, been a mainstay for successful digital mobilizations, and thus, is an appropriate place to begin. The second strategy, shaping preexisting digital platforms to enable members of the public to contribute focused and pointed user-generated content, emphasizes the need to direct the lifeblood of the social media organism. User-generated content can send a powerful message to those in power, but only if that message is clear. The third strategy, developing targeted content to be shared and distributed, helps to further the previous two strategies by both providing individuals with materials to share for the purpose of community-building, while also contributing to the strength of a focused and pointed message.

Indeed, by circumventing the Fourth Estate, and operating outside the framework of traditional systems of governance and engagement, the digitally mediated Fifth Estate presents the general public with a unique opportunity to reinvigorate the public watchdog function. The uprising that resulted in the response to Bill C-30 provides an important case study, and suggests that online activism, if strategically activated and coordinated, can, independent of traditional methods of public consultation, help to hold institutional power structures to account.[15]

NOTES

The authors would like to thank Andrew Clement (University of Toronto) for his help with the early formulation of this chapter. Leslie Shade also extends thanks to Steffen Schneider and the Association for Canadian Studies in the German-speaking Countries for their invitation to present an earlier version of this chapter at their 2012 conference in Grainau, Bavaria.

1. In the years preceding the introduction of Bill C-30 in February 2012, the Canadian government had provided opportunity for public input about cyber-surveillance legislation. That being said, C-30's proposed expansions to the government's cyber-surveillance capabilities, coupled with the aggressive (and disingenuous) nature of the bill's introduction emphasized the government's strong stance on the issue as of February 2012.

2. For an analysis of the networked public sphere during the SOPA debate, see, for example, Benkler et al. (2013).

3. Earlier legislation was introduced into the House of Commons in 2005 (the Modernization of Investigative Techniques Act), 2009 (two parallel bills, the Technical Assistance for Law Enforcement in the 21st Century Act and the Investigative Powers for the 21st Century Act), and 2010 (three corresponding bills, Improving Access to Investigative Tools for Serious Crimes Act; Investigative Powers for the 21st Century Act; and Investigating and Preventing Criminal Electronic Communications Act). All of the bills died on the order paper because of election calls.

4. These acts are the Competition Act and the Mutual Legal Assistance in Criminal Matters Act.

5. The title was changed from "An Act to Enact the Investigating and Preventing Criminal Electronic Communications Act and to Amend the Criminal Code and other Acts" to "The Protecting Children from Internet Predators Act."

6. See https://openmedia.ca/StopSpying.

7. The Stop Online Spying campaign's petition can be found at https://openmedia.ca/StopSpying; the Liberal government's petition can be found at http://petition.liberal.ca/online-privacy-surveillance-lawful-access-bill-c30-liberal-amendment/.

8. See https://openmedia.ca/blog/send-vic-toews-valentine.

9. See https://openmedia.ca/mp.

10. See https://openmedia.org/privacy/letter.

11. See www.unlawfulaccess.net/.

12. See www.youtube.com/watch?v=-I1WLfpD9Pk.

13. See http://bccla.org/our_work/moving-toward-a-surveillance-society-proposals-to-expand-lawful access-in-canada/.

14. US Declaration of Independence, www.archives.gov/exhibits/charters/declaration.html.

15. The defeat of Bill C-30 is heartening; however, there is a need for activists and citizens to be ever-vigilant about the reintroduction and intrusion of similar legislation. Indeed, Bill C-13, the Protecting Canadians from Online Crime Act, introduced by the government in late fall 2013, raised the ire of many Canadians especially concerned with its impact on surveillance and free expression for Canadian citizens. Closely related, Bill C-51, the Anti-Terrorism Act, received royal assent in June 2015. Intense criticism of the bill from civil society centered on expanded powers given to law enforcement and the Canadian Security Intelligence Agency, concerns over the bill's expanded definition of what constitutes a terrorist act, and the erosion of privacy rights. With the election of a new Liberal majority government in October 2015, renewed pressure to review and reform the bill has been initiated.

REFERENCES

Benkler, Y., H. Roberts, R. Faris, A. Solow-Niederman, and B. Etling. 2013. "Social Mobilization and the Networked Public Sphere: Mapping the SOPA-PIPA Debate." *Berkman Center Research Publication No. 2013–16* (July 19). http://cyber.law.harvard.edu/sites/cyber.law.harvard.edu/files/MediaCloud_Social_Mobilization_and_the_Networked_Public_Sphere_0.pdf.

Carlyle, T. 1905. *On Heroes: Hero Worship and the Heroic in History*. London: H. R. Allenson.

Carty, V. 2010. *Wired and Mobilizing: Social Movements, New Technology, and Electoral Politics*. New York: Routledge.

CBC News. 2012. "Online Surveillance Critics Accused of Supporting Child Porn." February 13. www.cbc.ca/news/technology/story/2012/02/13/technology-lawful-access-toews-pornographers.html.

Cooper, S. D. 2006. "The Blogosphere and the Public Sphere." In *Watching the Watchdog: Bloggers as the Fifth Estate*, 277–303. Spokane, WA: Marquette Books.

Donohue, G. A., P. J. Tichenor, and C. N. Olien. 1995. "A Guard Dog Perspective on the Role of Media." *Journal of Communication* 45, no. 2: 115–132.

Dubois, E., and W. H. Dutton. 2014. "Empowering Citizens of the Digital Age: The Role of a Fifth Estate." In *Society and the Internet*, edited by M. Graham and W. H. Dutton, 238–256. Oxford: Oxford University Press.

Dutton, W. H. 2008. "The Fifth Estate: Democratic Social Accountability Through the Emerging Network of Networks." *Understanding E-government in Europe: Issues and Challenges*, edited by P. G. Nixon, V. N. Koutrakou, and R. Rawal, 3–18. New York: Routledge.

———. 2009. "The Fifth Estate Emerging Through the Network of Networks." *Prometheus* 27, no. 1: 1–15.

Eizenstat, S. E. 2004. "Nongovernmental Organizations as the Fifth Estate." *Seton Hall Journal of Diplomacy and International Relations* 5, no. 2: 15–27.

Fitzpatrick, M. 2012. "Liberal Staffer Behind 'Vikileaks' Campaign." *CBC News*. February 27. www.cbc.ca/news/politics/story/2012/02/27/pol-liberals-vikileaks.html.

Geist, M. 2012. "How to Fix Canada's Online Surveillance Bill: A 12 Step to-Do List." Michael Geist blog. www.michaelgeist.ca/2012/02/12-fixes-on-C-30.

Hayes, A. S. 2008. *Press Critics Are the Fifth Estate: Media Watchdogs in America*. Portsmouth, NH: Greenwood Publishing Group.

House of Commons of Canada (HCC). 2012. Investigating and Preventing Criminal Electronic Communications Act (First Reading). First Session, Forty-First Parliament, Parliament of Canada. www.parl.gc.ca/HousePublications/Publication.aspx?DocId=5380965.

Kobrin, S. J. 1998. "The MAI and the Clash of Globalizations." *Foreign Policy*, no. 112: 97–109.

Little, A. D. 1924. "The Fifth Estate." *Industrial and Engineering Chemistry* 16, no. 11: 1105–1110.

Neumayer, E. 1999. "Multilateral Agreement on Investment: Lessons for the WTO from the Failed OECD-Negotiations." *Wirtschaftspolitische Blätter* 46, no. 6: 618–628.

Obar, J. A. 2010. "Democracy or Technocracy? An Analysis of Public and Expert Participation in FCC Policymaking." PhD diss., Pennsylvania State University.

———. 2014. "Canadian Advocacy 2.0: An Analysis of Social Media Adoption and Perceived Affordances by Advocacy Groups Looking to Advance Activism in Canada." *Canadian Journal of Communication* 39, no. 2: 211–233.

Obar, J. A., and A. Argast. 2014. "Adding Slacktivism to the Activist's Toolkit: Advocacy Group Perceptions of the Benefits and Drawbacks of Slacktivism." January 29. Working paper. http://ssrn.com/abstract=2387805.

Obar, J. A., and Schejter, A. M. 2010. "Inclusion or Illusion? An Analysis of the FCC's Public Hearings on Media Ownership 2006–2007." *Journal of Broadcasting and Electronic Media* 54, no. 2: 212–227.

Obar, J. A., P. Zube, and C. Lampe. 2012. "Advocacy 2.0: An Analysis of How Advocacy Groups in the United States Perceive and Use Social Media as Tools for Facilitating Civic Engagement and Collective Action." *Journal of Information Policy*, no. 2: 1–25.

OECD. 2001. "Citizens as Partners: Information, Consultation and Public Participation in Policy-Making." www.oecd-ilibrary.org/governance/citizens-as -partners_9789264195561-en.

Payton, L. 2013. "Government Killing Online Surveillance Bill: Justice Minister Rob Nicholson Says Controversial Bill C-30 Won't Go Ahead." *CBC News*. February 11. www.cbc.ca/m/touch/news/story/1.1336384.

Pollard, A. F. 1920. *The Evolution of Parliament*. London: Longmans, Green and Company.

Reilly, I. 2010. "Satirical Fake News and the Politics of the Fifth Estate." PhD diss., University of Guelph.

Schejter, A. M., and J. A. Obar. 2009. "Tell It Not in Harrisburg, Publish It Not in the Streets of Tampa: Media Ownership, the Public Interest and Local Television News." *Journalism Studies* 10, no. 5: 577–593.

Shirky, Clay. 2008. *Here Comes Everybody: The Power of Organizing without Organizations*. New York: Penguin.

Walker, J. L. 1966. "A Critique of the Elitist Theory of Democracy." *American Political Science Review* 60, no. 2: 285–295.

CHAPTER FOUR

WikiLeaks and "Indirect" Media Reform

CHRISTIAN CHRISTENSEN, *Stockholm University, Sweden*

MEDIA REFORM STRATEGY

The relatively short history of WikiLeaks (2006–2014) has offered numerous examples of how issues central to media reform (political economy, regulation, ownership, law) emerge from their leaks and activities. As I note, WikiLeaks is not a media reform group nor, for the most part, have WikiLeaks activities targeted media organizations or media-related policy. In terms of reform strategy, WikiLeaks—as perhaps the only whistleblowing site known across a broad cross-section of the global population—finds itself in an unusual, somewhat isolated position. My chapter on WikiLeaks is unlike the others in this book as it frames debates on media reform as emerging as a by-product of the actions of WikiLeaks, rather than addressing specific strategies for reform. I would argue that, given the nature of its activities, coalition-building around a single organization such as WikiLeaks would be, at best, difficult. However, other whistleblowing sites such as GlobalLeaks.org are starting to emerge, and larger-scale freedom of speech organizations such as the Electronic Frontier Foundation (EFF), the American Civil Liberties Union (ACLU), and Article 19 are evolving into a clear cluster of "digital defenders" advocating for the rights of whistle-blowers and users alike, and, thus, offer a valuable node for advocating media reform in relation to whistleblowing, transparency, and journalism.

In a special panel at the 2011 National Conference for Media Reform held in Boston, journalists Glenn Greenwald and Greg Mitchell, together with writer

Micah Sifry, were asked to consider the relationship between media reform and the rise of WikiLeaks. The discussion covered a wide range of issues, but what became clear was that WikiLeaks was considered both an important participant in a new, vibrant media ecology *and* a symptom of a deep malaise in US journalism. At the core of this malaise was a failure on the part of reporters and media organizations to engage in investigative journalism that questioned the fundamental powers of corporations and the state. WikiLeaks, the participants argued, had beaten hallowed organizations such as the New York Times and Washington Post at their own game, resulting in hostility toward the whistleblowing site on the part of the US press. Beyond competition, a more troubling explanation for the adversarial relationship between WikiLeaks and the press was that mainstream media—themselves large, profit-driven entities—were seen to be as much a part of the so-called establishment as the state and corporate entities they purported to oversee on the part of the general citizenry. As Emily Bell (2010, 5) put it: "this is the first real battleground between the political establishment and the open web . . . [which] forces journalists and news organisations to demonstrate to what extent they are now part of an establishment it is their duty to report."

There can be little doubt that, since 2010, the material leaked by WikiLeaks, the attention surrounding these leaks, the legal actions taken against Chelsea Manning and the threatened legal action against WikiLeaks founder Julian Assange (Pilkington 2011) have all contributed to a broad popular debate on the functioning of contemporary mainstream media systems (see Brevini et al. 2013). In an opinion piece on the relationship between WikiLeaks, Anonymous, and journalism, I posited the following regarding their rise to prominence:

> WikiLeaks and Anonymous are an expression, a crystallisation of a dissatisfaction with the extent to which primarily commercial, but also public service, news organisations have willingly absorbed elite discourses in relation to socio-economic, legal and military issues. Stories which expose political or corporate misconduct should not to be seen as the antithesis to these discourses. Often, such instances are simply defined as "the exceptions that prove the rule" while the greater meta-story of capitalism and western power remain unchallenged. (Christensen 2013)

The core of my argument was that groups such as WikiLeaks can be viewed through several lenses. The most logical perspective is to consider the concrete actions of these groups in the form of leaks and online activism, and the concrete impact(s) of these actions. In addition, however, I felt it was important to consider both what these two groups symbolized, why they emerged, and how they challenge and potentially reshape fundamental relationships: between citizens and the state (impacted by providing access to sensitive intelligence previously hidden from view); between citizens and the media (impacted by exposure of the shortcomings of an uncritical commercial media system); and, between media and governments (impacted by challenging the mantle of "watchdog" proudly trumpeted by major mainstream news outlets). All three of these relationships are fundamental within any discussion of media reform.

WIKILEAKS: FROM JOURNALISM TO LAW

Within this landscape of shifting relationships, it is my contention that an impact upon the broader media reform debate is an important by-product of the actions taken by groups such as WikiLeaks, as well as the legal, political, and economic consequences of those actions. As I have already stated, WikiLeaks is not a media reform activist organization (although it is not immune to providing harsh critiques of media systems or policy), and so this chapter does not provide an examination of a particular methodology for impacting reform. Rather, I would like to consider the place of WikiLeaks within media reform, and how the actions they have taken part in, and the responses to those actions in the media and politics, have important implications for these debates.

If we consider groups such as WikiLeaks to be players in media reform, then the first issue to be clarified is the role they fulfill. Michel Foucault made the following observation in a debate in 1974 with Noam Chomsky: "*The real political task in a society such as ours is to criticise the workings of institutions that appear to be both neutral and independent, to criticise and attack them in such a manner that the political violence that has always exercised itself obscurely through them will be unmasked, so that one can fight against them*" (Foucault 1974, 171).

Via their leaks and online activism, WikiLeaks and Anonymous have pitched, as their central *raisons d'etre*, the "unmasking" of the workings of power and, in particular, power structures—police, military, politics, corporations—to the general public as forces for, if not perhaps good, at least order and stability. What is interesting for the purposes of this chapter is that the efforts of WikiLeaks to expose the raw mechanics of power had, and continues to have, the crucial side-effect of throwing down the gauntlet in front of mainstream media outlets. I have also written that the rise of these WikiLeaks can, at least in part, be linked to what I called the failure of post-9/11 status quo journalism in the United States and Europe, leading to a desire on the part of many citizens for deeper analysis and critique of power. Thus, groups like WikiLeaks have filled a gap: "WikiLeaks and Anonymous are an expression, a crystallisation of a dissatisfaction with the extent to which primarily commercial, but also public service, news organisations have willingly absorbed elite discourses in relation to socio-economic, legal and military issues" (Christensen 2013).

With an organization such as WikiLeaks, it is important to distinguish between its leaks as a tool for instigating transparency and reform (military, political, corporate or otherwise), and media reform as a potential by-product of their acts of leaking or online activism. For the most part, WikiLeaks (with a few minor exceptions) has not placed media organizations or media policy in their activist cross-hairs, yet their activism has exposed a number of issues central to global media reform movements: an uncomfortably close relationship between governments/states and the media; the inadequacy of media-related policies in relation to protection of the public interest; an imbalance in mainstream media coverage between those in established positions of political-economic power and, for example, activist groups; and, importantly, the potentially antidemocratic impact of re-regulation, the commercialization of media, and related threats to public service systems.

While WikiLeaks had earned a reputation as a whistle-blowing site back in 2006, it wasn't until the release of the so-called Iraq and Afghanistan War Documents in 2010—and the high-impact "Collateral Murder" video—that the organization really broke through in the public, political, and journalistic imagination (WikiLeaks 2010a, 2010b, 2010c). While previous leaks on pharmaceutical companies, Swiss banks, and climate change negotiations were certainly important, they had a relatively minimal impact on the media. Revelations regarding US military activities in Afghanistan and Iraq, as well as the release of a massive number of US diplomatic cables (WikiLeaks 2010d), on the other hand, raised questions as to why it was WikiLeaks, and not news organizations such as the *New York Times* or *Washington Post,* that had obtained the material. Interestingly, at the trial of Chelsea Manning in 2013, Manning stated that she had initially tried to leak the material to both the *New York Times* and *Washington Post* (to no avail) before turning to WikiLeaks. For Benkler (2012, 330), organizations such as WikiLeaks force us to consider:

> the challenges that a radically decentralized global networked public sphere poses for those systems of control that developed in the second half of the twentieth century to tame the fourth estate, to make the press not only "free," but also "responsible." Doing so allows us to understand that the threat represented by Wikileaks was not any single cable, but the fraying of the relatively loyal and safe relationship between the United States Government and its watchdog. Nothing captures that threat more ironically than the spectacle of Judith Miller, the disgraced *New York Times* reporter who yoked that newspaper's credibility to the Bush Administration's propaganda campaign regarding Iraq's weapons of mass destruction in the run-up to the Iraq War, using Fox News as a platform to criticize Julian Assange for neglecting the journalist's duty of checking his sources and instead providing raw cables to the public.

McNair (2012, 77) notes that the "loyal and safe relationship" between government and media described by Benkler resulted in a flow of information controlled by political elites. This flow, however, was now being subverted by organizations such as WikiLeaks, with the possibility of reshaping the news agenda.

The work of Benkler (2012) on the legal implications of the WikiLeaks case for freedom of the press and freedom of speech in the United States is complemented by a significant volume of other legal scholarship on the issue, indicating the myriad media reform issues at the heart of the WikiLeaks case (e.g., Cannon 2013; Davidson 2011; Fenster 2012; Peters 2011; Rothe and Steinmetz 2013; Wells 2012). For Benkler, "the networked Fourth Estate" is one that will likely work not as a replacement for a strained preexisting media system, but rather one that will work *in conjunction* with that system. This interaction, Benkler (2012, 396) writes, will be "both constructive and destructive," but one that is necessary given that, "the traditional, managerial-professional sources of responsibility in a free press function imperfectly under present market conditions."

Beyond journalistic failure, scholars such as Tambini (2013) and Lynch (2010, 2013) have investigated the implications of the WikiLeaks releases for future regulation and censorship of related whistle-blowing sites in the United Kingdom and the United States. For Tambini (2013, 284), the huge volume of material released by WikiLeaks in the US diplomatic cable leak, "marked a watershed moment in the development of the legal and ethical frameworks for news publishing." The author notes that in order to ensure that future whistle-blowing organizations are afforded protection under the law, their ability to "self-regulate" and to "develop, codify, and implement ethical principles to govern their publication decisions" will be paramount. A failure to do so will result in the impression of an irresponsible information free-for-all, thus providing states with clear sufficient cause to pass restrictive laws geared toward censoring or even closing down their operations: "if rather than seeking protection through the law of free expression WikiLeaks continues to evade the law, this could result in a censorship backlash. . . . If mainstream media continue to conspire in the vilification of WikiLeaks, releases may provide a justification that undermines the contribution of the Internet to free speech" (285–286).

If, however, these groups succeed in establishing a transparent, ethical publication code, "there is a small chance they may be permitted to operate in a new extra-jurisdictional space of public interest journalism" (285). Lynch (2013, 321) makes a similar observation about WikiLeaks: namely that their impact on media law in the United States prior to what she calls the "megaleak" period was "salutary". After Cablegate, however, concern began to grow regarding the implications of a WikiLeaks prosecution in the United States on the mainstream press, and "the political backlash against WikiLeaks immediately proved disadvantageous to citizen journalists, as pending shield legislation in the United States that might have indemnified them from legal reprisal was sharply narrowed after the release of the Afghan War Logs" (321).

In the second half of the chapter I would like to consider the various concrete ways in which the actions of WikiLeaks have impacted issues of media reform. Although not a traditional media reform organization, the categorization of media strategies utilized by activists in relation to mainstream political cultures posited by Rucht (2004, 430) is a useful framework for considering the work of WikiLeaks. Based on a historical analysis of the media strategies of activist movements over several decades, Rucht proposed what he called the "Quadruple A" model: (1) Attack, (2) Adaptation, (3) Alternatives, and (4) Abstention. I would like to adjust the order of Rucht's strategies somewhat so that they better fit with the WikiLeaks chronology.

ALTERNATIVES TO THE MAINSTREAM

It is fair to say that the early years of WikiLeaks—from the first leak in December 2006 (a list of potential political assassinations in Somalia) up to the publication of the Iraq and Afghan war documents in early 2010—were marked by a degree of media fascination with the organization itself, but a relative paucity of articles on the content of the leaks. This is not to say that the stories leaked by the organization were insignificant. From 2008 to 2010, for example, WikiLeaks released documents showing potential money-laundering in the Cayman Islands by the major Swiss bank Julius Baer (WikiLeaks 2008a); the influence of large corporate pharmaceutical companies on the policy-making of the World Health Organization—including forcing developing nations to maintain high prices for drugs at the expense of poorer patients dying (WikiLeaks 2008b); the illegal dumping of toxic waste by corporations off of the Ivory Coast (WikiLeaks 2009b); the membership lists of the far-right British National Party (WikiLeaks 2009a); and, the use of intimidation tactics by the US government during the UN Climate Change Conference (COP15) in Copenhagen (Carrington 2010).

During these early years, WikiLeaks was somewhat of an anomaly. Journalists and politicians did not know what to make of the organization and, as already mentioned, media coverage was sparse. A strong argument could be made, however, that a number of the early leaks from this period—in particular those on WHO price-fixing and the UN Climate Change Conference—were stories with implications for the health and safety of billions around the globe. In response to the WHO documents, James Love (2011), the director of Knowledge Economy International, wrote the following:

> After reading these cables, it is difficult to stomach the defenses of US secrecy. Forcing developing countries to raise the price of drugs has predictable and well known consequences—it kills people, and increases suffering.

Many people could care less—including reporters and editors of newspapers. How much of this ends up in the *Washington Post*, the *New York Times* or the *Guardian* these days? But others who do care now have more access to information, and more credibility in their criticisms of government policy, because of the disclosures of the cables.

From a media reform perspective, and working from Love's arguments in relation to the WHO revelations, two issues emerge. First, that the stories, once leaked by WikiLeaks, did not receive significant media attention speaks to the issues raised by Benkler (and a myriad of critical political economists over the years) regarding the failure of the establishment corporate media. The second issue, however, runs deeper than just a lack of coverage of the leaks once they emerged, but has to do with the fact that the stories leaked by WikiLeaks about abuse of corporate and political power (particularly in relation to the subjugation of developing nations) are not addressed *regardless* of WikiLeaks. In other words, these early leaks illustrated a double failure on the part of mainstream journalism: that, historically, they have not covered the issues raised by WikiLeaks despite the massive international implications; and, second, their relative silence on the issues even after the material was released. It is perhaps only with the passage of time that perspective can be gained on both the relative importance of the leaks provided by WikiLeaks between 2006 and 2010—when, I would argue, WikiLeaks was acting as an alternative to the mainstream media—and the failure of mainstream media to engage with that material.

ADAPTATION TO THE MAINSTREAM

WikiLeaks witnessed a significant shift in both notoriety and operating procedures at the start of 2010. From early 2010 through 2011 the organization leaked a massive volume of material including the Iraq War documents, the Afghan War documents, the "Collateral Murder" video and the diplomatic cables. The impact of these leaks was greatly enhanced by the fact that WikiLeaks made the decision to collaborate with a number of mainstream media outlets for the collation, analysis, presentation, and release of the material (primarily the war documents). The strategy of releasing leaked documents to selected media outlets—the *Guardian*, *New York Times*, *Der Spiegel*, and *El Pais*—one month prior to a predetermined (and uniform) release date—was an indication that WikiLeaks was interested in maximizing the impact of these leaks. Some six months before these large releases, WikiLeaks founder Julian Assange outlined his thinking regarding the value of information: "You'd think the bigger and more important the document is, the more likely it will be reported on but that's absolutely not true. It's about supply and demand. Zero supply equals high demand, it has value. As soon as we release the material, the supply goes to infinity, so the perceived value goes to zero" (cited in Christensen 2011a, 2).

The leaked material proved to be a boon for both WikiLeaks and their main-stream media partners. As I have argued, given the political leanings of WikiLeaks, the collaboration may have appeared to be surprising, but it was one Assange was willing to make in exchange for exposure, particularly when taking into account the likely unwillingness of mainstream media outlets to engage with the material were it just to be dumped onto the WikiLeaks website for all to see:

> While WikiLeaks material is tailor-made for the critical eye of the alternative press, the political economy of most capitalist media systems means that these alternative outlets, and their contents, are de facto marginalized. While a deal with mainstream newspapers could be seen as a Faustian bargain for WikiLeaks, it was a deal that Assange was willing to make, probably because it would enable access to a sizeable chunk of citizens not part of the core of WikiLeaks' lovers (who follow the organization no matter what) or haters (who detest WikiLeaks no matter what). (Christensen 2011b)

The collaboration between these well-established international media outlets and WikiLeaks, the production of the influential "Collateral Murder" video (Christensen 2014), and the increased editorial role played by WikiLeaks and Assange in terms of material accumulation and release, also raised a fundamen-tal issue: the extent to which Assange could be seen as a journalist, and WikiLeaks a journalistic organization. Given the content of the WikiLeaks releases, the journalistic designation was central, as (at least in the US context) it could theoretically afford Assange a degree of protection under US law. The 2013 trial of Chelsea Manning brought this issue to the fore, with the charge of "aiding the enemy" levelled against the soldier who had leaked the material to WikiLeaks. The implications for journalism in the United States of a guilty verdict on this particular charge were enormous—potentially criminalizing any media outlet that published leaked material deemed to damage the national interest—yet the US media afforded the trial little attention. In the end, Man-ning was found not guilty' on the "aiding the enemy" charge, but still received a thirty-five-year prison sentence.

In addition to the legal implications of designating Assange a journalist, the WikiLeaks–mainstream media collaboration on the release of what Lynch (2013) calls "megaleaks" created a degree of tension among professional jour-nalists who saw Assange and WikiLeaks as providers of raw data, rather than journalists. In his analysis of how mainstream newspapers framed WikiLeaks in relation to professional journalistic boundaries, Coddington (2012, 378) found that "institutionality emerged as a key professional journalistic value . . . while source-based reporting routines and objectivity were largely bound within those contexts." In other words, WikiLeaks was seen as lacking the institu-tional, organizational and professional values and structures required of a true journalistic organization.

The reform issues raised during the WikiLeaks period of adaptation are myriad, but the question of "who is a journalist?" is undoubtedly fundamental (e.g., Handley and Rutigliano 2012). The attempt by large mainstream media organizations to downplay the role of WikiLeaks and Assange, and to construct professional and organizational walls between their organizations and WikiLeaks speaks not only to the commercial value of professional "boundary maintenance" discussed by Gieryn (1983), but also to the cultural capital associated with news work. Thus, while the material provided by WikiLeaks was important enough for major international news outlets to exploit over a matter of years, the whistle-blowing organization was not important enough to designate as a journalistic organization—even if the implications for such a barrier is the imprisonment of both Assange and Manning.

There is one final media reform issue in relation to the WikiLeaks–mainstream media collaboration that is important to note: the political-economic benefits afforded these commercial outlets as a result of the cooperation:

> It is clear that WikiLeaks have performed a crucial democratic function via their revelations, and it is also clear that this function was made possible— with the release of the Afghanistan and Iraq war documents—via a calculated collaboration with mainstream media. . . . [T]hrough this collaboration WikiLeaks sent a clear message to the broader public: whether we like it or not, large news outlets are still important in contemporary society. It is equally important to note that this collaboration was done with the clear side-effects of boosting the sales figures, advertising revenues and socio-cultural capital of these already powerful, Western media organizations. (Christensen 2011a, 2–3)

This is not to say that the message to the general public about the importance of mainstream outlets on the part of WikiLeaks was overt, but rather that it was a side-effect of their collaboration. WikiLeaks clearly challenged mainstream media power, yet, when we look at the issue of media reform using the lens of the networked Fourth Estate, we must also consider and analyze the extent to which such collaboration could end up boosting the influence and capital of already powerful corporate media actors.

ABSTENTION FROM AND ATTACK OF THE MAINSTREAM

I have merged two of Rucht's categories (abstention and attack) because I feel that the combination best explains the most recent phase of the WikiLeaks relationship with the international mainstream media in the post-Afghanistan/Iraq war dossier leak era. As noted in the previous section, the relationship between WikiLeaks and the newspapers with which they had collaborated became strained to the point of breaking (most famously with the *New York Times* and *Guardian*). Several pieces were written after 2011 in which harsh opinions about Julian Assange, and the extent to which the material leaked to WikiLeaks had, in fact, led to any positive change, were proffered:

> Former *New York Times* editor Bill Keller offered a particularly acid post-mortem when he wrote that "not all that much" had changed after the WikiLeaks releases, and that the leaks, "did not herald, as the documentarians yearn to believe, some new digital age of transparency. In fact, if there is a larger point, it is quite the contrary." In other words, according to Keller, WikiLeaks is actually responsible for the more aggressive stance taken by the US government in relation to security and surveillance. (Christensen 2012)

In addition, Bill Keller also (famously) described Assange as a news source and not a professional collaborator. In an interview for *Rolling Stone* magazine (Hastings 2012), Assange gave his interpretation of Keller's dismissal of himself and WikiLeaks:

> Keller was trying to save his own skin from the espionage investigation in two ways. First, on a legal technicality, by claiming that there was no collaboration, only a passive relationship between journalist and source. And second, by distancing themselves from us by attacking me personally, using all the standard tabloid character-assassination attacks. . . . Keller said, "Julian Assange may or may not be a journalist, but he's not my kind of journalist." My immediate reaction is, "Thank God I'm not Bill Keller's type of journalist."

The formal working relationship between WikiLeaks and major media outlets came to an end and from 2012 on, WikiLeaks returned to the position they occupied some three to five years earlier. In this case, however, I would define this latter era as one of *abstention* rather than *alternative*, as the organization had by now achieved a significant level of fame and influence. Interestingly, both of these newspapers, as well as many media outlets around the globe, continued to make use of material made available via WikiLeaks. During this period, WikiLeaks used social media (primarily Twitter, where the organization has over one million followers) to engage in debates with, and attacks upon, mainstream outlets the organization felt was misrepresenting their goals, actions or motives.

The media reform implications of the last few years of WikiLeaks, as well as the strained relationship between the mainstream media and the organization, cannot be assessed without also considering two factors: the crushing economic embargo placed upon WikiLeaks by Visa, MasterCard, and PayPal following pressure from the US government; and the threat of the indictment of WikiLeaks founder Assange by the US government. While the antagonistic relationship between WikiLeaks and their former collaborators has dampened the possibilities for productive cross-over journalism, the aggressive position taken by the US government in relation to WikiLeaks—made concrete by the thirty-five-year sentence handed down to Chelsea Manning—has sent a clear chilling message to whistle-blowers and whistle-blowing organizations. In addition, the position sends a message to media organizations considering collaboration with such individuals and organizations, as highlighted by the 2013 raid on the *Guardian* offices by British authorities to locate computer harddrives containing material obtained from Edward Snowden (Borger 2013). In short, the period of abstention and attack by WikiLeaks is marked by an increase in aggression from states over perceived threats to national security; a process that includes tactics such as financial strangulation and the threat of arrest. The result has been a marked decline in WikiLeaks activity over this period.

CONCLUSION

In this chapter I have attempted to outline how the relatively short history of WikiLeaks (2006–2014) has offered numerous examples of how issues central to media reform (political economy, regulation, ownership, law) emerge from their leaks and activities. As I noted at the start of this chapter, WikiLeaks is not a media reform group, nor, for the most part, have WikiLeaks leaks or activities targeted media organizations or media-related policy. In this conclusion, I would like to summarize what I consider to be the key media reform issues that have emerged via WikiLeaks, and to do so by considering the three stages on WikiLeaks relations discussed in the previous section.

First, the early WikiLeaks period as an "alternative" to mainstream output highlights not only the lack of original coverage of corporate and state malfeasance, but also an unwillingness of major media corporations to engage with important material even when it is released into the public domain. The important material obtained and released by WikiLeaks on, for example, climate change and global health points to the need for formalized support or protection for outlets such as WikiLeaks which expose criminal or unethical activities on the part of states and corporations—activities that are clearly in the public interest.

Second, the period of WikiLeaks–mainstream media collaboration around 2010, and the subsequent falling out, raised a number of key reform issues. The issues of what constitutes journalism, and who is a journalist, are obviously paramount as they connect to issues of law and legal protection and are thus in need of clarification. These issues also relate to the political economy of the media: Benkler was likely correct in pointing out that the networked Fourth Estate will be an amalgam of new media actors (such as WikiLeaks) and established media organizations. Yet Bill Keller's assertion that WikiLeaks was a source and not a collaborator has both legal and economic implications, as newspapers reap both profit and prestige from the free material provided to them by WikiLeaks. As the networked Fourth Estate evolves, a central question will be the extent to which (or if) such partnerships needs to be both funded and regulated so as to protect those taking the greatest risk from economic exploitation.

Finally, recent events connected to WikiLeaks—the Manning conviction, the economic embargo, and potential criminal charges against Assange—raise the specter of the use of the law by the state as a blunt instrument to intimidate those who challenge government or corporate power. At the time of writing, Edward Snowden sits in Russia, unable to return to the United States without being taken into custody and charged. Media reform is also about the protection of those who in good faith expose crimes or unethical behavior, as their information can constitute key components of media reports and news content. When news organizations that obtain information deemed to have been taken illegally—or in violation of the "national interest"—are also treated in a criminal fashion, it is the start of a slippery slope. What the WikiLeaks case has shown us is that it is possible to expose illegal state or corporate activity, and to distribute that information to a broad public. What remains to be seen, however, is the extent to which mainstream media organizations, corporations, and the state consider such an organization to be a threat to their professional, financial, or political interests, and, thus, not worthy of protection. The protection of whistle-blowers thus needs to be a central focus for media reform movements in the coming years.

REFERENCES

Bell, E. 2010. "How WikiLeaks Has Woken Up Journalism." Blog, December 7. http://emilybellwether.wordpress.com/2010/12/07/how-WikiLeaks-has-woken-up-journalism/.

Benkler, Y. 2012. "A Free Irresponsible Press: WikiLeaks and the Battle over the Soul of the Networked Fourth Estate." *Harvard Civil Rights-Civil Liberties Law Review* 47, no. 1: 311–397.

Borger, J. 2013. "NSA Files: Why the Guardian in London Destroyed Hard Drives of Leaked Files." *Guardian*, August 20. www.theguardian.com/world/2013/aug/20/nsa-snowden-files-drives-destroyed-london.

Brevini, B., A. Hintz, and P. McCurdy, eds. 2013. *Beyond WikiLeaks: Implications for the Future of Communications, Journalism and Society.* London: Palgrave Macmillan.

Cannon, S. C. 2013. "Terrorizing WikiLeaks: Why the Embargo Against WikiLeaks Will Fail." *Journal on Telecommunications and High Technology Law*, no. 11: 305–325.

Carrington, D. 2010. "US Goes to Basics Over Copenhagen Accord Tactics." *Guardian*, December 3. www.theguardian.com/environment/2010/dec/03/us-basics-copenhagen-accord-tactics?guni=Article:in%20body%20link.

Christensen, C. 2010. "WikiLeaks: Three Digital Myths." *Le Monde diplomatique*, August 1. http://mondediplo.com/blogs/three-digital-myths.

———. 2011a. "WikiLeaks and Celebrating the Power of the Mainstream Media: A Response to Christian Fuchs." *Global Media Journal—Australia* 5, no. 1. www.hca.uws.edu.au/gmjau/archive/v5_2011_2/rejoinder1.html.

———. 2011b. "WikiLeaks: Losing Suburbia." *Le Monde diplomatique*, September 1. http://mondediplo.com/blogs/wikileaks-losing-suburbia.

———. 2012. "We have Taken Our Eyes Off the Prize." *British Journalism Review* 23, no. 3: 48–53.

———. 2013. "WikiLeaks and Anonymous Respond to Status Quo Journalism." *Al Jazeera*, July 29. www.aljazeera.com/indepth/opinion/2013/07/20137289 1029632277.html.

———. 2014. "WikiLeaks, Collateral Murder and the After-Life of Activist Imagery." *International Journal of Communication*, no. 8: 2593–2602.

Coddington, M. 2012. "Defending a Paradigm by Patrolling a Boundary: Two Global Newspapers' Approach to WikiLeaks." *Journalism and Mass Communication Quarterly* 89, no. 3: 377–396.

Davidson, S. 2011. "Leaks, Leakers, and Journalists: Adding Historical Context to the Age of WikiLeaks." *Hastings Communication and Entertainment Law Journal*, no. 34: 27–91.

Fenster, M. 2012. "Disclosure's Effects: Wikileaks and Transparency." *Iowa Law Review* 97, no. 3: 753–807.

Foucault, M. 1974. "Human Nature: Justice versus Power." In *Reflexive Water: The Basic Concerns of Mankind*, edited A. Ayer and F. Elders, 133–198. London: Souvenir Press.

Gieryn, T. F. 1983. "Boundary-Work and the Demarcation of Science from Non-Science: Strains and Interests in Professional Ideologies of Scientists." *American Sociological Review* 48, no. 6: 781–795.

Handley, R. L., and L. Rutigliano. 2012. "Journalistic Field Wars: Defending and Attacking the National Narrative in a Diversifying Journalistic Field." *Media, Culture and Society* 34, no. 6: 744–760.

Hastings, M. 2012. "Julian Assange: The Rolling Stone Interview." *Rolling Stone*, February 2. www.rollingstone.com/politics/news/julian-assange-the-rolling -stone-interview–20120118.

Love, J. 2011. "Looking at WikiLeaks Cables on Pharmaceutical Drugs and Trade Pressures." *Le Monde Diplomatique*, September. http://mondediplo.com/ openpage/looking-at-wikileaks-cables-on-pharmaceutical.

Lynch, L. 2010. "'We're Going to Crack the World Open': WikiLeaks and the future of Investigative Reporting." *Journalism Practice* 4, no. 3: 309–318.

———. 2013. "WikiLeaks After Megaleaks: The Organization's Impact on Journalism and Journalism Studies." *Digital Journalism* 1, no. 3: 314–334.

McNair, B. 2012, August. WikiLeaks, Journalism And The Consequences Of Chaos. *Media International Australia*, 144, 77–86.

Peters, J. 2011. "Wikileaks Would Not Qualify to Claim Federal Reporter's Privilege in Any Form." *Federal Communications Law Journal* 63, no. 3: 667–696.

Pilkington, E. 2011. "WikiLeaks: US Opens Grand Jury Hearing." *Guardian*, May 11. www.theguardian.com/media/2011/may/11/us-opens-wikileaks -grand-jury-hearing.

———. "Manning Says He First Tried to Leak to *Washington Post* and *New York Times*." *Guardian*, February 28. www.theguardian.com/world/2013/feb/28/ manning-washington-post-new-york-times.

Rothe, D. L., and K. F. Steinmetz. 2013. "The Case of Bradley Manning: State Victimization, Realpolitik and WikiLeaks." *Contemporary Justice Review* 16, no. 2: 280–292.

Rucht, D. 2004. "The Quadruple 'A': Media Strategies of Protest Movements Since the 1960's." In *Cyperprotest: New Media, Citizens and Social Movements*, edited by W. van de Donk, B. D. Loader, P. G. Nixon, and D. Rucht, 29–56. London: Routledge.

Tambini, D. 2013. "Responsible Journalism? WikiLeaks, the Diplomatic Cables and Freedom of Expression in a UK Context." *Policy and Internet* 5, no. 3: 270–288.

Wells, C. E. 2012. "Contextualizing Disclosure's Effects: WikiLeaks, Balancing, and the First Amendment." *Iowa Law Review Bulletin* 97: 51–63.

WikiLeaks. 2008a. "Bank Julius Baer: The Baer Essentials Part 1." March 1. https://wikileaks.org/wiki/Bank_Julius_Baer#Swiss_bank_clients_exposed _by_Cayman_leak.

———. 2008b. "World Health Organization Avian Flu Draft 2008." August 14 https://wikileaks.org/wiki/World_Health_Organization_Avian_Flu_draft _2008.

———. 2009b. "Minton Report: Trafigura Toxic Dumping Along the Ivory Coast Broke EU Regulations." September 14. https://wikileaks.org/wiki/ Minton_report:_Trafigura_toxic_dumping_along_the_Ivory_Coast_broke _EU_regulations,_14_Sep_2006.

———. 2009a. "British National Party Membership List and Other Information, 15 Apr 2009." October 20. https://wikileaks.org/wiki/British_National _Party_membership_list_and_other_information,_15_Apr_2009.

———. 2010a. "Baghdad War Diary." October 22. www.wikileaks.org/irq/.

———. 2010b. "Collateral Murder." April 5. https://wikileaks.org/wiki/Collateral _Murder,_5_Apr_2010.

———. 2010c. "Kabul War Diary." July 25. www.wikileaks.org/afg/.

———. 2010d. "Secret US Embassy Cables." November 28. https://wikileaks .org/cablegate.html.

CHAPTER FIVE

Mobilizing for Net Rights

The IRPC Charter of Human Rights and Principles for the Internet

M. I. FRANKLIN, *Goldsmiths, University of London, United Kingdom*

MEDIA REFORM STRATEGY

The IRPC Charter of Human Rights and Principles for the Internet (IRPC Charter) is the outcome of a cross-sector collaboration between civil society organizations, human rights experts, scholars, and representatives from the intergovernmental and private sector to provide an authoritative, human rights–based legal framework for decisions around Internet design, access, and use. Essential to this project's success was an early decision to anchor the work in precursor civil society initiatives and international human rights law and norms. The coalition-building strategy that underpins the IRPC Charter brought a range of actors together, face-to-face and online, in the spirit of web-enabled collabowriting. *Since its launch in 2010/11 the IRPC Charter has grown steadily in stature, playing a formative role in key policy outcomes, landmark UN resolutions, and increasing awareness of the interrelationship between human rights issues and Internet policy-making at the national and global level (Hivos 2014, INDH 2013, Council of Europe 2014, NETmundial 2014, UNHRC 2012, Cultura Digital e Democracia 2014, Green Party 2014). A commitment to forging alliances and cooperation across diverse sectors in order to ensure human-centered Internet policy-making has been a key factor in the success of the IRPC Charter to articulate a viable framework for rights-based agenda-setting in a policy-making terrain dominated by powerful techno-commercial interests and competing political agendas.[1]*

The IRPC Charter emerged under the auspices of the Internet Rights and Principles Coalition (IRPC), one of the "dynamic coalitions" that constitutes the UN Internet Governance Forum (IGF), the 2006 successor to the ITU-hosted World Summit on the Information Society (WSIS). This chapter reconstructs key moments in the IRPC Charter process and points to the wider context that frames where and how Internet policy-making and agenda setting takes place. These are institutional and online spaces populated by official, and self-appointed bodies of experts, intergovernmental organizations, business interests, and emerging multilateral institutions where civil society participation is now a given rather than an aspiration. In this case the UN's brokering of "multistakeholder" consultations on Internet governance is one such space. Along with a brief reconstruction of the main points along the IRPC Charter's timeline the chapter looks at how the IRPC itself got going, how we completed the charter, promoted it, and then kept the project relevant and responsive to new arrivals and major developments in Internet governance discourses and the wider power relations in which this sort of coalition-building work has evolved. The chapter aims to show that having a concrete object to work with, in this case a legally recognizable human rights framework for the Internet, is just the start. Awareness, recognition, and implementation are different moments in the timeline of this long-term, coalition-building project. This longitudinal approach differs in principle and practice from a single-issue campaign even as it deploys similar sorts of established and newer media strategies to get off the ground and then stay not only viable but also visible.

WHERE, WHAT, HOW, AND WHO

The IRPC Charter as a project went hand in hand with the establishment of the IRPC in 2009 (IRPC 2013, 6; Franklin 2013a, 154 passim). As a rights-based strategy within the Internet Governance Forum the IRPC Charter project tapped into ongoing civil society advocacy on how media freedoms (of the press, the safety of journalists and bloggers) are also at stake online as well as individual rights and fundamental freedoms such as freedom of expression, or the right to privacy, education, and assembly. The intent of the early years (2009–2011) was to develop a document, hence the term *charter*, that could reframe disparate agendas and so revision a range of legal, technical, and political projects for the Internet's future development (e.g., social justice, sustainable development, and environmental considerations) into a more comprehensive, coherent, and recognizable legal framework anchored in international human rights law.

The underlying rationale for this ambitious approach was twofold. First, that existing international human rights law, although arguably sufficient to cover most online scenarios (debates on this continue), did need more explicit rendering for online situations, particularly in the relative absence of case law (see Benedek and Kettemann 2014). Second, that the historical and institutional disconnect between Internet policy-making constituencies, such as the IGF or ICANN,[2] and international human rights institutions and advocacy (e.g., at the UN Human Rights Council, Amnesty International) needed bridging. These aims underpinned the decision to make the IRPC Charter project part of a collaborative, writing, and thinking process in order to (1) articulate which aspects of existing human rights (e.g., freedom of expression) need to take fuller account of web-based, Internet facilitated scenarios (e.g., freedom of expression online in light of hate speech or cyber bullying in 24/7 communications); (2) raise public awareness about how our rights online, and on the ground, can be undermined through cumulative changes in Internet design, access, and use on the one hand and, on the other, the deployment of digital tools in illegal or disproportionate ways that threaten civil liberties. Taking international treaties and covenants as a working model was settled on, after some debate, as the most sensible way to focus the work. The argument being that this would generate a document with some legitimacy and legal clout and, thereby, one that had more chance of being accepted by policymakers working in fields where Internet services, human rights obligations, and public service commitments converge (e.g., Internet access in public libraries and schools). The final outcome after two years of drafting and feedback was version 1.1 of the IRPC Charter, a document modelled on the UN International Bill of Rights (UNGA 1948, 1966a, 1966b). The table of contents in Figure 5–1 shows how the charter reiterates human rights treaties but also points to how these articles relate to the online environment.

Alongside the full IRPC Charter two other projects have contributed to its development as a wider collaboration to change attitudes in technically arcane Internet policy-making circles around social issues, such as human rights. The first was the "Ten Punchy Principles" campaign in late 2010 just prior to the charter launch. This project, the work of a smaller working group, was set up to make the legally complex and high-level language of the charter more palatable to lay audiences and students, as well as help engage newer arrivals to the work more conversant with the idioms and methods of digital activism. The second was the IRPC Charter Booklet project from 2013. This brought the ten IRP principles together with the full charter and contextual information in booklet form, with a new design. The IRPC Charter booklet has been a great success, as a reference point and resource for the IRPC itself and those who the charter project has been aiming to reach. Figure 5–2 shows how the charter looks today, available in eight languages. In-between these two milestones, the IRPC contributed to successive Internet governance events such as the annual IGF, the European Dialogue on Internet Governance (EuroDIG), regional IGF meetings in the United Kingdom, Latin America and Asia-Pacific, ICANN meetings, and UNESCO-hosted reviews.

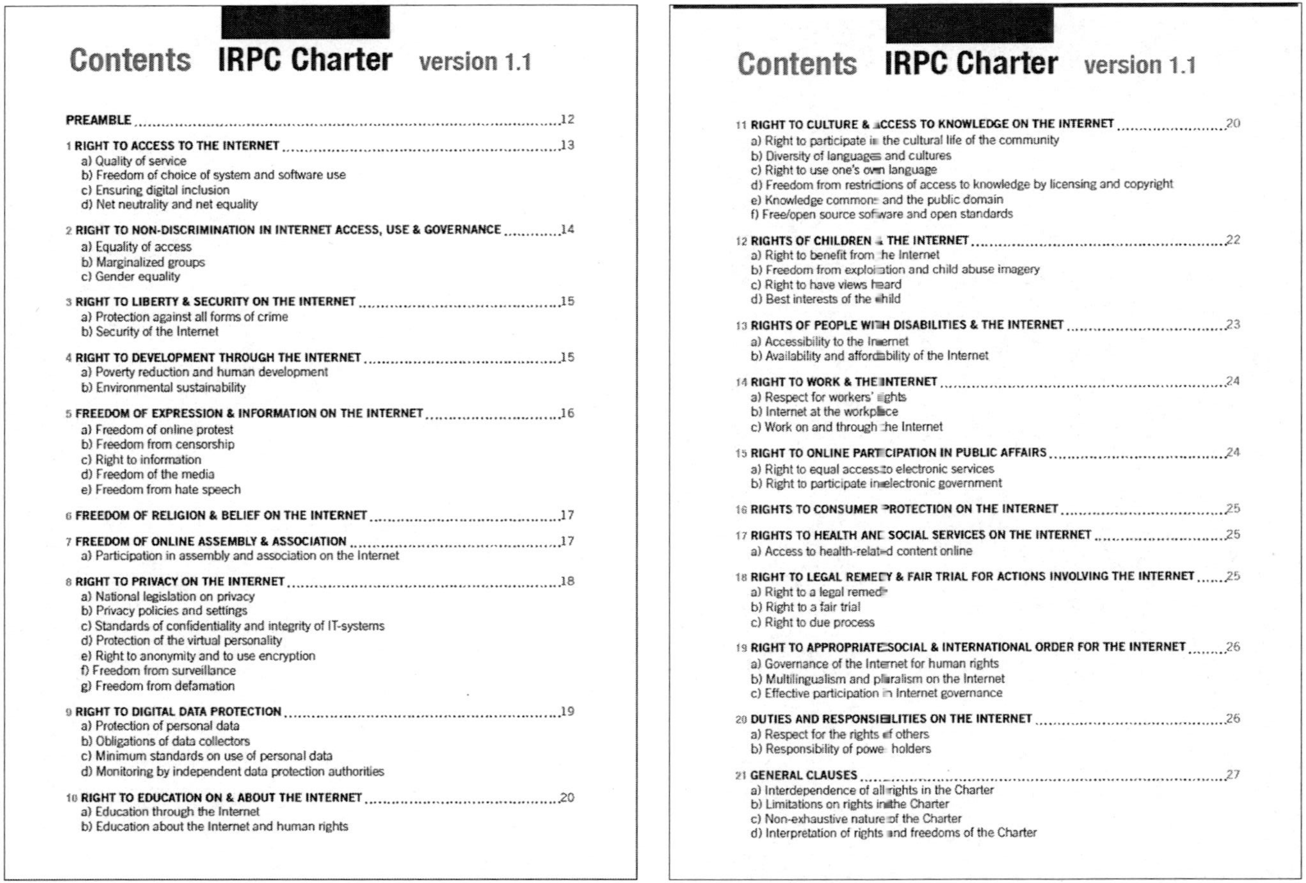

Contents IRPC Charter version 1.1

PREAMBLE 12

1 RIGHT TO ACCESS TO THE INTERNET 13
a) Quality of service
b) Freedom of choice of system and software use
c) Ensuring digital inclusion
d) Net neutrality and net equality

2 RIGHT TO NON-DISCRIMINATION IN INTERNET ACCESS, USE & GOVERNANCE 14
a) Equality of access
b) Marginalized groups
c) Gender equality

3 RIGHT TO LIBERTY & SECURITY ON THE INTERNET 15
a) Protection against all forms of crime
b) Security of the Internet

4 RIGHT TO DEVELOPMENT THROUGH THE INTERNET 15
a) Poverty reduction and human development
b) Environmental sustainability

5 FREEDOM OF EXPRESSION & INFORMATION ON THE INTERNET 16
a) Freedom of online protest
b) Freedom from censorship
c) Right to information
d) Freedom of the media
e) Freedom from hate speech

6 FREEDOM OF RELIGION & BELIEF ON THE INTERNET 17

7 FREEDOM OF ONLINE ASSEMBLY & ASSOCIATION 17
a) Participation in assembly and association on the Internet

8 RIGHT TO PRIVACY ON THE INTERNET 18
a) National legislation on privacy
b) Privacy policies and settings
c) Standards of confidentiality and integrity of IT-systems
d) Protection of the virtual personality
e) Right to anonymity and to use encryption
f) Freedom from surveillance
g) Freedom from defamation

9 RIGHT TO DIGITAL DATA PROTECTION 19
a) Protection of personal data
b) Obligations of data collectors
c) Minimum standards on use of personal data
d) Monitoring by independent data protection authorities

10 RIGHT TO EDUCATION ON & ABOUT THE INTERNET 20
a) Education through the Internet
b) Education about the Internet and human rights

Contents IRPC Charter version 1.1

11 RIGHT TO CULTURE & ACCESS TO KNOWLEDGE ON THE INTERNET 20
a) Right to participate in the cultural life of the community
b) Diversity of languages and cultures
c) Right to use one's own language
d) Freedom from restrictions of access to knowledge by licensing and copyright
e) Knowledge commons and the public domain
f) Free/open source software and open standards

12 RIGHTS OF CHILDREN & THE INTERNET 22
a) Right to benefit from the Internet
b) Freedom from exploitation and child abuse imagery
c) Right to have views heard
d) Best interests of the child

13 RIGHTS OF PEOPLE WITH DISABILITIES & THE INTERNET 23
a) Accessibility to the Internet
b) Availability and affordability of the Internet

14 RIGHT TO WORK & THE INTERNET 24
a) Respect for workers' rights
b) Internet at the workplace
c) Work on and through the Internet

15 RIGHT TO ONLINE PARTICIPATION IN PUBLIC AFFAIRS 24
a) Right to equal access to electronic services
b) Right to participate in electronic government

16 RIGHTS TO CONSUMER PROTECTION ON THE INTERNET 25

17 RIGHTS TO HEALTH AND SOCIAL SERVICES ON THE INTERNET 25
a) Access to health-related content online

18 RIGHT TO LEGAL REMEDY & FAIR TRIAL FOR ACTIONS INVOLVING THE INTERNET 25
a) Right to a legal remedy
b) Right to a fair trial
c) Right to due process

19 RIGHT TO APPROPRIATE SOCIAL & INTERNATIONAL ORDER FOR THE INTERNET 26
a) Governance of the Internet for human rights
b) Multilingualism and pluralism on the Internet
c) Effective participation in Internet governance

20 DUTIES AND RESPONSIBILITIES ON THE INTERNET 26
a) Respect for the rights of others
b) Responsibility of power holders

21 GENERAL CLAUSES 27
a) Interdependence of all rights in the Charter
b) Limitations on rights in the Charter
c) Non-exhaustive nature of the Charter
d) Interpretation of rights and freedoms of the Charter

FIGURE 5-1. Contents (excerpt) of the IRPC Charter.

FIGURE 5-2.   Cover of the IRPC Charter booklet.

HOW THE IRPC CHARTER CAME ABOUT

So how did a loose network of individuals and organizations, existing largely through a listserv, alongside a website and the usual social media accounts, manage to achieve these outcomes in a domain characterized by the spirit of previous attempts (Hamelink 1998), incumbent civil society campaigns (APC 2006) and intergovernmental initiatives (OECD 2011) and then maintain a foothold as rights-based projects mushroomed in the wake of the 2013 Snowden revelations? One reason is that the IRPC Charter as a whole goes a lot further and deeper than others. For despite earlier efforts from civil society, some stretching back to the 1970s (Frau-Meigs et al. 2012; Hamelink 1998; Jørgensen 2013; Murray 2010; Musiani 2009), the notion that human rights are integral— not just a "clip-on"—to how the Internet functions as well as to how it is run—had not yet been expressed in such a consistent or comprehensive way, nor in an authoritative legal idiom. The key to its success and durability in retrospect was an approach that generated the most internal debate at the time (Franklin 2013a, 164–169), namely by following the established, albeit old school, format of the 1948 Universal Declaration of Human Rights.

This approach paid off as the UDHR worked well as a working template. Its length and format allowed participants to take each clause in turn, along with those from ensuing international treaties and covenants such as the International Covenant on Civil and Political Rights (ICCPR) and Convention on the Rights of Persons with Disabilities (CRPD). In the spirit of a policy-hacking exercise, people divided themselves along each clause, debating not only language use and legal formulations but also substantive technical and legal points. For example, how "the development, promulgation and monitoring of minimum standards and guidelines for accessibility" (IRPC 2013, Articles 2 and 3, 14–15) for persons with disability imply research and development investment and governmental support for tailor-made technologies, training, and access to enable those with disabilities to fully enjoy the possibilities offered by the Internet. Figure 5–3 is a snapshot of the IRPC Charter's rendition of the aforementioned UN Convention on the Rights of Persons with Disabilities (CRPD). Moving this sort of collaborative effort from rough draft to meet international legal standards requires legal expertise, knowledge of the UN system, and writing skills. For some observers, the enterprise itself also implied some *chutzpah* to even consider policy-hacking one of the twentieth century's most iconic documents. Nonetheless, the IRPC Charter took shape through this collaborative effort between experts and lay people. The current version was refined with the input of an expert working group to craft the first full draft into an elegant, readable, and appropriate document for those working at the intersection of Internet policy and human rights.

A ROCKY ROAD WELL-TRAVELED:
A BRIEF TIMELINE

> The Internet is a global resource which should be managed in the public interest. . . . Rights that people have offline must also be protected online in accordance with international human rights legal obligations. . . . (NETmundial 2014)

> While the Internet has brought along substantial new possibilities for exercising and protecting human rights, the possibilities for human rights violations have also grown exponentially. (Benedek and Kettemann 2014, 18 and 19)

The above two quotes encapsulate the explicit role now being attributed to international human rights law and norms for policy-making on Internet design, access, and use, nationally and internationally. When the IRPC began work on formulating a comprehensive framework that could connect human rights advocacy with Internet policy agendas at the UN Internet Governance Forum, and in a way that could speak to government, business, and civil society interests, this was not the case. UN-brokered global Internet governance consultations, and other agenda-setting venues such as national legislatures or technical standard-making bodies, have habitually made sharp divisions between the techno-economic dimensions of contemporary media and ICT,

13 Rights of People with Disabilities and the Internet

People with disabilities are entitled to all of the rights in the present Charter. As enshrined in Article 4 of the United Nations Convention on the Rights of Persons with Disabilities (CRPD), "States Parties undertake to ensure and promote the full realisation of all human rights and fundamental freedoms for all persons with disabilities without discrimination of any kind on the basis of disability".

The Internet is important in enabling persons with disabilities to fully enjoy all human rights and fundamental freedoms. Special measures must be taken to ensure that the Internet is accessible, available and affordable.

On the Internet, the rights of people with disabilities include:

a) Accessibility to the Internet

Persons with disabilities have a right to access, on an equal basis with others, to the Internet.

Such access must be promoted through: the development, promulgation and monitoring of minimum standards and guidelines for accessibility; the provision of training on accessibility issues facing persons with disabilities; and the promotion of other appropriate forms of assistance to people with disabilities to ensure their access to information.

b) Availability and affordability of the Internet

Steps must be taken to ensure the availability and effective use of the Internet by people with disabilities.

Research and development must be undertaken to promote the availability of Information and Communications Technologies in a format suitable for persons with disabilities. Priority should be given to developing technologies at an affordable cost.

Persons with disabilities have the right to accessible information about assistive technologies, as well as other forms of assistance, support, services and facilities.

Visit us online for more information and to get involved internetrightsandprinciples.org

FIGURE 5-3.   IRPC Charter, Article 13.

concerns about Internet use and its role in society, and the respective "roles and responsibilities" of stakeholders in decision-making domains (ITU/WSIS 2005; WGIG 2005). Events in 2013 did much to crystallize increasing public awareness of human rights issues for the online environment, but work was well underway before then as the IRPC Charter work and its precursors show. The above statements also encapsulate a sea change in policy-making and social movement domains toward the Internet itself. In this case, the Internet (including here the World Wide Web and Web 2.0 services) is object, means, and medium for the project, which is distinct from how advertising or activist campaigns can and do use Internet media and technologies. The Snowden revelations of mass surveillance online in 2013 notwithstanding, the IRPC's contribution to this human rights turn in how public Internet policy is discussed is testament to how this work brings together NGOs, grassroots groups, socially engaged scholars, IT, and human rights lawyers, as well as intergovernmental organizations, to get human rights onto global Internet policy-making agendas.

When the idea of such a charter was first mooted at the 2008 IGF meeting in Hyderabad, India, it was conceived more as an aspirational and team-building exercise, to find some common ground and provide focus for those engaged in a plethora of crosscutting political, economic, and sociocultural agendas across the Internet policy-making spectrum. While there were already prominent civil society efforts in this area (see APC 2006), the argument this time was that a fully "multistakeholder" effort (i.e., not confined to one sector or another) could go further in bridging the gap between stakeholder groups and their respective viewpoints on the role and responsibility of governments on the one hand and, on the other, the private sector in how the Internet is run. The underlying rationale, diverse motivations, and political affiliations notwithstanding, was that, like the Internet itself, policy-making agendas straddle local, national, and international institutional settings (IRPC 2013, 6). By 2009, at the Sharm el Sheikh IGF meeting, the IRPC had been inaugurated, developed a governance structure, and devised a work plan.

When the initial drafting got underway in 2009, online discussions about content, phrasing, and organization (e.g., which article comes first) saw early caution predominate; experienced digital rights activists and experts noted that the charter was entering a deeply contentious and complex geopolitical and techno-economic domain (Franklin 2013; Vincent 2010, 32 passim). Taking care not to convey the impression that the aim was to create *new* rights in light of how fragile existing human rights on the ground remain, supporters of taking the IRPC Charter project up a level—that is, to revise the first very rough draft to have it comply with international human rights language—argued that this was no reason to hold back. For instance, contentiousness around freedom of expression and freedom of information had already come to the fore in light of national security arguments during the Wikileaks controversy, and the dangers that dissident bloggers and mobile-phone users ran during the Arab Uprisings was headline news in spring 2011. Meanwhile in the Internet's heartlands, media, and civil liberty watchdogs were also raising the alarm about the implications that changes in Internet functionalities (e.g., prioritizing transmission according to commercial or other priorities), or disproportionate levels of data-retention, have for fundamental freedoms such as freedom of information and privacy.

By mid-2010, a legally dense and ever-lengthening document had taken shape with three possibilities; to be an awareness-raising tool, legal instrument, or broader policy-making framework. One document was turning into three, not counting accompanying notes and umbrella statements. At this point the aforementioned side project took shape, to distil the larger charter's twenty-one articles into Ten Internet Rights and Principles that could be used for campaigning and educational purposes. While uncertainty over the purpose and eventual shape of the larger charter dominated discussions in 2011 and 2012 the "Ten Punchy Principles," launched online and in print as a flyer (see Figure 5–4) almost eclipsed the release of the full IRPC Charter at the time. This social media campaign, modest by today's standards, was an important learning point about the amount of time and effort—and financial resources—needed for online campaigning for those from the old media school of social movements. The ten principles remain an effective advocacy and educational tool (Hivos 2014) and key resource for subsequent high-level efforts to codify "global internet governance principles" (NETmundial 2014). The strategic point here being that these principles were derived from the charter, not the other way around.

The charter's first success was the role it played, substantively and through IRPC members' involvement, in a 2010 Council of Europe initiative to compile an overview of existing rights law for the Internet, one that has led to the *Guide on Human Rights for Internet Users* (Council of Europe, 2014) and ensuing analyses of human rights case law for the Internet (Benedek and Kettemann 2014; INDH 2013; Jørgensen 2013). This shift from being positioned as predominately a civil society-cum-academic project underscored moves at the UN level with the publishing of a landmark report by Frank La Rue, UN Special Rapporteur for Freedom of Expression (La Rue 2011) and an early supporter of the IRPC Charter project. La Rue's report stipulates, for the first time, that freedom of expression, and with that other human rights and fundamental freedoms, count online as well as on the ground. It paved the way for the UN Human Rights Council resolution the next year recognizing the connection between human rights law and the Internet (UNHRC 2012).

In hindsight, given the wake-up call of the Snowden revelations and current power shifts around who should call the shots in future Internet governance agendas, the above narrative sounds very neat and tidy, with consensus about the means and ends self-explanatory. This was—and is not—the case as divergent opinions about points of law as well as sociocultural and political affiliations characterize this sort of undertaking. Moreover, once completed, such a document needs to be communicated, translated, and taken up by a range of significant others. Neither was the ambition of including anyone who was interested, and doing so by moving the sort of nitty-gritty of drafting legally coherent and politically acceptable documents from relatively closed settings to open-access ones online. What did make it possible to get the IRPC Charter

The Internet offers unprecedented opportunities for the realisation of human rights, and plays an increasingly important role in our everyday lives. It is therefore essential that all actors, both public and private, respect and protect human rights on the Internet. Steps must also be taken to ensure that the Internet operates and evolves in ways that fulfil human rights to the greatest extent possible. To help realise this vision of a rights-based Internet environment, the 10 Rights and Principles are:

1 UNIVERSALITY AND EQUALITY: All humans are born free and equal in dignity and rights, which must be respected, protected and fulfilled in the online environment.

2 RIGHTS AND SOCIAL JUSTICE: The Internet is a space for the promotion, protection and fulfilment of human rights and the advancement of social justice. Everyone has the duty to respect the human rights of all others in the online environment.

3 ACCESSIBILITY: Everyone has an equal right to access and use a secure and open Internet.

4 EXPRESSION AND ASSOCIATION: Everyone has the right to seek, receive, and impart information freely on the Internet without censorship or other interference. Everyone also has the right to associate freely through and on the Internet, for social, political, cultural or other purposes.

5 PRIVACY AND DATA PROTECTION: Everyone has the right to privacy online. This includes freedom from surveillance, the right to use encryption, and the right to online anonymity. Everyone also has the right to data protection, including control over personal data collection, retention, processing, disposal and disclosure.

6 LIFE, LIBERTY AND SECURITY: The rights to life, liberty, and security must be respected, protected and fulfilled online. These rights must not be infringed upon, or used to infringe other rights, in the online environment.

7 DIVERSITY: Cultural and linguistic diversity on the Internet must be promoted, and technical and policy innovation should be encouraged to facilitate plurality of expression.

8 NETWORK EQUALITY: Everyone shall have universal and open access to the Internet's content, free from discriminatory prioritisation, filtering or traffic control on commercial, political or other grounds.

9 STANDARDS AND REGULATION: The Internet's architecture, communication systems, and document and data formats shall be based on open standards that ensure complete interoperability, inclusion and equal opportunity for all.

10 GOVERNANCE: Human rights and social justice must form the legal and normative foundations upon which the Internet operates and is governed. This shall happen in a transparent and multilateral manner, based on principles of openness, inclusive participation and accountability.

The 10 Internet Rights & Principles are available for download in 22 languages at http://internetrightsandprinciples.org/site/campaign

FIGURE 5-4.   IRPC flyer 2010 (*left*) and 2013 (*right*).

ready for public launching boiled down to several factors; first, an institutional setting that could provide legitimation and an offline space for getting participants, representing conflicting priorities and interests within these sectors let alone between them, to get together face to face. The Internet Governance Forum enabled this space. Second was the continual work achieved that kept civil society organizations keen to make their own mark in this terrain engaged, those others with expertise able to contribute, and those others motivated to undertake translations of the ten principles and the full charter into other languages. Third was a conscious decision to have the process be, as much as possible, an exercise in open discussion and participation that could take place online and offline across time zones and national borders. The decision to convene a group of legal experts to develop the final draft for a final round of consultation before the charter's public launch in 2011 diverged from this approach, but ensured a high quality outcome that lawmakers could recognize and engage with (CLD 2011; INDH 2013; Murray 2010,). In this sense, the IRPC Charter succeeded in applying a bottom-up, policy-hacking approach to getting online rights on the agenda within the opaque and expert-driven settings of UN-based consultations, thus moving past the dominant narrative of Internet governance as narrowly defined decisions around technical functionality and resources.

None of these efforts and outcomes was happening in a vacuum. Indeed, this whole project has taken shape during a crucial period in not only the Internet's history but also in the geopolitics of its incumbent and emerging decision-making institutions. The latter are uneasily and unevenly distributed between UN agencies such as the ITU and UNESCO, private corporations such as ICANN, standards-making bodies such as the Internet Engineering Task Force (IETF), and lobby groups and networks such as the Internet Society. Along with the political import of social media uses during regime change in the Middle East and North African region and the emergence of *cybersecurity* discourses and policy priorities in the wake of 9/11, other events have contributed to the change in mood toward human rights for the online environment.

The first was really two events in tandem; the first beginning with the globally organized mobilization against three bills in the US Congress with international repercussions for trading partnerships and digital copyright. The anti-ACTA, PIPA, and SOPA protests around the world in 2012 made a feature of social media and web campaigning tools, and did much to underscore just how integral the Internet had become to cross-border political and social mobilization, public commentary, and direct actions.[3] These protests saw major corporations join forces with a range of activist, consumer, and community groups, and new political parties based on free and open source software principles (e.g., the Swedish and German Pirate Parties) to successfully lobby against the bills. The second was taking place at the heart of the UN itself. ITU was hosting the World Conference on International Telecommunications (WCIT–12) at the end of 2012 in Dubai as this wave of mobilization around "net freedom" gathered steam. The agenda, to renew and update 1980s international telecommunications treaties, belied the historical and now technical interrelationship between imperial and Cold War–era telecommunications architecture and that of the Internet that now overlays this older architecture (Franklin 2013b, 195–199). While updating telecommunications treaties between UN member states looked like a done deal, the subtext was all about an emerging geopolitical standoff between the Internet's heartlands and non-Western powers with their own, national and unapologetically interventionist, approaches to Internet access and content provision to their citizenries: China and the Russian Federation in particular with India and Brazil staking a claim in deciding where and how to regulate, and build Internet provisions on their own terms. A campaign, spearheaded by groups self-identifying as digital natives and their fledgling global networks, was based on the slogan "The UN is trying to take over the Internet" (Access 2013) and with that the (hash) tag-line of "net freedom." Taking place shortly after the 2012 IGF meeting in Baku, Azerbaijan, marked by controversies over poor human rights and press freedom record of the Azeri government, the WCIT–12 meeting and its UN host came to epitomize all that is wrong with intergovernmental approaches to Internet policy-making (Goldsmith 2012; cf. Singh 2012). Sophisticated (social) media campaigns pushed the narrative of an Internet "born free" of state support and in doing so elided an ongoing debate about the historical role of government funding in the Internet's history, and that of the World Wide Web as well. This standoff between supporters of neoliberal models for the Internet's "self-regulation" and those advocating other variations of governmental regulation marks a move to pit multistakeholder models against multilateral ones for future decisions around the Internet. As UN member states make international law, this ongoing divergence of views has implications for human rights mobilization for the online environment that looks to states to also set the record straight.

Third, this positioning of the UN as "bad guy" and all governmental regulation as *ipso facto* antagonistic to Internet freedom was well under way by the time the news broke in 2013 of the US government and its allies' program of mass online surveillance that included US citizens and those from other countries. Ongoing revelations about the extent of first mass online surveillance and, second, other governments' complicity in the NSA affair have raised the profile, and the ante for human rights advocacy for the Internet. The IRPC Charter in effect now straddles a position that underscores states' legal obligations under international human rights law and the view of powerful (US-owned) ISPs and standard-setting authorities like ICANN (still US-based at time of writing) that the way forward is self-regulation. The political implications of the Snowden revelations for Internet governance as a UN-hosted concern have also heralded the emergence of powerful counter-narratives from other UN member states who have the techno-economic means and know-how to set policy agendas on their own terms. Here the BRICS powers of Brazil, Russia, India, China, and South Africa have been flexing their muscles in public meetings alongside the aforementioned ITU summit in Dubai, through to Seoul, Belgrade, and on to São Paulo. But this is not all. As the Internet becomes a part of US foreign and trade policy (Clinton 2010), a new silver bullet for the world's developmental woes and endemic poverty (UNGA 2000) it is also the linchpin of an increasingly digital arsenal against suspected cyberterrorists in a period still defined by the post-9/11 global war on terrorism.

These shifts in context have been a fillip to human rights mobilization for the Internet. At the same time, they raise new challenges as powerful actors—states and corporations—look to control the narrative. Taking a step back in history though for a moment, contemporary mobilization around human rights online has its forebears in a pre-Internet age. Rights-based aspirations and mobilizations have been integral to media reform movements throughout the last century (MacBride 1980). While the IRPC Charter is not the first initiative looking to bring the predominately techno-legal and economic approach to Internet design, access, and use in the same frame as sociocultural concerns and, with these, human rights (Hamelink 1998; Jørgensen 2006) it is easy to forget in the second decade of this century just how far-fetched it seemed in 2008, let alone earlier, to suggest that the online environment is a domain where the fundamental rights and freedoms encapsulated in international human rights law do need to be protected as well as enjoyed and that these protections implicate computer engineering and software design. The first point has been made. The second one is where rights-based coalitions are now starting to focus their efforts (IRPC 2014; NETmundial 2014).

Within this changing techno-historical and geopolitical context, the IRPC Charter work has come of age in what we could call a post-Snowden decade, one in which the Internet itself is under closer scrutiny along with incumbent governance agencies such as ICANN. While the IRPC Charter project has achieved its primary goals, the outlook is still far from clear. First, because cyberspace is being monitored and patrolled by nation-states who are also the guardians of human rights in disproportionate ways, without due process or rule of law, or respect for international human rights norms. Second, because breaches of certain high-profile rights (e.g., freedom of expression, privacy) belie the obstruction of others (e.g., rights of persons with disability, cultural diversity, of children online) that also occur through poor design, unchecked practices of digital monitoring, ad-tracking, and data retention by the world's major Internet service providers, as well as governments asserting national security priorities. These practices also require closer scrutiny and forms of legal redress, all of which the IRPC Charter presaged. Now being picked up by NGOs working in these regions where Internet access is booming and implemented as campaigns to raise awareness and educate new populations of Internet users about their rights online (Hivos 2014; INDH 2013), the IRPC Charter work has taken on a new impetus and relevance.

LESSONS LEARNT

This section covers some of the key lessons learnt during this project; a project that has to keep its bearings in the fast-changing and politically volatile context of human rights advocacy and networking for the Internet today. This is a project in which state actors are making a stand, private sector interests are holding their ground in the battle for the hearts and minds of the world's netizens as more revelations about state-sanctioned and corporate inroads into our online lives continue. It is also one in which the rights-based activism for Internet policy-making needs to consider how to implement such rights in a domain presaged on a post-Westphalian model of global and thereby Internet governance.

The first lesson is to recall that Internet-dependent communications have become the rule rather than the exception, and with that time spent online, in cyberspace, part of our everyday lives (Franklin 2004, 2013a, 2013b; Lovink 2012). Hence, the substantive content of human rights—in legal, moral, and political terms—have acquired online, cross-border, and computer-mediated dimensions. These developments also point to emerging new rights, such as that of access, because this "right underpins all other rights" (IRPC 2013, 13; see also Benedek and Kettemann 2014, 73 passim; Franklin 2013a, 169–171). But in these Internet-dependent contexts, governments are not the only power holders, nor the only arbiters of rights-based issues that occur in spaces that evade and confound national jurisdictions, rule of law, and media ownership models. The agenda-setting power of the global corporations, predominately registered and based in the USA, who own and control the lion's share of the Internet's critical infrastructure, consumer services and products, has added another layer of techno-economic and legal complexity to how governments fulfil, or fall short of their role in upholding human rights and fundamental freedoms in the online world. Further complicating things is the increasing activism and organizational skills of civil society organizations, large and small, professional NGOs and ad hoc networks working to be accorded a seat at the table where not only decisions about the current Internet's design, access, and use are made but future agendas are also set in motion. Cyberspace, where people and their avatars go when they access the Internet or use the Web, is now a means and an end in the flourishing of rights-based media advocacy. But this state of affairs is relatively recent, and increased awareness of how civil liberties can be enjoyed, yet also undermined, online has been even more recent; peaking at the time of writing as governments around the world stand accused of generating a "chilling effect" on the Internet as a free speech, democratic medium.[4]

The next lesson is about where to draw the line between rights and principles. As the inclusion of human rights *and* principles in the IRPC Charter title implies, there is a distinction how Internet design, access, and use has implications for individual human rights on the ground and online and identifying how these rights relate to the underlying principles that govern Internet decision-making to date, let alone in the future (IRPC 2013, 2). It is a fine line but a distinct one that has to be trod, for principles and rights are not interchangeable terms. There have been in the past, and could be in the future, operations such as terms of access and use that do not "support human rights" (ibid.). Lesson three is on practicalities once the point has been made. Much territory still needs to be covered between the aspirations of longstanding rights-based advocates for the Internet and implementation at the intergovernmental and corporate level. Moreover, official recognition that human rights do matter puts the legal role and responsibilities of governments in upholding human rights law squarely back in the picture. This "return of the state" (Franklin 2009, 2010; Giacomello and Erickson 2009) to Internet policy-making discourses and scholarship is all the more contentious given that the global corporations, taskforces, and think-tanks that oversee how the Internet functions all came of age under the aegis of neoliberal governments' commitments to (media) deregulation, privatization, and commercialization. It also separates US-based debates and mobilization around the relationship between regulation in the first instance and, in the second instance, human rights and fundamental freedoms online from those in Europe, the Global South, and the Asia-Pacific region.

CONCLUSION

The IRPC Charter staked its claim in a domain dominated at the time by an "if it ain't broke, don't fix it" rhetoric in the face of arguments about there being a need to address human rights online and thereby Internet policy-making. Since 2013 but presaged before then, the rhetoric has shifted to "the Internet is broken, and whose fault is that?" While the jury is still out on who is ultimately in charge, human rights are now on that agenda. Since its entry into the Internet Governance Forum and Human Rights advocacy domains the IRPC Charter project has gained traction in key intergovernmental and civil society venues where the future of the Internet is the focus. In these forums, but also further afield at grassroots level, the IRPC Charter and the IRPC behind the work have remained viable, beyond expectations and without structural funding or full-time staff. A major achievement has been the formal recognition of all three strategic aims of the IRPC Charter project (IRPC 2013, 3), in landmark UN and IGO reports and resolutions (Council of Europe 2014; La Rue 2011; UNHRC 2012) and evidence of human rights issues moving up Internet governance agendas and civil society campaigns (Hivos 2014; NETmundial 2014), into the mainstream media and, on going to press, the heart of democratic political processes (Cultura Digital e Democracia 2014; Green Party 2014). The IRPC Charter's sustainability has also accompanied shifts in national and global media policy discourses in the wake of the Internet's impact on national polities and legal norms. It has also echoed and looked to respond to increasing public awareness, and with that media panics over the threats to freedom of expression and privacy in light of recent events. Cognizant that from the outset the ownership and control of the Internet—that is, its design, access and use—have been politically, economically, and socioculturally skewed in favor of vested commercial interests and incumbent political power in a post–Cold War context, the spirit and intent of the charter from the outset was to re-vision legalistic and technically arcane discourses into more human-centered idioms. Its indebtedness to international human rights law and norms, criticized at the time for being overcautious, was a conscious tactic, based on the conviction that fundamental rights and freedoms, as both "hard" and "soft" law, need translating for their protection and enjoyment online.

The IRPC Charter has survived, and is now thriving precisely because of its legal idiom and UN-based institutional home in the first instance. In the second because it is based on a coalition that has forged a document that can speak across sectors and narrow interests in order to look to the future rather than focus on rectifying past wrongs. In addition, its recognition of precursor initiatives has helped it gain traction in domains that merge established and emerging institutions and organizational cultures where diverse state, and non-state actors converge—and diverge—on Internet policy-making agendas that by definition straddle the local, national, and transnational frames of reference and traditions of political legitimacy. These agendas concern an array of technical, economic, and political concerns that fall under the rubric of Internet governance but have been traditionally and strategically treated as separate, autonomous domains for decision-making and mobilization based on the premise that technology, social justice, and politics do not mix. The IRPC Charter has managed to articulate just where they do intersect and how even technical standards (e.g., Internet protocols, the domain name system), transmission architectures (server hubs, packet switching), and software codes impinge upon and are beholden to human rights standards.

Based at the newest of UN institutions, the Internet Governance Forum which is reaching the end of its second five-year mandate (up for a second renewal in 2015 having survived studied indifference from the corporate sector and getting by on a shoestring budget from the UN General Assembly), the Internet Rights and Principles Coalition has shepherded a document set that articulates the past, present, and future of the Internet from a rights-based sensibility, and does so in substantive moral, legal, and political terms. Voluntary labor, crowdsourced funding, growing endorsements, and increasing numbers of language editions from dedicated translation teams, with respective sources of funding have all sustained this project and its achievement as a sustainable, long-term strategy for reframing the terms of Internet governance debates once firmly framed by techno-economic rather than sociocultural prerogatives. Seeing the IRPC Charter become a source document for pioneering and innovative principle-based laws and remedies has confounded the sceptics and provided hope and sustenance for the optimists in the IRPC who have contributed to getting the charter this far and who continue to support the work, in spirit and in deed.

The question remains though; the IRPC Charter—like its predecessors and all the digital rights manifestos, declarations, and resolutions that have been mushrooming lately—consists after all only of words, aspirations rather than actions. This tension is borne out by the difficulty of enforcing human rights when those charged with upholding and protecting them—UN member states—are also perpetrators of human rights abuses. Nonetheless, before action or at least hand-in-hand with strategies that develop the means and tools to implement these rights online, there has been a need to articulate more coherently how Internet design, access, and use are not mutually exclusive domains to those covered by international human rights law and so relate these concerns directly to the responsibility of all power holders to uphold rights online as well as offline. This recognition and implementation of the IRPC Charter at the local and national levels (Green Party 2014; Hivos 2014) raises the ante in a world where the Internet has become "ordinary" (Franklin 2013a, 35 passim). Through the IRPC Charter the IRPC has succeeded in articulating a human-centered vision of Internet policy-making and continues to keep working with others to ensure that this vision is one that underscores agendas for a sustainable and equitable Internet that is accessible and affordable for all, now and in the future.

NOTES

1. This chapter was written in my role as co-chair of the Internet Rights and Principles Coalition (2012–2014) and member of its steering committee since the outset. A more detailed reconstruction of the IRPC Charter process is in Franklin (2013a, 138 passim). See also Franklin (2009, 2010, 2013a).

2. ICANN stands for the Internet Corporation of Assigned Names and Numbers, a private sector company registered in California (USA) to take charge of the domain name system that organizes where individuals and organizations are online, the epicenter of how the Internet works (Mueller 2002).

3. ACTA stands for the Anti-Counterfeiting Trade Agreement, PIPA for the Protection of Intellectual Property Act, and SOPA for the Stop Online Privacy Act.

4. This expression refers to how censorship practices online impinge directly on freedom of expression and other fundamental rights and freedoms.
See the definition on US Legal.com at http://definitions.uslegal.com/c/chilling-effect/, and the Electronic Frontier Foundation (EFF) Chilling Effects Clearinghouse website at www.chillingeffects.org.

REFERENCES

Access. 2013. "Tell the ITU: The Internet Belongs to Us!" *Access: Campaigns*. www.accessnow.org/page/s/itu.

Association of Progressive Communications (APC). 2006. "APC Internet Rights Charter." www.apc.org/en/node/5677/.

Benedek, W., and M. Kettemann. 2014. *Freedom of Expression and the Internet*. Strasbourg: Council of Europe Publishing.

Center for Law and Democracy (CLD). 2011. "Commentary on the Charter of Human Rights and Principles for the Internet." October. www.law-democracy.org/wp-content/uploads/2011/10/Charter-Commentary.pdf.

Clinton, H. 2010. Internet Freedom. Speech delivered at the Newseum in Washington, DC, January 21. www.foreignpolicy.com/articles/2010/01/21/internet_freedom?page=full.

Council of Europe. 2014. Recommendation CM/Rec(2014)6 of the Committee of Ministers to Member States on a Guide to Human Rights for Internet Users, Adopted by the Committee of Ministers at the 1197th Meeting of the Ministers' Deputies. 16 April. https://wcd.coe.int/ViewDoc.jsp?id =2184807.

Cultura Digital e Democracia. 2014. Marco Civil da Internet (Law 92.465)— Unofficial English Translation. https://thecdd.wordpress.com/2014/03/28/ marco-civil-da-internet-unofficial-english-translation/.

Franklin, M. I. 2004. *Postcolonial Politics, the Internet, and Everyday Life Online: Pacific Traversals.* London/New York: Routledge

———. 2009. "Who's Who in the 'Internet Governance Wars': Hail the Phantom Menace?" *International Studies Review* 11, no. 1: 221–226.

———. "Digital Dilemmas: Transnational Politics in the 21st Century." *Brown Journal of World Affairs* 16, no. 11: 67–85.

———. 2013a. *Digital Dilemmas: Power, Resistance and the Internet.* New York/ London: Oxford University Press.

———. 2013b. "How Does the Way We Use the Internet Make a Difference?" In *Global Politics: A New Introduction*, 2nd ed., edited by M. Zehfuss and J. Edkins, 176–199. London/New York: Routledge.

———. 2013c. "Like It or Not, We Are All Complicit in Online Snooping." *Conversation*, June 20. http://theconversation.com/like-it-or-not-we-are-all -complicit-in-online-snooping-15219.

Frau-Meigs, D., J. Nicey, M. Palmer, J. Pohle, and P. Tupper, eds. 2012. *From NWICO to WSIS: 30 Years of Communication Geopolitics*. Bristol, UK: Intellect.

Giacomello, G., and J. Eriksson, eds. 2009. "Who Controls the Internet? Beyond the Obstinacy or Obsoleteness of the State." Review Forum. *International Studies Review* 11, no. 1: 205–230.

Goldsmith, J. 2012. "WCIT 2012: An Opinionated Primer and Hysteria Debunker." *Lawfare* (blog). www.lawfareblog.com/wcit-12-opinionated-primer-and -hysteria-debunker.

Green Party of Aotearoa New Zealand. 2014. Internet Rights and Freedoms Bill. http://internetrightsbill.org.nz/.

Hamelink, C. 1998. "The People's Communication Charter." *Development in Practice* 8, no. 1: 68–74.

Hivos-IG MENA. 2014. "Click Rights Campaign." http://igmena.org/click-rights.

Instituto Nacional de Derechos Humanos (INDH). 2013. *Internet y Derechos Humanos.* Santiago: Chile. http://bibliotecadigital.indh.cl/bitstream/ handle/123456789/627/cuadernillo?sequence=1.

Internet Rights and Principles Coalition (IRPC). 2013. *Charter of Human Rights and Principles.* http://internetrightsandprinciples.org/site/wp-content/ uploads/2014/02/IRP_booklet_2nd-Edition14Nov2013.pdf.

———. 2014. *The IRPC Charter of Human Rights and Principles for the Internet.* Contribution to the Netmundial Global Multistakeholder Meeting on the

Future of Internet Governance, 23–24 April 2014. http://content.netmundial
.br/contribution/the-irpc-charter-of-human-rights-and-principles-for-the
-internet/161.

ITU/WSIS. 2005. Tunis Agenda for the Information Society. World Summit on
the Information Society, November 18. www.itu.int/wsis/docs2/tunis/off/
6rev1.html.

Jørgensen, R. F., ed. 2006. *Human Rights in the Global Information Society.* Cambridge MA: MIT Press.

———. 2013. *Framing the Net—The Internet and Human Rights.* Cheltenham, UK:
Edward Elgar Publishing.

La Rue, F. 2011. *Report of the Special Rapporteur on the Promotion and Protection of
the Right to Freedom of Opinion and Expression.* Human Rights Council:
UN General Assembly, May 16. www2.ohchr.org/english/bodies/hrcouncil/
docs/17session/A.HRC.17.27_en.pdf.

Lovink, G. 2012. *Networks without a Cause: A Critique of Social Media.* Oxford:
Polity Press.

MacBride, S., ed. 1980. *Many Voices, One World: Towards a New More Just and
More Efficient World Information and Communication Order.* Paris:
UNESCO.

Mueller, M. 2002. *Ruling the Root: Internet Governance and the Taming of Cyberspace.* Boston, MA: MIT Press.

Murray, A. 2010. "A Bill of Rights for the Internet." *IT Lawyer* (blog), October 21.
http://theitlawyer.blogspot.de/2010/10/bill-of-rights-for-internet.html.

Musiani, F. 2009. "The Internet Bill of Rights: A Way to Reconcile Natural Freedoms and Regulatory Needs?" *SCRIPTed—A Journal of Law, Technology
and Society* 6, no. 2. www.law.ed.ac.uk/ahrc/script-ed/vol6-2/musiani.asp.

NETmundial. 2014. Multistakeholder Statement of Sao Paulo. NETmundial
Global Multistakeholder Meeting on the Future of Internet Governance
April 24; http://netmundial.br/wp-content/uploads/2014/04/NETmundial
-Multistakeholder-Document.pdf.

OECD. 2011. *OECD Council Recommendation on Principles for Internet Policy
Making.* December 13. Paris: OECD. www.oecd.org/internet/ieconomy/
49258588.pdf.

Olesen, T. Ed. 2010. *Power and Transnational Activism.* London: Routledge.

Singh, P. 2012. "A False Consensus Is Broken." *Hindu,* December 21. www.the
hindu.com/opinion/lead/a-false-consensus-is-broken/article4222688.ece.

United Nations General Assembly (UNGA). 1948. *Universal Declaration of Human
Rights.* www.un.org/en/documents/udhr/.

———. 1966a. *International Covenant on Civil and Political Rights.* http://legal
.un.org/avl/ha/iccpr/iccpr.html.

———. 1966b. *International Covenant on Economic, Social and Cultural Rights.*
http://legal.un.org/avl/ha/icescr/icescr.html.

———. 2000. *Millennium Development Goals.* www.un.org/millenniumgoals/.

United Nations Human Rights Council (UNHRC). 2012. Resolution A/HRC/
RES/20/8: Promotion and Protection of All Human Rights, Civil, Political,

Economic, Social and Cultural Rights, Including the Right to Development, UN General Assembly: OHCHR. http://ap.ohchr.org/documents/dpage_e.aspx?si=A/HRC/RES/20/8.

Vincent, A. 2010. *The Politics of Human Rights*. Oxford: Oxford University Press.

Working Group on Internet Governance (WGIG). 2005. *Report of the Working Group on Internet Governance, Château de Bossey*. www.wgig.org/docs/WGIGREPORT.pdf.

CHAPTER SIX

Lessons from the SOPA Fight

RAINEY REITMAN, *Electronic Frontier Foundation, United States*

We won this fight because everyone made themselves the hero of their own story. Everyone took it as their job to save this crucial freedom. They threw themselves into it. They did whatever they could think of to do. They didn't stop to ask anyone for permission.

—*Aaron Swartz (1986–2013)*[1]

MEDIA REFORM STRATEGY

In 2012, free speech and access-to-information activists launched the largest digital protest in the history of the Internet to defeat SOPA, a copyright bill. While a range a tactics was used, several important factors proved instrumental in the success of the campaign. Specifically, organizations and activists relied on: (1) decentralization, that is, no single organization or coalition controlled messaging or campaigning; (2) speaking up quickly and frequently through blogging and social media in order to respond to breaking news; (3) powerful visual imagery that provided an entry point for journalists and the public; (4) engaging with Internet communities as a politically engaged force for action; and (5) crossing political lines.

In 2012, free speech activists orchestrated the largest digital protest to date. We fought back a bill called the Stop Online Piracy Act (SOPA), which was a ham-fisted proposal designed to give US law enforcement and corporations new tools to stop the unauthorized sharing of copyrighted materials. However, the bill was written in a way that could have censored whole domains for even a single piece of copyrighted content. It also would have threatened the domain name system (the technical backend that serves as a naming system for the Web), and it would have criminalized anticensorship tools used by activists worldwide.

The Electronic Frontier Foundation is a nonprofit civil liberties law and advocacy organization based in San Francisco, and we were one of a number of organizations that joined forces to stop SOPA. Together, we formed a coalition that crossed political lines, engaged groups who were not traditionally involved in Internet freedom fights, and successfully showed tech companies how this would impact them. The coalition was in many ways decentralized, no one was asking for or granting permission to get involved or launch an action. The organizations and individuals involved produced a steady barrage of educational and advocacy blog posts, a huge social media presence, and several digital protests that culminated in a highly visible digital blackout in which over one hundred thousand websites went dark to protest the bill. We demonstrated that modern Internet communities have serious political clout.

Not only was SOPA defeated, but the Internet community also sent a powerful message to lawmakers that it would vigilantly defend itself from future attempts at censorship.

A QUICK PRIMER ON THE COPYRIGHT FIGHTS

SOPA was only one battle in a war that had been waging for many years. On the one side, we have information freedom advocates pushing for sensible copyright policies that prioritize access to information worldwide through digital platforms while at the same time finding creative ways to ensure artists are paid. On the other side, we have Hollywood lobbyists fighting to impose brutish laws from the physical world onto digital spaces, often at the expense of free expression online.

Copyright does not and should not grant the same kind of property right as one would expect for a house or a car. According to the US Constitution, copyright isn't supposed to last forever; it's just supposed to last long enough to help artists get paid for their work so they'll be inspired to keep on writing and making music and such. However, a series of laws developed over many years have extended copyrights repeatedly so that new creative works produced in the United States are copyrighted for the entire life of the creator plus *an additional 70 years*. It's hard to imagine that these terms are the most effective way to ignite creative production, considering the artist in question is dead for decades before the copyright expires.

Technology today allows us to share data with other people around the world seamlessly and quickly, even content like books, movies, and music. Tech-savvy artists and musicians are adapting to the digital economy by selling access to content online, giving their fans new avenues to support them, and embracing technology as a fun new way to promote their work and connect with fans.

At the same time, industries like Hollywood are trying to create laws to restrict and criminalize sharing online. It's as if carriage makers, seeing the early rise of automobiles, tried to create laws preventing modes of transportation that weren't horse-driven: it's technophobia in response to innovation.

MEET SOPA, THE INTERNET BLACKLIST BILL

SOPA was an attempt to hamstring the Internet to prevent the sharing of unauthorized copies of digital works.[2] But it was more than that: it was a new, powerful way to censor the Internet. If SOPA had passed, whole websites could have been censored in the name of copyright enforcement.[3] The defenders of SOPA said it was supposed to help combat foreign websites dedicated to providing illegal content, but SOPA had provisions that would allow for the removal of enormous amounts of noninfringing content from the Web, including political speech.

It worked like this: if a website was reported for having copyrighted content, it could be added to a blacklist. The registrar of the website would be forced to rescind the domain name it had been granted, so that people who visited the site through the typical URL wouldn't be able to access the site; and search engines would be banned from directly linking to the site (Timm 2011). Technologies to circumvent the blacklist would also be criminalized (McCullagh 2011).

End result? A digital blacklist.

FIGHTING COICA, THE PREDECESSOR OF SOPA

Many people who read about the SOPA protest in the news didn't see the years of work that went into the fight. The blackout was the final chapter in a two-year advocacy effort.

The effort started with the Combating Online Infringements and Counterfeits Act (COICA), a predecessor to SOPA introduced in 2010 that had similarly dangerous provisions. Resistance to COICA was much less coordinated and more muted, but it served several important functions: free speech organizations that worked on copyright reform became well-versed in the bill; groups like the Electronic Frontier Foundation (EFF) began educating the public (Esguerra 2010) about the problems; tech journalists became familiar with the issues; and members of Congress began getting letters from constituents. It was small-scale digital advocacy—signing online petitions that resulted in letters to Congress. Ultimately Senator Ron Wyden promised to stop the bill from moving forward, and COICA was stalled for a year (Masnick 2010).

At the same time, the American public was learning more about censorship through copyright in the news. In November 2010, the US government seized more than eighty domain names of websites they claimed were engaged in the sale and distribution of counterfeit goods and illegal copyrighted works. Free speech defenders were quick to point out how legitimate websites were swept up in the digital raid (McSherry 2010), while illegal sites were propagating on new domains within hours. Technology and security experts also began speaking out against the problems associated with undermining the domain name system. This would be a theme throughout the SOPA fights: experts from the tech fields pushing back against legislation that sought to undermine the basic structure of the Internet. After COICA was halted, there was a lull that lasted a few months. But soon, PIPA and then SOPA were introduced.

ADVOCACY AGAINST SOPA AND PIPA HEATS UP

Advocacy against SOPA and PIPA took a few different forms, but in the earliest stages it involved a mailing list and bill (i.e., policy) analysis. This is the boring, foundational work of activism. Lawyers, technologists, and activists at EFF and similar organizations spent hours analyzing the text of SOPA and PIPA. We also reached out to other advocacy organizations to let them know we were concerned about this bill and urged them to get involved.

DC-based Public Knowledge created an open discussion e-mail list to help coordinate advocacy against SOPA. There were two fantastic things about this mailing list. First, anybody could join, provided they were in some way interested in stopping SOPA. This meant advocacy organizations, companies, journalists, and independent advocates were lumped together into one open discussion list. It was a politically diverse group, with libertarians responding to liberals and vice versa. Second, it was fueled with accurate, timely information about the bill movement. There was a constant stream of information about new cosponsors, hearings, political rumors, opposition, and much more. The mailing list became the place for discussing ideas, tactics, and analysis. It was part cheerleading squad, part law lecture, part gossip mill.

With this open communication structure, we were able to orchestrate a series of fast and fun campaigns. This included two initiatives that were spearheaded by the newly minted activist group Fight for the Future. The first was a joke website that showed singer Justin Bieber in jail and proclaimed that the new copyright bill would put artists like Bieber in prison for copyright infringement. This adorable and relatable site was shared widely on social media, helping to educate the public. Not long after, Fight for the Future worked together with EFF and other groups in launching American Censorship Day, which was a prequel to the SOPA blackout that encouraged websites to post a banner that protested Internet censorship through SOPA. Tens of thousands of websites participated, including large sites like Tumblr.

We also successfully engaged the technical community, which was key to our success. Organized by EFF, a group of eighty-three prominent Internet engineers spoke out in an open letter to the US Congress opposing SOPA (Eckersley and Higgins 2011). We also saw tech companies increasingly begin to speak out against the bill, including through a full-page ad in the *New York Times* (Doctorow 2011).

We also saw public attacks on SOPA supporters. For example, Go Daddy, a prominent Internet registrar, suffered a boycott as a result of its support for SOPA. The boycott started on Reddit but advocacy groups, including Fight for the Future, quickly championed the boycott. Namecheap, a Go Daddy competitor, capitalized on the backlash by promising donations to EFF for all domains transferred to them. Go Daddy caved to the pressure and withdrew its support for the bill.

Blogging and social media proved especially powerful resources for advocates working on the issue.

THE BLACKOUT

The advocacy organizations that had been working to stop SOPA and PIPA didn't call for a blackout. In fact, the origins of the idea are hard to trace. There was an early suggestion from Reddit, and another from Jimmy Wales at Wikipedia that cited an earlier blackout of the Italian version of the site. Regardless of its true origins, once the idea was out there, it caught fire—and that was in large part because of the online network that had already developed and the advocacy groups spreading the word.

On January 28, 2012, over 115,000 websites are believed to have shut down or gone dark in protest. This included major websites like Wikipedia and Google. Many of these websites urged visitors to contact Congress to protest SOPA. Over 8,000,000 calls were attempted through Wikipedia's call look-up tool. Over 10,000,0000 signatures were gathered on petitions opposing SOPA, and over 4,000,000 e-mails were sent to members of Congress via activist platforms run by groups like EFF, Fight for the Future, and Demand Progress.[4] The protest was featured in major news outlets around the world. That same day, cosponsors of both SOPA and PIPA began withdrawing their support.

The day before the protest, 80 members of the US Congress supported SOPA, 31 opposed. Two days later, there were 63 supporters and 122 opposed (Geist

2012). The bill's author announced that he was withdrawing the bill. We had succeeded.

FIVE LESSONS FROM THE SOPA CAMPAIGN

Lesson 1: Embrace decentralization. This is the number one lesson from the SOPA fight. We never tried to control things. Instead, the role of advocacy organizations was to educate and inspire action, and then reframe and promote the actions of others. Much of this was accomplished through blogging and social media, whether explaining SOPA's impact or highlighting the work of an anti-SOPA advocacy campaign or urging people to join an upcoming action. It was never telling people what to do and then approving it; that does not scale. Instead, the real work was educating people about the issues, inspiring them to action, then framing and promoting those actions.

Lesson 2: Speak up quickly and often. One of the strongest tools we had in fighting SOPA was blogging and tweeting. The power of blogging cannot be emphasized enough. EFF alone was publishing blog posts about SOPA day and night, responding to breaking news, exploring new reasons to oppose the bills, and reacting to public statements and actions by members of Congress. These posts, in turn, fueled conversations on Hacker News, Reddit, Twitter, and other Internet communities. These blog posts also provided endless new hooks for technology journalists, which themselves provided stories for more mainstream publications; drove traffic to petitions and other actions; were useful in engaging users on social media; and helped bring in more advocacy groups.

Lesson 3: Be visible. In the same way a physical protest or a banner drop can create a memorable image that is forever emblematic of a movement, the SOPA protest added a powerful visual element to a political fight. The visibility helped to educate people who had not been following the issue, and helped penetrate major news outlets.

Lesson 4: Use Internet communities as a political force. The SOPA fight demonstrated clearly that the Internet is a political force. The communities of online users who connect and collaborate on platforms like Reddit, Twitter, and Hacker News are feeding thoughtful political discussion, educating one another about complex technical and political issues, and engaging in activism. These communities, plus many others, spoke up because they felt threatened by proposals that would censor the Internet. We must now understand that digital communities can be powerful political forces that can and will fight for their freedoms.

Lesson 5: Cross political lines. SOPA went beyond traditional left and right political lines. It was an innovation issue, a free speech issue, and an issue that impacted both big companies and everyday users. Groups put aside their political differences to stop the bill. We couldn't have stopped it otherwise.

This chapter, unlike the rest of the book, is licensed under a Creative Commons Attribution 4.0 International License.

1. Aaron Swartz, a leading activist in the fight against SOPA, in his 2012 "How We Stopped SOPA" speech, available on YouTube, www.youtube.com/watch?v=Fgh2dFngFsg.

2. Note that SOPA was amended significantly during the course of the debate. In general, this article refers to SOPA as it was originally introduced to the US House of Representatives. Since this article is focused on the activism that stopped SOPA, I haven't explained in full how the bill changed from its original form to its later amended version. For a better understanding of this topic, refer to the posts collected under "SOPA/PIPA: The Internet Blacklist Legislation," *Deeplinks Blog*, Electronic Frontier Foundation, www.eff.org/issues/coica-internet-censorship-and-copyright-bill.

3. The Stop Online Piracy Act (SOPA) and the PROTECT IP Act (PIPA) were sister bills in the House and Senate that attempted to codify a procedure for creating and censoring a blacklist of websites.

4. For complete numbers, see "The January 18 Blackout/Strike: In Numbers and Screenshots," SOPA Strike, Fight for the Future, http://sopastrike.com/numbers/.

REFERENCES

Doctorow, C. 2011. "Internet Giants Place Full-Page, Anti-SOPA Ad in NYT." *Boingboing*, November 16. http://boingboing.net/2011/11/16/internet-giants-place-full-pag.html.

Eckersley, P., and P. Higgins. 2011. "An Open Letter from Internet Engineers to the U.S. Congress." Deeplinks Blog, December 15. Electronic Frontier Foundation. www.eff.org/deeplinks/2011/12/internet-inventors-warn-against-sopa-and-pipa.

Esguerra, R. 2010. "Censorship of the Internet Takes Center Stage in 'Online Infringement' Bill." Deeplinks Blog, December 15. Electronic Frontier Foundation. www.eff.org/deeplinks/2010/09/censorship-internet-takes-center-stage-online.

Geist, M. 2012. "Geist: The Day the Internet Fought Back." *thestar.com*. www.thestar.com/article/1119151-geist-the-day-the-internet-fought-back.

McCullagh, D. 2011. "How SOPA's Circumvention Ban Could Put a Target on Tor." *CNET*, December 21. www.cnet.com/news/how-sopas-circumvention-ban-could-put-a-target-on-tor.

McSherry, C. 2010. "U.S. Government Seizes 82 Websites: A Glimpse of the Draconian Future of Copyright Enforcement?" Deeplinks Blog, November 29. Electronic Frontier Foundation. www.eff.org/deeplinks/2010/11/us-government-seizes–82–websites-draconian-future.

Masnick, M. 2010. "Senator Wyden Says He'll Block COICA Censorship Bill." *Techdirt*, November 19. www.techdirt.com/articles/20101119/05102211946/ senator-wyden-says-hell-block-coica-censorship-bill.shtml.

Timm, T. 2011. "The Stop Online Piracy Act: A Blacklist by Any Other Name Is Still a Blacklist." Deeplinks Blog, November 7. Electronic Frontier Foundation. www.eff.org/deeplinks/2011/11/stop-online-piracy-act-blacklist -any-other-name-still-blacklist.

CHAPTER SEVEN

Internet Freedom from the Outside In

CRAIG AARON *and* TIMOTHY KARR, *Free Press, United States*

MEDIA REFORM STRATEGY

Progress on any important issue will be impossible without changes to the policies, laws, and politics that created the media we have now. The only way to make such changes is by creating a broad, popular movement for media reform. To mobilize this movement, Free Press uses an outside-in approach. For its net neutrality campaign, which culminated with a historic victory at the Federal Communications Commission (FCC) in February 2015, Free Press mobilized hundreds of organizational allies and millions of "outside" activists to sign petitions, call Congress, and attend FCC hearings and local rallies. Free Press combined its robust field strategy with "inside" expertise—including a Washington-based team of lawyers and lobbyists—to compel policy-makers to side with the public interest. Whether working for Internet freedom in the United States or abroad, advocates won't win the right policies without a symbiosis of fieldwork and policy intelligence. While the fight for net neutrality is far from over, the best long-term strategy for saving the Internet will be to work from the outside in.

On a cold Thursday morning in January 2014, a small group of advocates gathered outside the imposing edifice of the FCC in Washington, DC. They opened the trunk of a red Ford Fusion parked nearby and began unloading about twenty white banker's boxes. Within minutes, they had assembled a makeshift cardboard podium. Inside the boxes were more than a million signatures collected in just two weeks from people across the country.

Each person had signed on to an online letter demanding the FCC Chairman Tom Wheeler protect net neutrality, the principle that ensures that the Internet remains open and free from discrimination by the large phone and cable companies that control broadband access for most Americans.

These petition deliveries were just the beginning of a year of Internet activism that resulted in unprecedented public involvement in an issue before the FCC, unexpected popular attention on the once obscure issue of net neutrality, and a remarkable reversal that would eventually lead both the president of the United States and the chairman of the FCC to endorse the strongest public interest protections.

Net neutrality as an idea is but one part of a larger global movement of people fighting for Internet freedom. It is a movement that includes democracy activists in Eastern Europe, Arab Spring protesters in the Middle East and North Africa, and dissident bloggers and "hacktivists" across Asia. In early 2012, more than ten million people mobilized online and off to defeat the SOPA/PIPA Web censorship legislation in the United States (Lee 2012). Activists are using the open Internet to fight unchecked spying and surveillance by the NSA and demand online privacy and free speech rights.

The Internet was designed to be an engine of disintermediation, free speech, and inclusion—a means by which anyone could route around the gatekeepers, build online communities, and share information. It opened the door for new forms of grassroots political organizing and gave smart online activists an impressive means to influence policies and shame bad actors.

With only a tiny fraction of financial resources of our opponents, Internet freedom advocates struggle every day to preserve this openness. In response to the Internet freedom movement's successes, industry has hired a phalanx of front groups to paint activists as extremists and doomsayers (Gustin 2010; Powell 2014). The fate of the Internet is best left in the hands of "experts," they say, those think-tank analysts who are themselves often on the dole of powerful cable and telephone companies (Ekran 2014).

For their part, the empires of companies like AT&T, Verizon, Comcast, and Time Warner Cable were built not on innovation, free speech, and inclusion, but on control of markets and squelching new competition. Despite what their PR offices would like us to believe, their armies of lobbyists earn their keep by winning anticompetitive legislation and exclusive government franchises while stifling the entry of new applications and services on their networks.

And it's quite an army. The three largest Internet service providers—Verizon, AT&T, and Comcast—have spent more than $500 million dollars on lobbyists and an additional $185 million in campaign contributions over the past twenty-five years.[1] This corporate largesse has delivered a return on investment. The past three decades of media policymaking is a story of growing consolidation of media power into the hands of a very few.

Essential to their success is the corruption of the political process and an ever-revolving door between the regulators and the regulated. Of the twenty-six FCC commissioners and chairs to have served at the agency between 1980 and the end of 2012, at least twenty-one have gone on to work—as an employee, consultant, lobbyist, lawyer, or paid board member—for corporations in industries they were in office to regulate. Michael Powell who served as FCC chairman from 2001 until 2005 is now the top lobbyist for the National Cable and Telecommunications Association. In 2011, Meredith Atwell Baker left her job as FCC commissioner for a top lobbying position at Comcast. She announced her transition just four months after voting to approve the cable giant's takeover of NBC (Stearns 2011).

This is the challenge facing any Internet freedom fighter, While we are far from pinning down this many-tentacled beast, we believe that outside-in organizing is the best strategy to undo the damage caused by a generation of insidious corporate influence over media policy. It is a strategy that can yield results not just in Washington but also with any government that is structured—at least nominally—to be responsive to public input. Outside-in organizing is an approach that has allowed us to use the Internet to save the Internet. This confluence of the medium and the message give new leverage to the fight against seemingly insurmountable odds.

OPPOSING FORCES

The history of the Internet has been a history of openness and innovation as well as a history of steady encroachment on the fundamental ideas underlying the Internet's evolution.

It has been twenty-five years since Sir Tim Berners-Lee designed an open protocol for the World Wide Web. It is this innovation—and the end-to-end network engineering principle put in place by other Internet founders like Vint Cerf—that served as catalysts for the Internet's explosive growth (Jakma 2011). These foundations had far-reaching political implications, favoring systems that are more decentralized and democratic, promoting universal access to information and amplifying dissident voices. Without the inherent protections an open Internet offers, political voices that lack a big-money megaphone will be drowned out.

It is an approach that has also given us a truly free marketplace of ideas where even the smallest entrepreneur can theoretically compete with giant corporations. (Would we have many of the Internet's most innovative businesses—like Twitter, YouTube, or Etsy—had they been unable to enter the market on a level playing field?).

The alternate history of the Internet is one of steady encroachment on these fundamental ideas by gatekeeper media. In the United States, companies like AT&T, Comcast, and Verizon have managed to chip away at the rules that were put in place to protect openness. Their goal is to impose a gatekeeper business model on content—anything that allows Internet service providers to squeeze more profit out of the rush of data.

At the center of the tug-of-war over the Internet is the issue of net neutrality. Should Internet users' rights to communicate without interference be protected under the law? Or should Internet access providers be able to treat the network and the content that flows across it as theirs to censor and exploit?

THE OUTSIDE-INSIDE CHALLENGE

Since its founding ten years ago, Free Press has taken up the cause of media users of every stripe. The net neutrality battle has taken center stage in our work. Our challenge in the face of such institutional corruption is to shame those in Washington who sell their influence to the highest bidder while amplifying the public voice in key policy debates.

To accomplish this we had to frame the debate correctly. The battle for net neutrality was not a clash of corporate titans and their lobbyists—even if the mainstream media often painted it as a standoff between Google and AT&T. The real story was one of a new political movement pitting the interests of all Internet users against the narrow concerns of a handful of wealthy and politically influential corporate elites (see, for example, Labaton [2007] for a journalist using the *Clash of Titans* frame). Although net neutrality advocates were vastly outspent and outmanned in the halls of power, industry could not kill the power of an engaged, passionate, and connected public.

Our outside-in approach on net neutrality relied upon the expertise and credibility of Free Press "inside" experts—lawyers, researchers and advocates—while simultaneously mobilizing hundreds of organizational allies and millions of "outside" activists to sign petitions, call Congress, and attend FCC hearings and local rallies from Minneapolis to Mountain View.

Employing this strategy for net neutrality seems so obvious in some respects that it's amazing how often it is not used. Insiders regularly fail to engage the public on key issues and then wonder why people never seem to care. And popular ideas fail to take root in Washington because there's nobody on the inside making the case—or worse, the appointed insiders are actively working against the interest of their constituents.

The key is establishing a symbiosis. The insiders need the threat of a public outcry to get a seat at the table. But hundreds of thousands of calls and letters to Congress and the FCC are worthless without the inside knowledge to target the right decision-makers at the right time. You need to strike a balance between the credibility of the policy experts and the creativity of the field.

Though vastly outnumbered on Capitol Hill—and largely absent at industry cocktail parties at campaign fundraisers—Free Press has relied on a small group of advocates and allies to represent us in Washington. We work with a network of social justice and civil liberties allies to connect with key congressional staff-ers and educate "grasstops" leaders, and conduct independent research and analysis. We collaborated closely with legal scholars such as Columbia's Tim Wu, and Harvard University's Susan Crawford, and Lawrence Lessig, whose widely influential writing on open networks, universal access and free culture have captivated a generation of Internet advocates.[2]

The campaign insiders also reached out to potential industry allies. Internet "edge" companies—including Google, Amazon, and Netflix—had formed the Open Internet Coalition in 2006, but its corporate members often got cold feet when facing pressure from Capitol Hill. While at times working on legisla-tive strategy alongside the Internet companies that supported net neutrality, Free Press never took their money. As tempting as it might have been to tap into the corporate coffers, it was more important to remain independent. One needed only to look at our opponents to see what happened when legitimate groups got coopted by corporate cash.

Insiders are often skittish about grassroots efforts—largely because they can't control them. The public outreach—spearheaded by Free Press, MoveOn.org, CREDO Action, Demand Progress, Fight for the Future, and other groups—has helped find remarkable and genuine local stories that showed how taking away net neutrality would harm everyday people and small businesses.[3] By letting the grassroots speak for themselves we managed to overcome the reluc-tance of some Washington, DC, decision-makers, who view the larger, public-interest advocacy groups as just another interested party.

SETTING THE STAGE FOR 2015

The story of 2014 was of all these overlapping networks coalescing into a remarkable campaign. Online, a massive Internet slowdown was joined by forty thousand websites, including such major platforms as Netflix, Tumblr, and Etsy. All told, more than four million people filed official comments with the FCC—more than on any other issue in the agency's history. Offline, the highlights included repeated rallies outside the FCC and at Obama fundrais-ers, people camping out on the FCC's doorstep, enormous video billboards erected in Washington, DC, packed public hearings across the country, and creative street theater. This upsurge in activism attracted the attention of the press and entered popular culture through the viral-video rant of HBO's John Oliver.

It was the activists who moved the companies and the politicians, not the other way around. The activists created the political space in which President Obama could step into the bully pulpit to push for the strongest net neutral-ity protections. And that gave FCC Chairman Tom Wheeler the room to change his mind. On February 26, 2015, the FCC voted to protect net neutrality and reclassify broadband under Title II of the Communications Act. Wheeler, who was interrupted by multiple standing ovations, called it "the proudest day of my public policy life" (Kang and Fung 2015).

This moment was ten years in the making and marks the biggest win ever for the public interest at the FCC. Net neutrality as an issue has become the gateway to Internet freedom activism on several related fronts, including opposition to industry mergers like the scuttled deal between Comcast and Time Warner Cable, the fight for universal, affordable Internet access, online privacy, and free speech both domestically and on a global scale. Our goal has never been about getting faster movie downloads or new gadgets. The fight for Internet freedom at its core is about democracy—it is about ensuring that citizens have the information they need to hold their leaders accountable and the tools to have an impact on the decisions shaping their lives. That is why saving the Internet is so important: It might be our best—and last—chance to make an end-run around the mainstream media gatekeepers. But the Internet won't reach this promise if we don't demand a seat at the table where decisions about its future are being made. And the only way to get that seat is from the outside-in.

NOTES

1. Figures from Influence-Explorer, an online resource of the Sunlight Foundation, drawing on data provided by the Center for Responsive Politics, the National Institute for Money in State Politics, Taxpayers for Common Sense, the Project on Government Oversight, the EPA, and USASpending.gov.

2. Tim Wu (2003, 143) described this threat as network operators' "unfortunate tendency to want to ban new or emerging applications or network attachments, like Wi-Fi devices or Virtual Private Networks, perhaps out of suspicion or an (often futile) interest in price discrimination." In an August 2003 submission to the FCC, Tim Wu and Lawrence Lessig (2003) urged the agency to pay closer attention to possible deviations from a neutral Internet regime and warned of the impact such discrimination would have on online innovation.

3. To read more about the coalition's grassroots organizing, see Candace Clement and Mary Alice Crim (2015).

REFERENCES

Clement C., and M. A. Crim. 2015. "A 'Political Miracle': What We Learned from the Net Neutrality and Comcast Fights." Free Press, June 24. http://www .freepress.net/blog/2015/06/24/'political-miracle'-what-we-learned-net -neutrality-and-comcast-fights.
Ekran, G. 2014. "Net Neutrality Struck Down: Techpolicydaily Experts Respond." American Enterprise Institute. January 14. www.aei-ideas.org/2014/01/ net-neutrality-struck-down-techpolicydaily-experts-respond/.

Gustin, S. 2010. "FCC Passes Compromise Net Neutrality Rules." *Wired*, December 21. www.wired.com/business/2010/12/fcc-order/.

Jakma, P. 2011. "Cerf and Kahn on Why You Want to Keep IP Fragmentation." June 28. http://paul.jakma.org/2011/06/28/cerf-and-kahn-on-why-you-want -to-keep-ip-fragmentation/.

Kang, C., and B. Fung. 2015. "FCC Approves Strong Net Neutrality Rules." *Washington Post*, February 26. www.washingtonpost.com/blogs/the-switch/ wp/2015/02/26/the-fcc-set-to-approve-strong-net-neutrality-rules/.

Karr, T. 2006. "National Outpouring of Support for Net Neutrality." *MediaCitizen*, August 31. http://mediacitizen.blogspot.com/2006/08/national-outpouring -of-support-for-net.html.

Labaton, S. 2007. "Congress to Take Up Net's Future." *New York Times*. 9 January. www.nytimes.com/2007/01/10/washington/10net.html.

Lee, T. 2012. "SOPA Protesters by the Numbers: 162m Page Views, 7 Million Signatures." *Arstechnica*, January 19. http://arstechnica.com/tech-policy/ 2012/01/sopa-protest-by-the-numbers-162m-pageviews-7-million -signatures/.

Powell, M. 2014. "Internet Doesn't Need Phone-Era Rules: Opposing View." *USA Today*, January 14. www.usatoday.com/story/opinion/2014/01/16/internet -net-neutrality-fcc-broadband-editorials-debates/4542665/.

Stearns, J. 2011. "Pressure Builds after Baker Leaves FCC post for Comcast." *Free Press* (blog), May 13. www.freepress.net/blog/11/05/13/pressure-builds-after -baker-leaves-fcc-post-comcast.

Wu, T. 2003. "Broadband Discrimination, Network Neutrality," *Journal of Tele-communications and High Technology Law* 2, no. 1 (Fall): 141–175. http:// www.jthtl.org/content/articles/V2I1/JTHTLv2i1_Wu.PDF.

Wu, T., and L. Lessig. 2003. "Ex Parte Submission in CS Docket No. 02–52," August 22. www.timwu.org/wu_lessig_fcc.pdf.

CHAPTER EIGHT

A Victory for Digital Justice

JOSHUA BREITBART, *Formerly New America's Open Technology Institute, United States*

MEDIA REFORM STRATEGY

When Congress created the Broadband Technology Opportunities Program (BTOP) in 2009, the media reform movement marked it as a major, albeit partial victory. The nonprofit Allied Media Projects in Detroit and Media Mobilizing Project in Philadelphia responded to this opportunity by building digital justice coalitions with groups focused on housing, workers' rights, youth, education, and environmental justice that had not worked on the digital divide in the past. The Open Technology Institute supported these innovative, visionary efforts with technical and policy expertise, a network of researchers, and a commitment to enduring partnerships. As a program, BTOP provided an opportunity for an enduring impact on broadband in the United States. In Philadelphia and Detroit, the coalitions used the grant-seeking process to craft visions for how the Internet could better serve their respective communities, and turned these visions into successful proposals for the newly available BTOP grant funds. With the grants, the local communities formed relationships, developed skills, and built infrastructure with which to take on new challenges. The digital justice coalitions proved a highly effective strategy for transforming the local digital ecosystem in response to the conditions created by the BTOP.

In 2009, digital justice coalitions in Detroit and Philadelphia seized an opportunity to turn a temporary federal program into a lasting transformation of the Internet's role in their local communities.

When Congress included $7.2 billion for broadband access and adoption in the American Reinvestment and Recovery Act of 2009 (ARRA), the media reform movement marked it as a major, but partial victory. It was a victory because ARRA funded a Broadband Technology Opportunities Program (BTOP) that marked a dramatic shift marked a dramatic shift in policy from a reliance on the private market to a more active role for the government in extending the benefits of the Internet to everyone across the United States; only a partial victory, though, because the hard work of devising and enacting a new approach to the Internet remained.

No longer were we limited to constraining the harm the federal government might cause; we could spend public dollars directly on making the Internet more useful and accessible for those who had previously been shut out from its benefits. To do this, we had to engage groups that had never before worked on media policy issues. While past media reform campaigns called on people to make a small investment of time in the form of a call to Congress or a comment to the FCC, we needed to generate multiple, high quality, innovative federal grant proposals, which analysts at the Open Technology Institute (OTI, then known as the Open Technology Initiative) estimated would take a skilled team of four or five people approximately 240 to 250 hours to complete (New America Foundation 2009).

FORMING DIGITAL JUSTICE COALITIONS

Allied Media Projects (AMP) in Detroit and Media Mobilizing Project (MMP) in Philadelphia responded to this opportunity by building digital justice coalitions with groups focused on housing, workers' rights, youth, education, and environmental justice that had not worked on the digital divide in the past. The coalitions crafted visions for how the Internet could better serve their respective communities, and planned to turn these visions into successful proposals for the newly available federal funds (MMP 2009).

OTI was well positioned to support these innovative, visionary efforts with technical and policy expertise, a network of researchers and a commitment to enduring partnerships. OTI provided an organizing roadmap for Philadelphia based on an analysis of the city's previous major attempt at expanding broadband connectivity and then carried the idea of starting a digital justice coalition (DJC) to Detroit (Breitbart, Lakshmipathy, and Meinrath 2007). AMP and MMP used the annual Allied Media Conference in Detroit to share lessons directly and connect with other BTOP hopefuls.

As we translated these visions into the immediate context of the new federal policy, OTI and our partners wanted the National Telecommunications and Information Administration (NTIA) to affirm multimedia production as an essential component of digital literacy, in addition to basic skills like search and e-mail. We also sought to have nontraditional anchor institutions included as hosts of computer centers, along with the public libraries and community colleges that were referenced in the Recovery Act (US Congress 2009).

When NTIA approved our grant proposals and implemented programs along these lines, it set an important precedent of government support for this type of community media production as an economic development strategy and as a new vision of public media. More importantly, we saw how it allowed people to engage as full authors from places where they already had community support. Instead of just adopting broadband as individuals, whole communities could go online together so the Internet would be a welcoming enhancement to their offline lives.

This media reform victory offers lessons beyond Internet policy for anyone hoping to build cross-sector coalitions and reach traditionally underserved groups with new technology. There were four phases to this campaign: (1) shaping the funding program; (2) winning funds for exemplary, innovative projects, (3) implementation, including documentation and evaluation; and (4) communicating and building on the results of the implementation.

As Congress began consideration of the Recovery Act, OTI worked with other public interest groups in Washington, DC, to shape the broadband component of the bill to support adoption and public computer centers, and ensure that community organizations could be eligible for grant funding. The statutory language left a lot of discretion to the implementing agencies: the Rural Utility Service, a part of the US Department of Agriculture, and NTIA, a division of the Department of Commerce.

The NTIA's program, known as the Broadband Technology Opportunities Program (BTOP), included funds for mapping current broadband availability and for building new infrastructure, and for two other areas focused on encouraging use of the technology through "sustainable broadband adoption" activities and "public computer centers." Once the legislation was passed, we focused our attention on these grant programs as the more accessible opportunity for smaller organizations, and the only one for groups in urban areas.

PURSUING A BROAD VISION

As we moved into the second phase of outreach to potential applicants, we gathered many ideas from people as to how BTOP could most benefit their communities. After the first grant opening, the NTIA issued a request for comments on how it should administer subsequent rounds of the program. We submitted written comments in November 2009 with more than two dozen recommendations endorsed by more than forty organizations. We proposed an expansive vision of what the program could support. In our comments, we encouraged a broader definition of community anchor institution, cited the importance of digital media and social media skills for broadband adoption, and proposed the use of publicly available data like unemployment and poverty rates to determine eligibility for BTOP funds, rather than hard-to-find broadband adoption rates as NTIA had required.

We faced another challenge in pursuing a broad, community-based vision of broadband expansion: Many hard-to-reach groups did not know the program existed. We set up a broadband stimulus resource library where we offered guides to each funding stream as well as a Primer on Federal Internet Funding in July 2009 that we produced with People's Production House, a community media organization (OTI and People's Production House 2009). The single page, printed front and back in black and white and folded into a small booklet, included bullet points with background information and basic guidance for groups to determine whether this opportunity was right for them. We used this as a handout at a series of strategic discussions titled "A Healthy Digital Ecology: Creating a Community Vision for Federal Internet Funding" across multiple national conferences that OTI organized with a range of media reform and media justice organizations throughout summer 2009.

MMP and AMP, both of which emphasize media production in their missions, organized digital justice coalitions in their cities, connecting media education and technology groups with a range of social justice organizations and community anchor institutions. The coalitions bridged a gulf in their respective communities between those who understood the potential of broadband as an organizing tool and the organizers of communities that did not use broadband. These coalitions had already identified BTOP as an organizing opportunity. OTI provided expertise in the federal policy guidelines and how the program was playing out across the country.

INVOLVING THE COMMUNITY

The digital justice coalitions were grassroots and composed of experts in *not* having access to technology, either through their personal experience or from their immediate constituents. The coalitions formulated principles to guide their work and fashioned visions for the role of broadband in their communities that put a potential BTOP proposal in a broader context (Detroit Digital Justice Coalition 2009). To inform their proposals, the digital justice coalitions adopted participatory techniques from community organizing and urban planning. Public fora were discussion based, with small breakouts. The Detroit Digital Justice Coalition held neighborhood technology fairs called DiscoTechs (short for "discovering technology") that served as both a teaching opportunity and a form of participatory research (Breitbart 2012). Moreover, even if the had not led to a winning proposal, the public forums and DiscoTechs expanded the coalitions and enhanced the shared community vision.

The coalitions in Philadelphia and Detroit took distinct paths, as shaped by their local conditions. In Philadelphia, the coalition spurred the city government and large institutions to action, and with its members ultimately taking part in applications from the City of Philadelphia and the Greater Philadelphia Urban Affairs Coalition. The Detroit group, operating in a city where local government and most major institutions had failed residents, retained its identity as the Detroit Digital Justice Coalition and was able to carve out the components for Detroit in twin statewide proposals from Michigan State University.

Both coalitions were successful in winning funds for their visions of broadband adoption and public computer centers. The Philadelphia Freedom Rings Partnership (FRP) was awarded $11.8 million for broadband adoption and $6.4 million for computer centers. Michigan State University, whose statewide proposals included the Detroit Digital Justice Coalition, received $5.2 million for adoption and $6.1 for centers.

Winning the grants brought us both funding and new challenges as we moved into the implementation phase. These projects—cross-sector coalitions with novel teaching programs and new computer labs—were inherently ambitious. The coalitions had built themselves up through long hours and a lot of grit, through which a relatively small group of leaders came to know each other rather well. Receiving the grants and scaling up meant moving from hard-won relationships and no money to new relationships and a lot of money. Having a prominent lead organization on the proposals was essential to win the grant and manage the federal reporting requirements, but in practice this also meant translating the horizontal relationships of the organizing coalitions into hierarchical relationships of prime grant recipients, subrecipients, and subcontractors.

OTI moved from being policy advocates to instigators to grant writers to contracted evaluators, which required balancing independence, subordination, and evangelizing for the shared (yet evolving) vision. Once the money was awarded, we moved quickly to implement the projects. Grant recipients had to proceed with hiring, contracting, procurement, refinements to the program design and the development of an evaluation framework in parallel. Some of the partners were doubling or tripling their budgets and staff sizes almost overnight. Finding staff with the requisite skills, cultural competence and shared vision was itself a challenge to implementation. The Recovery Act was designated as economic stimulus and there was both political pressure and requirements in the grant to act fast. These pressures also made it hard to build relationships across grant programs in different locations.

The vision of these programs as locally defined—eschewing a formulaic approach to teaching basic digital literacy and placing computer centers in traditional institutions—meant local partners had to figure a lot of things out for themselves. In practice, this ambition meant triple the work: the baseline set of organizational and educational activities that one would expect, plus a steady cycle of evaluation and iteration, and regularly having to make the case for your methods to the funder and other traditional actors in the field.

OTI and partners had to figure out how to document and communicate the impact of the digital justice coalitions' innovative approaches. We did not want to rely on the accepted indicators of skill acquisition, new broadband subscriptions, and computer center use, which are in and of themselves rather difficult to track across multiple programs using separate curricula to tackle various barriers among different populations. Those metrics did not fully reflect our goals or values. We developed new indicators to document the important role of personal and institutional networks in facilitating broadband adoption and how these relationships could sustain these impacts in the form of "durable social infrastructure" (Breitbart 2013).

In Philadelphia, OTI's evaluation identified "the FRP's successful approach to embedding digital literacy and public computer access in existing community-based and social service networks" (Gangadharan, Carolan, and Chan 2013). We showed how the people and institutions in these networks built relationships that led to "individuals [using] broadband technologies in settings that they find most appropriate to their personal needs and contexts" (ibid.). We collected a wealth of data on how this approach and these relationships make the Internet relevant and useful for those it had previously underserved.

The Detroit Digital Justice Coalition built networks of teachers, youth, and artists and trained them to use media production and web development for organizing, education, and economic development. The lead groups in Detroit promulgated the use of the #detroitfuture hashtag on Twitter among all program participants as a way to collaboratively develop an online narrative about Detroit that would be attractive for new broadband users. OTI archived this crowdsourced program documentation to analyze the ripple effects of the trainings and identify the expanding social networks of the program (Breitbart, Byrum, and Bullen 2014).

OTI was able to move these insights into the mainstream of academic and policy discourse on broadband adoption. OTI convened a consortium of researchers to discuss "meaningful metrics for broadband adoption" in April 2012 and published the resulting papers in a peer reviewed special section of the *International Journal of Communications* (Gangadharan and Byrum 2012). NTIA drew heavily on this framework in a discussion about establishing common metrics for broadband adoption at the close of BTOP in 2014 (National Telecommunications and Information Administration 2014).

The Detroit Digital Justice Coalition continues to work with DiscoTechs, focusing on housing and other pressing issues for the city. OTI and AMP developed a concept of digital stewardship and a related training program that emphasizes community control of communications infrastructure, teaching neighborhood residents to build their own wireless networks.

The BTOP money has come and gone. As a program, it was fleeting, but it provided an opportunity for an enduring impact on broadband in the United States. In Philadelphia and Detroit, we were able to use the grant-seeking process as a vehicle for visioning and organizing, and for bringing new voices and audiences into the conversation about our shared digital future. With the grants, the local communities formed relationships, developed skills and built infrastructure (in the form of the computer centers) with which to take on new challenges, whether transforming the Internet or using the technology to improve their cities in other vital ways. The digital justice coalitions proved a highly effective strategy for transforming the local digital ecosystem in response to the conditions created by the Broadband Technology Opportunities Program.

REFERENCES

Breitbart, J. 2012. "The DiscoTech Is an Opportunity to Teach—and Learn." *Detroit Digital Justice Coalition*, no. 4: 21–23. https://www.alliedmedia.org/ files/ddjc_zine_4.pdf.
———. 2013. Strategies to Sustain Your PCC/SBA BTOP Grant. Workshop. Schools, Health, and Libraries Broadband Coalition annual conference.
Breitbart, J., G. Byrum, and G. Bullen. 2014. *A Network Model of Broadband Adoption: Using Twitter to Document #DetroitFuture Programs*. Preliminary analysis by the Open Technology Institute. New America Foundation.
Breitbart, J., N. Lakshmipathy, and S. D. Meinrath. 2007. *The Philadelphia Story: Learning From a Municipal Wireless Pioneer*. New America Foundation. www.newamerica.net/files/NAF_PhilWireless_report.pdf.
Detroit Digital Justice Coalition. 2009. "Introducing the Detroit Digital Justice Coalition." Allied Media Projects. http://alliedmedia.org/#!/news/2009/11/ 22/introducing-detroit-digital-justice-coalition.
Gangadharan, S. P., and G. Byrum. 2012. "Defining and Measuring Meaningful Broadband Adoption." *International Journal of Communication* 6. http:// ijoc.org/index.php/ijoc/article/view/1836.
Gangadharan, S. P., K. Carolan, and K. Chan. 2013. "The KEYSPOT Model: A Home away from Home. An Evaluation of the Philadelphia Freedom Rings Partnership." Open Technology Institute, December 10. www.newamerica .org/oti/the-keyspot-model-a-home-away-from-home/.
Media Mobilizing Project (MMP). 2009. "Digital Philadelphia Initiative Summit." *YouTube*. www.youtube.com/watch?v=pz4Pt4mDNgY.
National Telecommunications and Information Administration (NTIA). 2014. Research and Evaluation: Meaningful Metrics. Webinar.
New America Foundation. 2009. *Application Guide for Public Computer Center Funding from the Broadband Technology Opportunities Program (BTOP)*. https://static.newamerica.org/attachments/3994-public-computer-centers -program-application-guide/Public_Computer_Centers_Program_ Application_Guide.23a568c438384c8ba942f6e0bc8c125a.pdf

Open Technology Initiative and People's Production House. 2009. "Primer on Federal Internet Funding." New America Foundation. https://static.newamerica .org/attachments/3891-primer-on-federal-broadband-funding/BTOP _primer_zine.b163183770d44c069fc07478ddof01d2.pdf.

US Congress. 2009. Senate and House of Representatives. American Recovery and Reinvestment Act of 2009. H.R. 1. 111th Congress, 1st Session. https:// www.govtrack.us/congress/bills/111/hr1/text.

CHAPTER NINE

Working Toward an Open Connected Future

DAVID CHRISTOPHER, *OpenMedia, Canada*

MEDIA REFORM STRATEGY

The Internet won't save itself. And no single organization can save it either. That is why everything OpenMedia does is about placing citizens at the heart of digital policy. Our digital future is just too important to be left to industry lobbyists and government bureaucrats. OpenMedia engages, educates, and empowers people through long-term citizen-driven campaigns. For this, it is important to ensure that citizen voices are at the heart of the debate. We ensure citizens can have their say through accessible online actions, innovative participatory tools, and through Internet Town Halls and other events.

Having gathered citizen input, we do our utmost to make sure those voices are heard at the heart of government. We do so by appearing at regulatory proceedings, meeting with politicians, creating online tools that make it easy for citizens to take part in regulatory processes, and by leveraging media coverage to encourage decision-makers to listen to citizens. Working with Canadians, OpenMedia has developed crowd-sourced, informed, in-depth research that shapes our policy plans. This approach is having a real impact and has led to a significant improvement in recent years in terms of how the Canadian Radio-television and Telecommunications Commission (CRTC) listens to Canadians.

Imagine it is the year 2020 and you've been having a frustrating day. You've just been stopped at the border—the US guard wouldn't tell you why, but you guess it is because of something you said or did online recently. They collect so much data on your everyday life these days that it's difficult to know what, or why they red-flagged you. It's been happening to a lot of your friends recently too (Bridge 2011).

Returning home, you log on to your Internet account. It's costing you over two hundred Canadian dollars a month for this, and some days you wonder why you bother. You still enjoy connecting with people, but so much content is off limits nowadays that your Internet is feeling more and more like cable TV (Isaacson 2014). That extra twenty-dollar monthly fee to access Facebook still annoys you—and you just can't find the money needed to access online news content other than from your sensationalist local TV station. At least that station comes as part of your Internet package because it is owned by your provider—but to access the BBC would be an extra twenty-five dollars a month, and your provider blocks access to other online news portals completely.

Checking your e-mail, you discover a threatening letter (Brown 2013) from a US media giant. It seems to have something to do with a song your six-year-old niece accidentally downloaded when she visited a few months ago. The e-mail warns that under copyright rules negotiated under a previous Trans Pacific Partnership, you will be kicked off the Internet entirely if it happens again (Desjardins 2013), and now you've got to pay an extortionate fine to avoid getting taken to court.

At the same time, you are acutely aware that storing everything everyone says or does online in giant government databases comes at a cost to taxpayers. And you're constantly worried that you'll be the victim of a massive leak of your private data (Bennett 2014)—it seems the government just can't keep anything secure.

You fondly remember the good old days, where you could read and connect privately with whoever you wanted online. It hasn't been like that for years—not since those old net neutrality rules were struck down and your Internet provider started blocking or charging extra for content it didn't own. And you have no option outside the high-cost Big Telecom Internet providers since they succeeded in blocking you from more affordable, independent services. You wonder why you didn't make more of a fuss about that at the time.

Does all this sound terrifying? Thankfully, we are still some way off from this dystopian vision of the future. But we are already on a very slippery slope and this is exactly what the Internet could look like in a few years if citizens step back and leave digital policy to telecom lobbyists and spy agency bureaucrats.

OpenMedia exists to stop this from happening—it is an award-winning community-based organization that safeguards the possibilities of the open Internet. We work toward informed and participatory digital policy by engaging hundreds of thousands of people in protecting our online rights.

We do this work because digital policy affects each and every one of us. Our high cell phone bills are holding back our economy and Canada as a whole. Mass surveillance of law-abiding citizens has grown secretive, expensive, and out-of-control—stifling free speech in Canada and overseas. And supposed trade deals negotiated behind closed doors contain extreme Internet censorship rules that would seriously undermine citizens' ability to communicate freely online.

If we stood back and did nothing then at best we'd end up with an Internet that was far more expensive, locked down, censored, and policed. And, if recent challenges in the United States to net neutrality are anything to go by, we may not even have an Internet at all—at least not as we know it today. That's why, at OpenMedia, we work for digital policies that make sense for all users of the Internet—policies that help build the open, connected society we all deserve.

CASE STUDY OF A SUCCESSFUL REFORM STRATEGY

Our campaign on reshaping Canadian wireless policy offers a compelling example of how the OpenMedia community works to bring about positive change. Canadians pay some of the highest prices in the industrialized world for often patchy and unreliable wireless service. This has been confirmed time and again by independent reports (Christopher 2013). Many Canadians have no choice but to pay the high prices on offer by the Big Three: Bell, Rogers, and Telus. These three providers control over 90 percent of the market, meaning many Canadians have no independent options whatsoever.

In January 2012, we launched the "Stop the Squeeze" campaign, which urged Canadians to speak out against the Big Three's near-oligopoly. As part of this campaign we launched an online petition—along with videos, posters, and viral share images—to equip Canadians with the tools needed to speak up and pressure decision-makers. Over sixty-five thousand Canadians took part and in October 2012 the CRTC announced a public consultation to create new national rules for wireless services. At OpenMedia we wanted to make sure that these reflected the lived realities of Canadian cell phone users. We created a new online tool to enable Canadians to share their cell phone horror stories about the effect high wireless prices and poor service had on their daily lives.

Over three thousand submissions were received through this campaign—some of which consisted of multiple pages detailing remarkable mistreatment at the hands of the Big Three. A common complaint was from Canadians returning home after trips to the United States or overseas to receive staggering bills for roaming charges. This citizen input was used to create a crowdsourced report entitled *Time for an Upgrade* (OpenMedia 2013) that reflected how Canadians wanted to see the wireless market reformed. During a crowdsourced presentation to the CRTC, the key themes of this report were reinforced: lower prices, greater choice, and better service. Canadians were kept informed via live tweets and online updates from the CRTC hearing.

OpenMedia focused on making the case that any new wireless code of conduct must be grounded in the evidence-based realities that our campaign had highlighted. In June 2013 this work paid off—the CRTC announced a broadly positive new set of customer protection rules (Rennie 2013) that capped data roaming rates, shortened contracts, and made it easier to switch to a new provider.

By this time, it seemed that decision-makers were starting to listen to Canadians but there was still a long way to go. Despite these new customer protections, independent reports revealed that wireless prices were still increasing at thirteen times the rate of inflation (CNW 2013). Canadians were also concerned about attempts by large telecom providers such as Rogers and Telus to take over scarce wireless spectrum resources that had been set aside for independent providers.

As focus shifted from customer protections to the need for greater wireless choice, over forty leading Canadian innovators and entrepreneurs spoke out to tell Industry Canada that high prices were stifling job creation and holding Canada's economy back. Responding to this pressure, the government announced new rules (Industry Canada 2013) to help prevent the Big Three from taking over independent providers.

With a new industry minister, James Moore, taking office we encouraged Canadians to send a message asking him to stop the Big Three acting as gatekeepers to resources that were essential for Canada's economy. We underlined the need for the government to listen to Canadians, whose crowdsourced roadmap called for opening the wireless networks to deliver real affordability and choice. Once again, citizen pressure delivered results, with the new industry minister promising to take modest steps to ensure independent providers could access wireless networks at an affordable rate (Market Wired 2013), along with financial penalties for telecom giants who break the rules. Notably, the government has effectively staked its reputation on delivering "more choice and lower prices"—even launching a government website and a TV ad campaign to underline this goal. This would never have happened without citizens speaking out en masse and working together to provide a positive roadmap for policy change.

THE INTERNET IS KEY TO BUILDING A PARTICIPATORY SHARED SOCIETY

OpenMedia has seen again and again that when people speak up en masse, it really does make a difference—delivering real results and positive change. The common theme with all our successful campaigns is our emphasis on citizen participation. Our community is placed at the heart of everything OpenMedia does—whether that is crowdsourcing reports, policy plans, presentations, and even meetings with government ministers.

When designed well, OpenMedia's campaigns effectively act as platforms for broad-based citizen engagement and collaboration. It is important to ensure that citizens play the key role in shaping campaigns, from messaging and tactics all the way to long-term policy solutions. Time after time, it is citizens' decisions to participate in campaigns that has ultimately led to their success. Together, our growing community has accomplished some remarkable things in a short period of time—whether overturning invasive government spying schemes or stopping pans to install a pay meter on the Internet.

Sadly, the need for OpenMedia's work has not diminished. Efforts to spy on private online activity, to censor the Internet, and to price-gouge Canadians on their monthly Internet bills make clear that the Internet continues to need people who are willing to work to defend it. By coming together online, citizens are demonstrating why the Internet is key to a future where collaboration and participation are hallmarks of our democracy. We are really just beginning to explore the Internet's potential for building a democracy that truly responds to citizens' needs. With threats to the Internet intensifying, it seems that we are at a tipping point: Either we stand back and let leaders, bureaucrats, and lobbyists tip us back into disempowerment, or we step forward to build a participatory shared future that works for all of us.

REFERENCES

Bennett, D. 2014. "Minister 'Outraged' Over Stolen Laptop Holding 620,000 Albertans' Health Data." *Globe and Mail*, January 22. www.theglobeandmail .com/news/national/alberta-seeking-stolen-laptop-holding-620000–peoples -personal-health-data/article16459295/.

Bridge, S. 2011. "Canadians with Mental Illnesses Denied U.S. Entry." *CBC News*, September 9. www.cbc.ca/news/canada/canadians-with-mental-illnesses -denied-u-s-entry-1.1034903.

Brown, J. 2013. "Welcome No More in U.S. Courts, Copyright Trolls Looks to Canada." *Macleans*, May 13. www.macleans.ca/society/technology/welcome -no-more-in-u-s-courts-copyright-trolls-look-to-canada/.

Christopher, D. 2013. "Canadians Pay Some of the Highest Prices for Some of the Worst Telecom Service in the World." OpenMedia.ca, July 1. https:// openmedia.ca/blog/confirmed-canadians-pay-some-highest-prices-some -worst-telecom-service-industrialized-world.

CNW. 2013. "J. D. Power and Associates Reports: Satisfaction Is Highest Among Wireless Customers Who Use Their Carriers' Online Service and Sales Channels." May 9. www.jdpower.com/sites/default/files/2013064-canadian _wireless_2013.pdf.

Desjardins, J. 2013. "Harpers TTP Free-Trade Deal Threatens All Canadians." *Bulletin*, November 18. http://thebulletin.ca/harpers-ttp-free-trade-deal -threatens-all-canadians/.

Industry Canada. 2013. Framework Relating to Transfers, Divisions and Subordinate Licensing of Spectrum Licenses for Commercial Mobile Spectrum, DGSO-003-13. www.ic.gc.ca/eic/site/smt-gst.nsf/eng/sf10653.html.

Isaacson, B. 2014. "One Frightening Chart Shows What You Might Pay for Internet Once Net Neutrality Is Gone." *Huffington Post*, January 17. www.huffingtonpost.com/2014/01/17/net-neutrality-gone_n_4611477.html.

Market Wired. 2013. "Harper Government Taking Action to Stand Up for Consumer Choice and Affordable Prices in Canada's Wireless Sector." December 18. www.marketwired.com/press-release/-1863548.htm.

OpenMedia. 2013. "Time for an Upgrade: Demanding Choice in Canada's Cell Phone Market." https://openmedia.ca/sites/openmedia.ca/files/TimeForAnUpgrade_OpenMedia_130419.pdf.

Rennie, S. 2013. "CRTC Code of Conduct: New Rules Include Cellphone Contract Changes." *Huffington Post*, June 3. www.huffingtonpost.ca/2013/06/03/crtc-code-conduct-cellphone-contracts_n_3376970.html?just_reloaded=1.

PART THREE

The Power of the Media Reform Movement

CHAPTER TEN

A Perfect Storm for Media Reform

Telecommunications Reforms in Mexico

ALEJANDRO ABRAHAM-HAMANOIEL, *Goldsmiths, University of London, United Kingdom*

MEDIA REFORM STRATEGY

The Mexican Association for the Right to Information (AMEDI to use its Spanish acronym), the most prominent media reform group in the country, has for the last decade conducted a permanent campaign for the democratization of media in Mexico. During this time AMEDI has developed a number of effective strategies, which include: constant advocacy, *by organizing conferences and symposiums, providing relevant information through their website and public communiques and cementing an eclectic but unified front against media power;* permanent monitoring, *by identifying in National Congress all MPs with close personal or professional connections to big media and evaluating the performance of media regulators and key civil servants;* strategic litigation, *by fighting in court commercial transactions carried out by media corporations which could further entrench media concentration and contesting in the Supreme Court of Justice the constitutionality of particular pieces of legislation; and* research and policy-making exercises, *by publishing academic books and drafting media law to be submitted for legislative consideration. Alongside an advantageous political environment and the pressure asserted by a dynamic student movement, these strategies have been responsible for the 2013 telecommunications reform and the transformation of the legal framework regulating media in Mexico.*

For the first time in half a century, the legal framework regulating media in Mexico has been comprehensively reformed. A number of far-reaching constitutional amendments approved in June 2013 will radically change the way radio and television are regulated in the country, opening up a highly monopolized market and raising the possibilities for the creation of a national public broadcasting service. Among the most relevant changes brought up by this reform was the creation of the Federal Institute of Telecommunications (IFT to use its Spanish acronym), a new regulator which, unlike its predecessor, is constitutionally independent from both the Presidency and Congress. The IFT will have the authority to implement asymmetric regulation and the tools to impose multi-million peso fines. IFT will have both financial and administrative independence and their commissioners are now selected through a publicly sanctioned process. In a country in which radio and television was born out of the intimate connections between a small number of affluent families and politicians, these reforms represent an attempt to sever the murky and, at times, illegal connections between media moguls and government officials.

Perhaps more importantly, the 2013 telecommunications reform recognize the duty of the state in the creation of a diverse and plural media environment: both the right to information and the right to access to information and communication technologies (ICT) are now constitutional prerogatives. In addition, a number of previously inoperable laws, from the right to reply to the laws protecting personal privacy, are to be harmonized in new secondary legislation. Although the laws necessary to implement these far-reaching reforms have yet to be developed, most social commentators and academics recognize the unprecedented possibilities these reforms will bring to the media ecology of the country. This chapter, however, does not explore the particularities of these constitutional reforms, but focuses instead on the sociopolitical circumstances that made them possible and, in particular, on the strategies used by Mexican Association for the Right to Information (AMEDI), the largest and most energetic media reform group in the country.

As Silvio Waisbord (2000, 50) has rightly pointed out, like in most other Latin American countries, radio and television in Mexico was for many years caught between "the hard rock of the state and the hard place of the market." At the dawn of broadcasting technologies, the countries of the region—ruled by military regimes or by authoritarian political elites—adopted a liberal media model which left to the market the introduction and operation of these new services. In fact, political control over radio and television was a key element in the longevity of these administrations; this influence, however, was not exercised through direct ownership or management. In postrevolutionary Mexico, broadcasting was controlled through a complex system of patronage and commercial links, in which media owners were allowed to conduct their commercial affairs undisrupted as long as they showed unquestionable loyalty to the hegemonic party and in particular to the president in turn. When comparing media systems, Hallin and Papathanassopoulos (2002) use the term *clientelism* to describe the way countries in southern Europe and Latin America organize their media industries; they describe media systems in which "access to social resources is controlled by patrons and delivered to clients in exchange for deference and various kinds of support" (ibid., 184–185).

The combination of the unflinching backing from authoritarian regimes and the financial resources generated by thinly regulated markets produced a number of extremely wealthy and powerful corporations. By the mid-1970s, a handful of groups in Mexico owned most radio stations in the country and the Azcárraga family had total control over most television stations. In 1990 Emilio Azcárraga Milmo, father of Emilio Azcárraga Jean, current CEO of Televisa, famously declared himself to be "a solider of the party" (Fernandez and Paxman 2013). All through the twentieth century, the Azcárraga family maintained strong connections to the leadership of the Institutional Revolutionary Party (PRI to use the Spanish abbreviation), the hegemonic organization which ruled Mexico for 70 years. These links allowed the company to fend off any attempt to modify the archaic laws which supposedly regulated their industry. In fact, as María de la Luz Casas Pérez (2006, 96) has suggested, media regulation almost always "appeared by request [of the media corporations] in order to formalize already existing favourable conditions or to ensure their future stability." Politicians and academics (Esteinou and de la Silva 2009; Trejo Delarbre 2009) have even implied that pieces of media legislation were often written by the legal department of Televisa rather than by elected MPs.

In Mexico, unlike other countries in the region, the slow process toward democracy which began in the late 1980s and accelerated during the 1990s, was not wholeheartedly supported by any major media corporation. As Hallin (2000, 103) has stated, Televisa actively resisted democratization and most "commercial broadcasters, far from leading the charge to pluralism, had to be dragged kicking and screaming by the state." Media companies dreaded a decline in influence a different political party could bring and kept their options open. However, the adoption of a multiparty system in which electoral results were respected and the balance of power effectively shifted did very little to curtail their influence on the political and social life of the country. The fact they never fully adopted an ideological position and pursued instead purely commercial interests, allowed most of the largest corporations to be flexible enough to cozy up to all parties, including opposition forces which relied on publicity to be elected.

Rather than auguring media plurality and the decentralization of the media market, the 2000 presidential election in which the PRI was finally ousted, tragically increased the power of these corporations. In pursuit of the media exposure considered vital for electoral victory, politicians of all parties tried to outbid each other in the purchase of airtime. This dynamic, alongside hefty contracts for political advertising, gave media lobbyists huge political leverage. In the following decade, media corporations were not afraid to use this influence, pushing outrageous legislation through their proxies in Congress; for example, they forced the president to slash their taxes by 1,000 percent (Trejo Delarbre 2010; Esteinou and de la Silva 2009). By the 2006 presidential election the entire political class was effectively being blackmailed by Televisa, the largest Spanish-speaking media cooperation in the world.

THE SOCIOPOLITICAL CONTEXT BEHIND MEDIA REFORM

In the last couple of years, a number of Latin America countries, from Argentina to Nicaragua, have fundamentally changed the way they regulate and organize their media. Just like in Mexico, many of these changes were, at least in part, the result of a sustained campaign by media reform groups. Activists, academics and university students began to create organizations that strived to democratize the media and make the connections between authorities and media owners publicly accountable. The Latin American media observatories, which Maria del Mar Rodríguez Rosell and Beatriz Correyero Ruiz (2008, 23) describe as "organizations which deliver a critical look at the global media system and have as a common objective its democratization," differ from country to country (Herrera 2005; Rey 2003) but the great majority of them shared a "reformist agenda" (Herrera 2006). These organizations have been at the forefront of all major media reform movements in the region and are a rich environment in which alternative models of media production and distribution are tested. However, unlike other countries in South and Central America (Waisbord 2010), no Mexican media observatory has been fully endorsed by a political party and comprehensive media reform has never been part of any government's agenda. Contrary to countries like Argentina, Colombia, Ecuador, Nicaragua, and Venezuela, in which populist governments "dovetail with longstanding demands by civic groups, namely, the need to restrict the power of large media corporations and diversify media ownership" (ibid., 99), Mexican reform groups had only limited influence within the established political system. It was therefore somewhat surprising to witness the scope and progressive outlook of the 2013 telecommunications reform. Before concentrating on the strategies employed by the most prominent media reform group in the country, this section explores two sociopolitical events that might help to explain such a dramatic change. Alongside the strategies used by AMEDI, the #YoSoy132 (#IAm132) student movement and the return of the Institutional Revolutionary Party to the presidency are key elements that help to explain the recent legal transformations.

THE CONNECTIVE LOGIC OF #YOSOY132

#YoSoy132 is a student movement that erupted two months before the 2012 presidential election. The result of an incident which took place at a private university, the movement was a popular reaction to a political snub: a group of students who demonstrated against the visit of Enrique Peña Nieto, the then front-runner in the presidential race and today's Mexican president, were publicly called hooligans and political pawns by members of his campaign. In what would become very relevant, not only did the major television channels distort their grievances and opinions but they questioned their identities as students. Using mobile phone recordings, the students at the Iberoamerican University (Ibero) contested the official version of events by posting a number of videos on YouTube and flooding social media with their own account. On May 14, three days after the incident, 131 protesters posted a video in which they showed their student IDs and confirmed their presence in the original protest. This video not only went viral, but was soon picked up by radio and television stations, if not by the largest corporations. This video "would end up shaking the electoral atmosphere like nothing else before, in a presidential campaign which had given the impression to be immobile" (Galindo Cáceres and Gonzáles-Acosta 2013, 83). When students at other universities began to be aware of the movement and identify with it, they considered themselves to be the 132nd protester, even if they had only been *virtually* there. They used the #YoSoy132 hashtag to make their struggle public through the use of social media but also to identify themselves as *joveNets*, or networked-youngsters (Candón Mena 2013), the young "digital natives" who are proficient in the use of digital technologies and are not passive consumers of "old media."

Although the movement soon took to the streets and organized regional rallies and assemblies, #YoSoy132 followed what Lance Bennett and Alexandra Segerberg (2012) have described as the *logic of connective action*. By adopting horizontal methods of decision-making and heavily relying on social media for making their struggle public, the movement escalated exponentially and gained central stage in the political agenda. With its use of "easily personalized ideas" (ibid., 744) and its rejection of hierarchies, the students were able to generate an incredible momentum which took politicians and social commentators by surprise. By May 23, just over a week after the first incident, tens of thousands of young protesters marched through Mexico City and held a massive assembly near the city center. Not all protesters identified themselves as members of the #YoSoy132 movement (Meneses Rocha, Ortega Gutiérrez, and Urbina Cortés 2013), but it was precisely the ideological flexibility of the movement and its lack of formal membership which allowed the formation of an eclectic group, able to accommodate diverse and dissimilar demands: from calling for an end to the War on Drugs to demanding respect and legal protection for unionized labor.

Nevertheless, and despite having no formal leadership to script a manifesto or a coherent set of demands to present to Congress, *media democratization* soon became the core exigency of the movement; it became the only idea that bonded all protesters together. Rather than directing their outrage against a particular political party or candidate, most students identified mainstream media, and in particular Televisa and TV Azteca, as the main target and the biggest threat to the democratic process. As Gabriel Sosa Plata (2012, 103) has clearly described: "Why do young people talk about the need to democratize media? Because, as a matter of fact, television is an overwhelming presence in all Mexican homes, in particular the two companies [which control most of the airwaves]. No other media has these levels of penetration, coverage and national reach."

The students considered media corporations a vestige of an undemocratic and authoritarian past which they no longer recognized and began to employ digital technologies as organizational tools while at the same time using social media as *performative acts of resistance*. During demonstrations students hold banners with slogans such as: "They have the TV, we have social networks" and "Televisa: Welcome to the Digital Age!" In an interesting twist of events, the demographic who consumed traditional broadcasting content the least—the young, metropolitan, tech-savvy university student—was the sector of society that recognized radio and television stations as important centers of political power. Even without setting up a detailed action plan or developing a clear political agenda, #YoSoy132 was crucial in bringing media reform to the fore. Nonetheless, without the endorsement of the political elite, far-reaching amendments to the Constitution would have never been possible, even after a popular protest of such a scale. Taking this into consideration, the election of Enrique Peña Nieto and his subsequent political maneuvering are crucial to understand the circumstances which made media reform possible.

THE RETURN OF THE PRI

One of the most intriguing aspects of the 2013 telecommunications reform is the fact it was not only encouraged but actively supported by Peña Nieto. The changes to the Constitution, which might seriously jeopardize the interests of media owners, were sponsored by a politician widely considered to be in cahoots with the media, particularly with Televisa. Even before running for the presidency, Peña Nieto made use of radio, television, and print media as key elements in his political strategy, spending millions in advertising during his time as governor of the State of Mexico. He relied very heavily on public relations and fostered close links to the owners of radio and television stations (Villamil 2009). In November 2010 these connections became personal, when Peña Nieto married Angélica Rivera, a successful soap opera actress with a very long history of working for Televisa. Arguably, one of the reasons which made Peña Nieto decide to push for media reform during its first year in office was an attempt to clean up his image and distance himself from the corporations that helped him to get elected. During the seventy years the PRI controlled the country, it was often the case that when a new president took office, an element of the old administration had to be sacrificed in order to maintain legitimacy. The recent reforms have been seen by some as an act of legitimization (Calleja 2013; Sosa Plata 2013).

A more direct factor in the political support shown for these reforms was the Pact for Mexico (Pacto por México), a political agreement signed by leaders of the three main political parties days after Peña Nieto took office. The pact seeks "to accomplish reforms that have been obstructed by political gridlock" (Sada 2013) and listed a number of areas in which all parties agreed to work together. The pact aimed at speeding up the law-making process by establishing a committee in charge of drafting legislation with cross-bench support and giving them legislative priority. Although fraught with difficulties, the agreement allowed for a relatively smooth transition through Congress and prompted the 2013 Telecommunications Reforms to be approved in record time. It is to expect that without this political environment, all the efforts of an engaged civil society would still have not been enough to produce the changes recently experienced in the country. Nevertheless, it is vital to recognize the work done by academics and media activists who for the last twenty years have worked for such reforms. Without their input, the perfect storm which encouraged this change would have never produced such a promising transformation.

THE MEXICAN ASSOCIATION FOR THE RIGHT TO INFORMATION

Instead of enforcing control through coercion, the PRI was skilled in subsuming dissent through collusion. During most of its time in power, the party could accommodate diverse and contrasting interests, from peasant's movements to chambers of commerce, within its flexible hierarchy. It diffused conflict by granting positions within its structure to every potential opponent, while strongly prosecuting those who rejected. Although there were moments of violent repression, most famously the 1968 student massacre at the Plaza de las Tres Culturas, Mexico escaped most of the killings and disappearances common to other regimes in the region. This culture of collusion made the development of civil society organizations particularly difficult, since it undermined possibilities for organized political resistance. Nevertheless, since the mid-1990s civic participation and public engagement have started to play a larger role in the process of policy making. Successive Mexican governments, slowly and sometimes grudgingly, have begun to open "windows of opportunity" (Merino 2013) for the engagement of civil society in matters of public concern. It is in this new environment that media reform groups have begun to operate.

For over a decade, AMEDI has been the most effective media observatory in the country, calling politicians and media owners to account and promoting media plurality as an essential element in the construction of democracy. With many of the characteristics shared by other Latin American observatories (Beas

Oropeza 2007; Correyero 2008), the association has been able to combine academic research with political advocacy, drawing together scholars, politicians, and journalists. AMEDI inherited a long history of research and radical action, rooted in the democratic struggles of the late 1980s and early 1990s. In 1999, after yet another failed attempt to push through Congress a comprehensive set of media reforms, a small group of academics, activists, and legislators decided to set up an organization that would "promote total respect, by government as well as by the media, to the citizen's rights to information, which must ensure that all Mexicans have access to truthful and objective information regarding matters of public concern" (AMEDI 2010). In a recent interview, two of its founding members, Senator Javier Corral and Professor Beatriz Solis, admitted creating AMEDI as an antidote to the influence the media lobby were able to exert on government. Until then, every attempt to modernize the laws regulating radio and television in the country had been defeated by the coordinated efforts of the media corporations. Not only was every bill publicly and gratuitously attacked, often without even superficial analysis, but the individuals sponsoring them soon became open targets. Televisa and TV Azteca were particularly vicious against any politician floating the idea of changing the media framework that, for half a century, had guaranteed their influence and wealth. Although commercial rivals, these two corporations have proven to be, time and time again, staunch political allies every time their interests are threatened.

SOME AMEDI STRATEGIES

Since its inception, AMEDI has made use of a variety of methods; I will focus my attention only on four of its most successful strategies: its advocacy, its work monitoring Congress and the civil service, its use of strategic litigation, and its research and policy-making efforts. These tactics are often merged for specific campaigns; nonetheless they are sufficiently distinct as to be described separately. Alongside the popular student movement and the favorable political atmosphere experienced after the 2012 presidential election, it is my contention that these strategies are responsible for the shape and scope of the 2013 telecommunications reform.

Perhaps the most important policy employed by AMEDI is its total commitment to public advocacy; no other organization has been able to promote media reform to such an extent. It has long been able to build high levels of social awareness and provide a clear alternative to the current paradigm. Part of this success is due to its diverse and committed membership, drawn from the entire political spectrum, who have been able to maintain pressure and a unified front even during the most adverse political times. AMEDI has become an ideological and political safe haven in which MPs, activists, and academics can openly discuss ideas without fear of censorship or peer pressure; consequently, it has become a fertile ground for cross-party cooperation. In addition, its members have developed a strong sense of identity which allows them to coordinate their efforts effectively.

AMEDI's members have also a permanent presence in the media, with regular columns in major newspapers and constant appearances in radio and television programs. The association produces and broadcasts *Espacio AMEDI* (2014), a weekly radio program which works as a platform for their campaigns while introducing topics such as freedom of information, participative democracy, and the right to information. It compiles a weekly newsletter in which all major media topics are discussed and which includes links to newspaper articles. In a country in which media issues are scattered around and difficult to find, this is a vital resource for anybody interested in the subject. The public recognition and trajectory of many of its members and supporters—including well-known actors, respected filmmakers, and famous writers—has given AMEDI considerable reputational resources. It is generally acknowledged as a serious, committed, and plural organization and its members possess considerable influence over key decision-makers, both in the business community and in Congress. However, its strategies go beyond inflicting "reputational costs" (Hevia de la Jara 2013), and many of their ideas are now central in debates around media reform.

A second strategy used by AMEDI is the constant monitoring of major decision makers, key policy players and relevant civil servants. Although it employs only a couple of full-time staff and most of its members freely give their time and expertise, the association has been able to maintain regular oversight of Congress and all institutions with a stake in the legal framework of the media. Although it has been described as a media observatory, content analysis has yet to become an important element in its overall strategy. Instead, AMEDI has traditionally relied on quantitative research conducted independently by its members or on work done by other organizations, such as the Citizen Observatory for Gender Equality in the Media (Observatorio Ciudadano por la Equidad de Género en los Medios), which invites audiences to monitor sexist content on Mexican media. For a detailed analysis of state expenditure on official publicity, AMEDI has relied on the Centre for Analysis and Research (Fundar). The association can therefore be better described as a "political media observatory" or, following Consuelo Beas Oropeza's (2007) classification, as an observatory for supervision (*observatorio de fiscalización*).

Interestingly, AMEDI asks its supporters, and Mexican society in general, to take part in their monitoring efforts. After the 2012 general election, the association launched the Telecracy is not Democracy! (*¡Telecracia no es democracia!*) campaign. This program identified members of Congress with close connections to the media corporations and analyzed their individual legislative output as an exercise in accountability. The campaign identified a distinct and heterogeneous group of MPs, from relatives of media owners and lawyers previously employed by the corporations to former daytime presenters (MultimediA-MEDI 2012). AMEDI christened this group *Telebancada* (the Television Bench) and used this campaign to inform citizens about these connections and invite anybody interested to follow their legislative record. The campaign also showed how, in most cases, these MPs were not directly elected but accessed Congress through rules of electoral proportionality.

Similarly, AMEDI monitored the appointment of all the new members of the Federal Institute of Telecommunication. It not only closely followed the process itself but also led a public scrutiny of the successful candidates. Unfortunately, on its first assessment, the association has concluded that four out of the seven commissioners did "not comply with the required experience and/or independence established by the Constitution" (AMEDI 2013). Although this type of monitoring does not always stop wrongdoings or bring the changes it advocates, it represents a method of accountability and a challenge to the decisions taken away from public scrutiny. The work done by AMEDI is an example of how the value of accountability can be institutionalized within a national media system.

Strategic litigation has become another very successful strategy against attempts to further concentrate the media and telecommunication market in Mexico. AMEDI has filed lawsuits and undertaken legal actions against the largest media corporations as a disruptive technique, forcing them to publicly defend some of their most dubious deals and sometimes even demonstrating the collusion between judges and media owners. The association has even challenged the constitutionality of a number of bills introduced to Congress by their proxies: AMEDI's most significant legal victory to date has been the overturning of a number of reforms rushed through Parliament in 2006 in the midst of a hotly contested presidential election. The Televisa Law (*Ley Televisa*), as these reforms were dubbed at the time, attempted to modify "the Federal Law on Radio and Television" and would have expanded the dominance of Televisa and TV Azteca into areas of telecommunication and Internet provision, granting them an unfair advantage in the process of digitalization (Esteinou and de la Silva 2009). Among other concessions, *Ley Televisa* would have allowed for an automatic renewal of all media licenses and extended their validity from fifteen to twenty years. It would have given existing corporations the possibility to fully exploit the dividends of digital switch-over, exempting them from additional taxation and making the appearance of new competitors extremely difficult. In a highly monopolized market these concessions would have further entrenched the influence of the media duopoly.

In 2007, alongside other organizations and a number of MPs, AMEDI successfully challenged an act of unconstitutionality at the Mexican Supreme Court, nullifying *Ley Televisa* and paving the way for the 2013 telecommunications reform. In addition, the publicity surrounding the litigation was a powerful vehicle for the socialization of media reform and demonstrated how the courts were a viable space in which to challenge the corporations and their political allies. Since then, AMEDI has continued to use lawsuits to advance its agenda. It has, for example, recently challenged the purchase by Televisa of 50 percent of mobile phone company Iusacell, on the basis that TV Azteca, its sole competitor in free-to-air commercial television, already owns the other 50 percent. This alliance would have created a dangerous and unconstitutional connection between the two major producers of media content in the country. Although the lawsuit was admitted by the courts, it was summarily dismissed due to legal technicalities; the Supreme Court has unfortunately shown no interest in further pursuing the case (AMEDI 2012), but there is now a precedent to follow. Aleida Calleja, who vigorously employed these tactics during her time as AMEDI's president, is very clear on effectiveness of litigation as a strategy to further the cause of media reform:

> I am convinced that to acquire new rights you have to litigate. Denouncing is not enough. I do not trust Congress. You have to win [these new rights] through facts on the ground. You need to make these topics public and then assert a legal precedent in court; this would inhibit illegal conduct by the corporations. You could shout and yell outside their offices and nothing will happen. If we manage to succeed with these lawsuits, we will have strengthened our position. (Calleja 2012)

An additional element in AMEDI's strength resides in the fact that many media scholars in the country—alongside reputed legal experts, historians, and economists—are either members or supporters. The intellectual rigor of their contributions have been captured in the scholarly books AMEDI has published and the international conferences it has organized. Most publications (Ávila Pietrasanta, Calleja, and Solís Leree 2001; Bravo 2011; Solís Leree 2000; Trejo Delarbre et al. 2009; Trejo Delarbre and Montiel 2011) are edited volumes in which policy is carefully discussed and possible alternatives are delineated. As done by other media observatories in Latin America, AMEDI has positioned itself half way between a pressure group and a policy think-tank. The loyalty of its members is in part due to the clarity of its objective and the perceived illegality of the actions of its opponents.

Through their members and supporters in Congress, the association has gained working knowledge not only on policy making but also of the politics behind constitutional reform. All these skills have been put to practice in the drafting of a bill, which in turn has been submitted to Congress for consideration. The Citizens' Bill for a Convergent Law on Broadcasting and Telecommunication (Iniciativa Ciudadana de Ley Convergente en Materia de Telecomunicaciones y Radiodifusión) is a comprehensive text which reflects years of research and academic discussion. It aims to implement the constitutional amendments approved by the 2013 telecommunications reform, creating a new convergent law that could jointly regulate broadcasting and telecommunication services. The Citizens' Bill (Iniciativa Ciudadana de Ley 2013) has a corpus of over 391 articles and represents an attempt to harmonize the new requirements established by the Constitution. The project has already been adopted by a good number of MPs and will soon be discussed in Congress.

Two fundamental elements on this Citizen's Bill are its recognition of broadcasting as a social service, paying particular attention to the rights of audiences, and the reassertion of the state as the sole administrator of the airwaves. In contrast to the current legal framework, this bill does not consider broadcasting simply another industry within the capitalist market, but sees it as a vital building block in the construction of citizenship.

CONCLUSION

By the end of the 1970s, the culture of *subordination* which saw the birth of commercial broadcasting in Mexico was replaced by a culture of *collusion*, in which media owners and politicians haggle for concessions but work together for their mutual benefit and the protection of a profitable status quo. This period saw an unprecedented concentration of radio and television frequencies in the hands of a few party faithfuls and created the corporations that now monopolize the media market. Paradoxically, the transition to democracy experienced by the country in the last two decades, rather than diversifying the media environment and triggering an era of content plurality, has ushered in a culture of *submission* in which the entire political class could be blackmailed by the corporations. As Manuel Alejandro Guerrero (2010, 274) has pointed out, after the ousting of the PRI, radio and television stations "gained an enormous amount of symbolic power, facing a fragmented political class with an enormous need for airtime."

The approval of the 2013 telecommunications reform hints at a shift in the balance of power and, for the first time, it has introduced a new element in the process of media policy-making: an organized and capable civil society. The amendments to the Constitution approved by Congress still need to be implemented by secondary legislation and there is a real danger of a counter-reform promoted by the media lobby (Corral Jurado 2014). The winds have nonetheless changed. The growing public awareness of the role played by the media in the democratic life of the country will make difficult readopting cultures of subordination, collusion or submission, which have for so long characterized the relationship between politicians and media owners in Mexico. The strategies employed by AMEDI, from their commitment to advocacy to their strategic use of litigation, have ensured that media power is no longer unaccountable in the country.

REFERENCES

AMEDI. 2010. *Estatutos.* February 9. www.amedi.org.mx/18-amedi/305-estatutos.

————. 2012. "Televisa-Iusacell presionan al Poder Judicial para que deseche el amparo de la AMEDI." September 10. www.amedi.org.mx/index.php/ campanias/comunicados/812–televisa-iusacell-presionan-al-poder-judicial -para-que-deseche-el-amparo-de-la-amedi.

————. 2013. "Errónea decisión del Senado en designación de comisionados para Ifetel: AMEDI." *Homozapping.com,* 11 September. http://homozapping.com .mx/2013/09/erronea-decision-del-senado-en-designacion-de-comisionados -para-ifetel-amedi/.

Ávila Pietrasanta, I., A. Calleja, and B. Solís Leree, eds. 2001. *No más medios a media, Participación ciudadana en la revisión integral de la legislación de los medios electrónicos.* México: Fundación Friederich Eber-México/AMEDI.

Beas Oropeza, C. 2007. "Comunicación y participación ciudadana. Los observatorios de medios de comunicación en Ame'rica Latina." In *Comunicación y Educación. Enfoques desde la Alteridad,* edited by Mauricio Andión Gamboa et al., 91–114. México: Porrúa-UAM.

Bennett, L., and A. Segerberg. 2012. "The Logic of Connective Action." *Information, Communication and Society* 15, no. 5: 739–768.

Bravo, J., ed. 2011. *Panorama de la Comunicación en México. Desafíos para la Calidad y Diversidad.* Mexico: AMEDI/Camara de Diputados

Calleja, A. 2012. Interview by author. December 9. Mexico City.

————. 2013. "La expectante reforma." *La Silla Rota,* March 5. https://aleidacalleja .wordpress.com/2013/03/05/la-expectante-reforma/.

Candón Mena, J. 2013. "Movimientos por la democratización de la comunicación: Los Casos del 15'M y #YoSoy132," *Razón y Palabra* 3, no. 82 (March). www .razonypalabra.org.mx/N/N82/V82/32_Candon_V82.pdf.

Casas Pérez, M. 2006. *Políticas públicas de comunicación en América del Norte.* México: TEC—LIMUSA Editores.

Corral Jurado, J. 2014. "¿Honrara' Peña Nieto su palabra en Telecom?" *El Universal,* February 4. http://m.eluniversal.com.mx/notas/articulistas/2014/02/ 68576.html.

Espacio AMEDI. 2014. 660 AM. http://espacioamedi.blogspot.co.uk/.

Esteinou, J., and A. de la Silva. 2009. *La Ley Televisa y la lucha por el poder en México.* Mexico: Universidad Autónoma Metropolitana.

Fernández, C., and A. Paxman. 2000. *El Tigre. Emilio Azcárraga y su imperio Televisa.* Mexico: Raya en el Agua-Grijalbo.

Galindo Cáceres, J., and J. Gonzáles-Acosta. 2013. *#YoSoy132. La Primera Erupción Visible.* Mexico: Global Talent University Press.

Guerrero, M. A. 2010. "Los medios de comunicación y el régimen político." In *Los grandes problemas de México,* edited by S. Loaeza and J. Prud'homme, 231–302. Mexico: El Colegio de México.

Hallin, D. 2000. "Media, Political Power, and Democratization in Mexico." In *De-Westernizing Media Studies*, edited by J. Curran and M. J. Park, 97–110. London: Routledge.

Hallin, D., and S. Papathanassopoulos. 2002. "Political Clientelism and the Media: Southern Europe and Latin America in Comparative Perspectives." *Media, Culture and Society* 24, no. 2: 175–195.

Herrera, S. 2005. "Retrato en diez rasgos de los observatorios de medios en América Latina." *Sala de Prensa* 7, no. 3. www.saladeprensa.org/art638.htm.

———. 2006. "Los observatorios de medios en Latinoamérica: Elementos comunes y rasgos diferenciales." *Razón y Palabra* 51, no. 1. www.razony palabra.org.mx/anteriores/n51/sherrera.html.

Hevia de la Jara, F. 2013. Introducción. In *Rendición de cuentas social en México. Evaluación y control desde la sociedad civil*, edited by M. Merino et al. 18–35. México: Gobierno del Estado de Oaxaca. www.oaxtransparente.oaxaca.gob .mx/rdcsocialenmx/rdc.pdf.

Iniciativa Ciudadana de Ley. 2013. Asociación Mexicana de Derecho a la Información. www.amedi.org.mx/documentos/proyecto-AMEDI-ley-secundaria -octubre-2013.pdf.

Meneses Rocha, M., E. Ortega Gutiérrez, and G. Urbina Cortés. 2013. "Jóvenes, participación político ciudadana y redes sociales en México 2012." In *La Libertad de Expresión en el proceso electoral de 2012*, edited by A. López Montiel and E. Tamés. México: Porrúa/PNUD/Tecnológico de Monterrey/ COPARMEX. www.amedi.org.mx/documentos/proyecto-AMEDI-ley -secundaria-octubre–2013.pdf.

Merino, M. 2013. "La otra mirada a la democracia." In *Rendición de cuentas social en México. Evaluación y control desde la sociedad civil*, edited by M. Merino et al., 12–15. México: Gobierno del Estado de Oaxaca. www.oaxtransparente .oaxaca.gob.mx/rdcsocialenmx/rdc.pdf.

MultimediAMEDI. 2012. "La Telebancada." *YouTube*. June 5. www.youtube.com/ watch?v=oYmTM3XOZjc.

Rey, G. 2003. *Ver desde la ciudadanía. Observatorios y veedurías de medios de comunicación en América Latina.* http://library.fes.de/pdf-files/bueros/kolumbien/ 04198.pdf.

Rodríguez, M., and B. Correyero. 2008. "Los observatorios como agentes mediadores en la responsabilidad social de los medios de comunicación: panorama internaciona." *Sphera Publica* 8, no. 1: 15–40.

Sada, A. 2013. "Explainer: What Is the Pacto por México?" *Americas Society/Council of the Americas*, March 11. www.as-coa.org/articles/explainer-what-pacto -por-m%C3%A9xico.

Solís Leree, B., ed. 2000. *Medios de Comunicación y Procesos Elecotrales. Un Comrpomiso para el Futuro.* Mexico: Camara de Diputados/AMEDI.

———. 2009. "De cómo llegamos hasta aquí. Los antecedentes de la 'Ley Televisa.'" In *La Ley Televisa y la lucha por el poder en México*, edited by J. Esteinou et al., 27–53. Mexico: Universidad Autónoma Metropolitana.

Sosa Plata, G. 2012. "#YoSoy132: jóvenes frente a las redes sociales y la democratización de los medios de comunicación." In *Esfera pública y tecnologías de la información y la comunicación*, edited by C. Arango et al., 81–119. Mexico: Instituto Electoral del Distrito Federal/Sinergia.

———. 2013. "Ma's para ver." *El Universal*, March 5. http://blogs.eluniversal.com .mx/weblogs_detalle17980.html.

Trejo Delarbre, R. 2009. "Los diputados ante la 'Ley Televisa.'" In *La Ley Televisa y la lucha por el poder en México*, edited by J. Esteinou et al. 76–101. Mexico: Universidad Autónoma Metropolitana:.

———. 2010. *Simpatía por el rating. La política deslumbrada por los medios*. México: Cal y Arena.

Trejo Delarbre, R., and A. Montiel, eds. 2011. *Diversidad y Calidad para los medios de comunicación. Diagnósticos y propuestas, una agenda ciudadana*. Mexico: AMEDI/Camara de Diputados.

Trejo Delarbre, R., et al., eds. 2009. *¿Qué Legislación hace Falta para los Medios de Comunicación en México?* Mexico: AMEDI/Camara de Diputados.

Villamil, J. 2009. *Si Yo Fuera Presidente. El Reality Show de Peña Nieto*. Mexico: Grijalbo.

Waisbord, S. 2000. "Media in South America. Between the Hard Rock of the State and the Hard Place of the Market." In *De-Westernizing Media Studies*, edited by J. Curran and M. J. Park, 50–62. London: Routledge.

———. 2010. "The Pragmatic Politics of Media Reform: Media Movements and Coalition Building in Latin America." *Global Media and Communication* 6, no. 2: 133–153.

CHAPTER ELEVEN

Between Philosophy and Action

The Story of the Media Reform Coalition

BENEDETTA BREVINI, *University of Sydney, Australia*
JUSTIN SCHLOSBERG, *Birkbeck, University of London, United Kingdom*

MEDIA REFORM STRATEGY

Media reform strategy needs to straddle the intersection of engagement and resistance and to negotiate the myriad trade-offs that confront media reform groups. For example, there is a need to strike a balance between devising solutions and developing ideas, and between building consensus and protecting core principles. Above all, we advocate engagement with hegemonic media and policy-making institutions as public platforms, but resistance to the terms and definitions of policy problems imposed by media and political elites, and to the "logic" of dominant media narratives that serve to reinforce communication power. This approach recognizes that even amidst unprecedented, critical junctures in media policy, the political and media establishment is unlikely to reform itself in a substantive, meaningful, or long-lasting way. But critical junctures can offer unparalleled opportunities for media activists and reformers to get their messages out, particularly on occasions when the mainstream media are compelled to cover issues ordinarily shunned from the agenda. Equally, however, we should not lose sight of the reality that hegemonic institutions can distort public debate and they should not be relied upon exclusively as vehicles of outreach. Ultimately, we believe that media reform groups must embrace media activism while avoiding totalizing perspectives that see any kind of engagement with hegemonic institutions as intrinsically counter-productive.

The notion of *critical junctures* has been used by sociologists to explain the windows of opportunities for social change offered at specific times in history. In the United Kingdom, Lord Justice Leveson's inquiry into the ethics and practices of the press announced in 2011 was certainly represented as an important juncture that media reform groups sought to capitalize on. The phone hacking scandal made the news media themselves a headline story to a degree unparalleled in recent times, and the inquiry—as well as subsequent criminal trials—promised to sustain the spotlight.

This engendered unique opportunities as well as challenges for media reform movements that we will reflect on in this chapter. We will use the Media Reform Coalition (MRC) in the United Kingdom as a case study to elucidate the discursive struggles that tend to foreshadow media activism.[1] Indeed, it is within the sphere of media policy-making that the actions of the MRC were concentrated and where its successes and failures to date can be duly (if somewhat preemptively) assessed.

First, we will summarize the brief history of the MRC's formation, actions and challenges. Second, we will theoretically problematize and define the domain of media policy-making in which the MRC had to operate. Third, we will conceptualize the struggle of the MRC beyond traditional framings of media reforms (Napoli 2007) by linking them to wider battles for social justice and social inclusion. Contemporaneous to events reflected on here, the Occupy movement was critiqued for eschewing engagement with existing political institutions and structures, and evading concrete policy solutions to the problem of concentrated wealth (Roberts 2012). In respect of concentrated media power, there has been no such lack of engagement or defined proposals for change on the part of media reform groups. But media activism in the United Kingdom and elsewhere has suffered from a lack of congruence in ideas and the kind of grassroots identity of resistance that enabled the Occupy movement to make a meaningful challenge to dominant discursive frameworks around inequality (Juris and Razsa 2012).

Ultimately, we argue that strategies of engagement and resistance should not be seen as mutually exclusive in media reform movements; that there is a particular need to reclaim the notion of *resistance* in the context of concentrated media power from its cooption by neoliberal discourse around press freedom; and that critical media scholars, journalists, and activists have an integral role to play in documenting media failure and imbalances in the coverage of social justice issues at large, as a vehicle for wider social movement integration.

FOUNDING PRINCIPLES OF THE MRC

The MRC was founded by media academics at Goldsmiths, University of London during summer 2011. Organizational goals were fixed around the concept of a coordinating committee intended to mobilize and build consensus among civil society groups, academics, and activists in the context of the phone-hacking crisis. From the outset, the focus was on policy engagement and coordinating civil society responses to the myriad public inquiries that were set to follow.

These founding goals endowed the MRC with certain unique characteristics as a reform organization from the outset. First, it began from a set of broad principles rather than its own explicit manifesto for change or policy reform proposals. Media reform advocacy groups were already well-established in the United Kingdom, including the Media Standards Trust, Media Wise, the Campaign for Press and Broadcasting Freedom, and the more recently formed Hacked Off (backed by and representing the victims of phone hacking).

There seemed little point in adding another voice to this mix. Rather, the vision from the outset was to engender some semblance of an alliance that could overcome the fractures that media policy issues provoke within progressive politics. For instance, as has been demonstrated in the North American context (Hackett and Carroll 2006b), media activism in the United Kingdom is typically split between a focus on engagement with hegemonic institutions (seeking to influence the practices and output of professional media groups and advocating for reform at the policy level) and an oppositional resistance toward them (fostering independent "third sector" media and developing audience frameworks for mainstream media criticism).

In regards to the Leveson Inquiry, the latter position was articulated strongly by investigative journalist and radical media critic John Pilger (2012) who regarded it from the outset as a show trial ultimately directed toward "the preservation of the system." Such comments were typical of critiques emanating from the radical Left and implied that engagement would merely confer legitimacy on a process that was structurally rigged in favor of established media interests.

But this view was countered by Des Freedman, among others, who argued that to dismiss the significance of the hearings all together was to ignore "the possibility that the exposure of media power during the course of the Inquiry might help to radicalise victims' groups and other media reform campaigners." This, in turn, could lay the basis "for a more sustained challenge to the hegemony of corporate media" (Freedman 2014, 66–67).

This latter position, adopted by MRC from the outset, was not based on a naïve faith in the efficacy of the process. It was based rather on a conviction that Leveson marked a rupture in the dominant discourses that have long served to legitimize the status quo of media policy and regulatory frameworks. As such, the hearings presented two unique opportunities for media reform activism in the United Kingdom. First, they provided unprecedented levels of disclosure in regards to the mechanics of media power which could fuel reform campaigns and activist research in a wide range of contexts. Second, they offered a platform for the expression of alternative—even radical—views about the ownership, structure and regulation of the press which, however marginalized from the coverage of Leveson itself, offered a rare opportunity for such views to be heard outside the usual confines of media activist circles.

No matter how fleeting and ephemeral, the MRC's founding philosophy was based on a perceived need to exploit such opportunity, ultimately in a bid to help energize and sustain media reform activism in Britain. In a tone reminiscent of Chomsky's (1967) thesis on the "responsibility of intellectuals," Freedman (2014, 67) argued that "the left has a responsibility in this context to amplify these arguments about the flaws of an entrenched media power as part of a broader argument about the operation of the capitalist state."

But there was another aspect to Pilger's critique which did share common ground with the intellectual foundations of the MRC. This was the view that Leveson's focus on the ethics and practices of the press did not come close to capturing the scope of structural problems impacting on media performance. Though the MRC's position was not to dismiss the significance of press regulatory reform and the demands of phone-hacking victims, it did share with more radical critiques a conviction that these were insufficient conditions for meaningful change. In particular, so long as media ownership issues were consigned to the margins, Leveson would fall prey to the relatively narrow terms of debate set by the mainstream press. Even within this framework, gaps and omissions skewed the press coverage in ways that undermined the voice of victims. But they also lent further weight to demands for more root-and-branch reform that extended beyond the complaints handling system of the press (Bennett and Townend 2012).

This holistic approach was reflected in the MRC's initial three working groups covering media ethics, plurality, and funding. Although the latter two were subsequently merged, the output of the working groups reflected an ongoing focus on plurality and ownership issues as fundamental rather than tangential to the campaign.

NECESSARY TENSIONS

With a strong emphasis on inclusivity, it was inevitable that tensions in the campaign would emerge. They surfaced most acutely in the areas of policy deliberation, campaign strategies and internal governance. With regards to the first, the MRC had to negotiate political and ideological fault lines extending beyond that between engagement and resistance outlined above. In the UK context, media reform activism has long been structured around positions that either endorse public service media as an engine of the public sphere and counter-weight to concentrated private media power (Curran 2002; Garnham 1986; Scannell 1989) or perceive all corporate media (both public and private) as inherent threats to the independent performance of journalism (Keane 1991).

From the beginning then, the MRC straddled the intersect of conflicting realms of discourse even within "friendly" stables. But this was not so much a weakness as its *raison d'etre*: consensus building was perceived as an essential component of its strategy to exploit the apparent gap that had been opened in the media policy paradigm. A united front seemed the only way to at least offer a challenge to the dominant narratives of the press, which were rapidly closing ranks around a position of cosmetic rather than substantive reform.

Early casualties highlighted the difficulties of this task. The Voice of the Listener and Viewer, an advocacy group representing the consumer and citizen interest in public broadcasting, pulled out at the first hurdle citing concerns that initial policy proposals were not sufficiently explicit in regards to protecting the scope and funding of the BBC. At the other end of the scale, attempts to engage with press freedom groups bore little fruit given their resistance to any suggestion of state involvement in the regulation of the press.

Part of the challenge was an intellectual one. In order to engage key stakeholders such as the National Union of Journalists and Hacked Off, the MRC had to find a way to speak in support of both journalists *and* the victims of phone hacking. Indeed, such a holistic voice seemed to be precisely what was missing from the media reform landscape at large. This further sharpened the focus on media owners as the ultimate bearers of responsibility not only for ethical misconduct in their newsrooms, but also the erosion of journalist autonomy that was a crucial (although less talked about) backdrop to the scandal.

Engagement with stakeholders also gave rise to tensions in governance. On the one hand, an initial steering committee had to be all but abandoned due to a failure to sustain the active involvement of group representatives in the production of collaborative evidence. On the other hand, individuals who did maintain a close involvement expressed concern over what was perceived as a retreat from collaboration by the MRC's Goldsmiths-based organizers.

Tensions derived in part from the MRC's founding ambition to play a coordinating role among existing civil society groups rather than create an academic think tank. But those groups had distinct priorities, agendas and ways of operating ranging from grassroots member-driven organizations like Avaaz to issue-specific advocacy groups like Hacked Off. As such, each group was understandably conscious of the need to ensure that its own voice was heard above the pulpit and that effective representation of its constituents was not diluted by allegiance to a somewhat abstracted coalition. In this regard, any media reform alliance faces a much steeper challenge in outreach compared to single-issue campaigns such as the Stop the War Coalition.

Underlying all of this was the question of representation, which became explicit by the time the MRC founding principles had developed into concrete proposals for reform. Although very much the product of collaborative research and deliberation, the proposals inevitably gave the MRC a distinct organizational identity that in some ways conflicted with its founding principles. Clearly, there would have been little sense in simply collating and representing the views and perspectives of various civil society groups. A core rationale behind the very notion of a coordinating committee was not just to identify common ground among reform groups, but to push for deliberation on a host of emergent issues around which many groups had yet to adopt a definitive position. Contentious questions regarding the framework of media regulatory reform included whether membership of a new press complaints group should be mandatory or voluntary; whether a comprehensive system of fixed media ownership limits should be imposed within key news markets; and how new sources of public funding for the media should be sourced and allocated. Striking the right balance between advocating concrete proposals for reform while leaving room for consensus building proved to be a key strategic challenge.

Strategic challenges were also related to practical considerations of how best to prioritize and allocate the organization's minimal resources. On the one hand, public engagement as well as media strategies seemed to be an essential component of any broad-based coalition for reform. But MRC organizers lacked the funds and network building blocks to make significant in-roads into the public consciousness on their own accord. Added to this was a virtual impasse in regards to achieving even neutral press coverage.

In this respect, it is significant that MRC's most high profile event to date, a celebrity-fronted public rally in May 2012, received virtually no contemporaneous press coverage. Yet nearly a year later, the event was described in detail by Andrew Gilligan in a piece for the *Daily Telegraph* in which he attempted to portray Leveson's modest recommendations for press regulatory reform as the outcome of a left-wing conspiracy. Somewhat ironically, the fact that the MRC "received virtually no publicity in the mainstream media" was used to invoke this sense of conspiracy (Gilligan 2013).

DEMANDING THE IMPOSSIBLE

By the fourth module of the Leveson hearings (addressing policy solutions), deliberation had given way to advocacy as the central thrust of the MRC's work, though a strong instinct for collaboration and consensus building continued to drive the organization's advocacy efforts.

It was at this point that a fully mobilized mainstream press began a relentless effort to discredit civil society reform proposals. But in spite of this, the MRC and its partner organizations succeeded in reframing policy debates and the Leveson recommendations—at least in respect of the press complaints system— bore the hallmarks of a civil society as well as subsequent cross-party consensus. For a brief moment, what had eluded all of Leveson's long line of predecessors began to look like a realistic possibility: a system of self-regulation of the press with broad representation, statutory underpinning, in-built protections for journalist autonomy, and meaningful powers of sanction and enforcement (Leveson 2012).

Although the press lobby succeeded in deterring Parliament from legislating, the Hacked Off lobby managed to pull off a cross-party deal on a royal charter that embodied the bulk of Leveson's recommendations. What's more, the royal charter was to be uniquely entrenched to the extent that ministers would be unable to alter it without Parliamentary approval, effectively giving it the same status as legal statute.

A rival charter proposal backed by a majority of the press endorsed ministerial discretion over parliamentary oversight. This seemed to lay bare the hollow rhetoric of libertarian defiance that had dominated the opinion editorial pages for months. In any case, the cross-party charter won out and was approved by the Privy Council in the autumn of 2013.

Perhaps in an attempt to hedge his bets, Leveson opted not to address the ownership and plurality question in a substantive way, despite the inquiry's unprecedented disclosures of excessive intimacy and institutional corruption between media and political elites. But buoyed by the success of the cross-party agreement, the Labour party front bench began to pick up and run with ownership reform proposals. Shadow culture secretary and deputy Labour leader Harriet Harman spoke of ownership concentration as having fostered "a culture of invincibility" among dominant media and that "it's wrong for democracy for there to be a lack of plurality in newspapers."[2]

But red lines on reform began to be drawn almost as soon as the prospect of real change glimmered on the horizon. For a political establishment still gripped in the discursive throes of austerity, funding solutions were a policy nonstarter. And in a matter of months, the gap had narrowed further as *caps* rapidly became a dirty word and even discussion of ownership became all but displaced by the less effacing issue of plurality.

At the same time, the royal charter began to look like a hollow victory as newspapers collectively opted out and pressed ahead with their own plans for a new complaints-handling body. By the close of 2013, the Independent Press Standards Organisation (IPSO) had emerged as the favorite to replace the discredited Press Complaints Commission. Newspapers seemed content with the threat of exemplary damages if complaints could not be resolved by IPSO, which would not be recognized under the terms of the royal charter.

Within a matter of months then, what had looked like unprecedented achievement by media reform groups appeared to unravel into something more like spectacular failure. But it is perhaps too early to draw meaningful lessons. Much of the historical impetus for the MRC derives from the Channel 4 legacy: it took the best part of three decades to reach a political compromise over Britain's third public broadcaster in 1982. But it nevertheless resulted in a comparatively radical mandate to represent minority interests and to cover political issues in-depth without impartiality constraints within particular programs (Curran and Seaton 2009).

The battle over Channel 4, however, was not equivalent to the struggle for a democratic and accountable media system and the fundamental reform that would entail. In this context, the word *reform* seems inappropriately modest when matched against an end-goal that demands a root-and-branch overhaul of media policy and regulatory frameworks.

UNDERSTANDING MEDIA POLICY-MAKING
AS ORDER OF DISCOURSE

Because the MRC had to function and operate predominantly within the domain of policy-making (given inadequate resources to support substantive public engagement campaigns), it is useful to ponder over the meaning of policy and its relevance for the struggle of media reforms.

The literature tends to understand policy and policy making in different ways (Colebatch 2002; Ham and Hill 1993). For some, policy is "what governments do, why they do it, and what difference it makes" (Dye 1976, 1). However, governments are not the only bodies that produce policy, and there are also policies developed by independent regulatory bodies, public institutions, and trade organizations (Galperin 2004). A common model applied to the study of policy is the "cycle model" (Colebatch 2002, 55) that sees policy-making as a sequence of stages: "agenda setting (problem recognition), policy formulation (proposal of a solution), decision-making (choice of a solution), policy implementation (delivery), and policy evaluation (checking the results)" (Howlett and Ramesh 2003, 10).

Of course, in reality, the policy-making process is more convoluted than the cycle model assumes. Stages may not follow the predicted order or may be omitted.

A particularly fruitful line of thought aimed at analyzing policy-making and which has attracted considerable attention in the study of media reforms (Napoli 2007) has been the "interest group approach to policy" that explains policy as "the outcome of competing powerful interest groups" preferences that can span from NGOs to trade union organizations to business organizations and so on. As Galperin (2004, 160) explains: "Policy outcomes are typically explicated by the organization and the resources available to interest groups and their support coalitions. In other words, policy outcomes are a function of the power that each interest group is able to amass and wield in support of its preferred outcome."

However, this configuration shares with the cycle model approach a tendency to overlook the complexities built into the policy-making process and neglects an analysis of social and cultural power at play. Policy outcome becomes the result of a predictable war between competing groups with different goals, thus ignoring that "much political activity is concerned with maintaining the status quo and resisting challenges to the existing allocation of values" (Ham and Hill 1993, 12).

For this reason, the work of Michel Foucault on governmentality and discourse (1981 [1970], 1991) might facilitate a better understanding of the realm of policy-making and provide fruitful observations for activists seeking to build strong and successful reform movements. Foucault's work on governmentality draws on the belief that "the how of power is as important as who has the power" (1991, 103). Foucault's study on the *production of discourse* is also particularly useful in that it problematizes the practices of government and how public policy is developed, shaped, and changed. According to this thesis, relations of power "permeate, characterise and constitute the social body, and these relations of power cannot themselves be established, consolidated nor implemented without the production, accumulation, circulation and functioning of a discourse" (1980, 93).

In "The Order of Discourse" (1981 [1970]), Foucault investigates the social practices of disciplines and shows how discourses are embedded in social practices (and disciplines) through rituals, values, and routines. Foucault's ideas are extremely valuable in an analysis of policy-making, because they challenge the belief that media policy is a rational process. On the contrary, Foucault warns that the rules of government are not themselves defined by the rule of law, but rather by the rules of knowledge, power, and discourse formation. So what is presented to us as incontrovertible truth or evidence in policy-making is rather the product of "the will to truth": "Thus all that appears to our eyes is a truth conceived as a richness, a fecundity, a gentle and insidiously universal force, and in contrast we are unaware of the will to truth, that prodigious machinery designed to exclude" (ibid., 56).

Drawing on Foucault, the work of Norman Fairclough (1989) is extremely helpful in highlighting the strategic importance of an "order of discourse" that embodies specific ideologies in policy: "Since discourse is the favored vehicle of ideology, and therefore of control by consent, it may be that we should expect a quantitative change in the role of discourse in achieving social control" (ibid., 37). Thus, policy becomes not the product of policy-making institutions but rather an order of discourse structuring what can and cannot be said as well as who has the power to speak. But as Gramsci (1996) has argued, social control in modern complex societies is not exerted through coercion but through consent, therefore the ideologies conveyed through discourse are not imposed, but simply accepted.

Drawing on this approach, the order of discourse that the coalition had to challenge becomes apparent: on the one hand, the hegemonic mantra of market liberalism, on the other hand, the cultural background of an established commercial media system.

For example, a dominant discourse of market liberals' policy-makers might interpret media democratization as deregulation (Hackett and Carroll 2006a), while a change of media systems might seem unthinkable because it becomes detrimental to a freedom of speech argument, especially in the United States where it is often evoked as the principle upholding liberal individualism.

This can explain, for example, why in policy discourse in (and beyond) the United Kingdom, there is a common belief that regulation of media ownership should not be a priority for policy-makers but rather that plurality as an instrument of market competition should be the main concern. Indeed, plurality is becoming an increasingly popular term of reference for policy-makers developing reforms in the United Kingdom, Europe, and Australia.

FROM MEDIA REFORMS TO SOCIAL JUSTICE

So how then can a dominant policy discourse be transformed by a media activist campaign and an order of discourse be subverted? What are the tools? And what is then the role of activist groups? For years, as Napoli (2007, 21) rightly argues, early work on media reform lacked a solid theoretical perspective, while the focus seemed to be consistently "legalistic." When later studies embraced different frameworks, they seemed to adhere to interest group theories of policy-making discussed above. More recently however, scholarship has found more fruitful ground through conceptualizing the battles of groups advocating media reforms as the actions of an emerging social movement (Atton 2003; Calabrese 2004; Hackett and Carroll 2006a).

Hackett and Carroll (2006a) elaborated a thorough sketch of the most important frames employed by scholarship to understand media reform movements. Out of the main frames, perhaps one of the less developed is the idea of media reform as "a first issue that can draw new people into public life, citizen activism, and wider struggles for social justice" (Jansen 2011, 7). Indeed, according to McChesney (2007, 220), "fates of media reform and social justice research are intertwined. They will rise or fall together." Jensen (2011, 2) also notes that Amartya Sen's (2009) theory of justice dovetails with McChesney's perspective, as "the removal of barriers to free and open discussion, the development of the right ('capability') to communicate, and the institutionalization of an 'unrestrained and healthy media' (to give 'voice to the neglected and disadvantaged')" becomes fundamental "to the pursuit of human justice and security."

According to Mouffe (1999, 756), "every consensus exists as a temporary result of a provisional hegemony, as a stabilization of power and that always entails some form of exclusion." If hegemony is provisional, then media reform movements should concentrate on building a permanent campaign far beyond a single-issue crusade. Given that it took thirty years to introduce and develop a successful policy discourse to launch Britain's third public broadcaster, Channel

4, in 1982, a more overreaching media reform campaign would need to create a wider base, a consistent message, and a long-term agenda.

If we consider that the key challenge for any media reform movement lies not so much in building consensus among existing stakeholders, but in articulating a narrative that can resonate with wider struggles for social justice, then the MRC experience offers some useful lessons. First, it seems clear that the organization was somewhat constrained by a degree of unavoidable short-termism in its outlook and approach. To understand why this was the case, we need to recall the peculiar circumstances and political context into which it was born. It was the prospect of a public inquiry into the media with a remit unprecedented in scope which provided the impetus for the founding of the MRC in 2011. It was also the basis on which it secured limited short-term funding from the Open Society Foundations with the explicit objective of coordinating civil society responses to the inquiry. Along with these specificities enshrined in the MRC's founding agenda and funding terms, the scope for action was further constrained by the imminent time pressures imposed by the various deadlines for submission of evidence to the Leveson and concomitant inquiries.

This also contributed to situating the MRC firmly within the policymaking community. It was a position that was further anchored by the MRC's collaboration with Hacked Off, an organization with an explicit lobbying focus. As we have seen, this provided the MRC, for a brief moment at least, with an opportunity to intervene and to some extent reframe media policy debates. But it also constrained the organization to the extent that the press had an incomparably more fundamental role in defining and circumscribing the terms of that debate. Thus the notion of press freedom was always exclusively tied to the threat of statutory underpinning of press regulation. Media reform groups were forced to articulate defenses on the basis that statutory underpinning would *not* threaten press freedom, with little room to demonstrate the real threat to press freedom posed by *ownership*.

Maintaining such a close proximity to the heart of the policymaking community also limited the MRC's capacity to reach out to groups that were not already engaged in media policy debates but might nevertheless align with a progressive agenda for media reform. As already mentioned, the MRC was constrained in this respect also by its limited resources. But one resource in plentiful supply was its capacity to present new evidence of media failure through critical scholarship, new studies and new data. By tying this critical scholarship directly to the concerns of other social justice movements, media reform groups can provide a basis for wider engagement and thus a more assured footing in their resistance to dominant neoliberal discursive frameworks. As Hackett and Carroll (2004, 18) succinctly observe: "This frame repositions the project as one of social justice in a world organized around global capitalism, racism and patriarchy, and directly connotes the need for alliances, even integration, with other social movements."

Of course, this is not a new idea. There are plenty of historical precedents demonstrating success in linking media reform with other social movements. For example, Horwitz (1997, 344) explains that the success of US media reforms movements of the 1970s was due to a "connection to the broader social movement of Civil Rights." The key challenge is to think of how to build these connections in the contemporary socio-political landscape and how best to link media reform movements to campaigns such as climate change activism, with a view to making fair, accountable and accurate coverage of climate change issues a key concern of campaigning groups. This would in turn draw meaningful support for media reform from a wider layer of other groups. We could perhaps also draw lessons from the way in which climate change activism has evolved into a movement for climate *justice*, encompassing a network of organizations committed to social, gender as well as ecological iniquities.[3]

CONCLUSION

A stronger discursive focus on resistance might also make it easier for media activist and advocacy movements to engage a broader spectrum of groups already active in their struggles against poverty, racial and gender discrimination, ecological risks, and violations of human rights; to foster and build solidarity across different social justice movements in an effort to change and reshape dominant media policy discourses. Most importantly, a focus on resistance poses a direct challenge to the dominant discursive order that defines resistance exclusively in the context of the threat of state interference and the struggle to safeguard press freedom (or more accurately, press *power*).

Such a focus need not necessitate disengagement from the media policy process or from established stakeholders, provided that the target of resistance is framed clearly and exclusively as media owners who use their assets in a bid to control the political and public agenda. Such a platform could even be used to further engage with professional journalists themselves with a view to redressing the balance of power between capital and labor within newsrooms (Sparks 2007).

Resistance to corporate media ownership also provides a basis upon which to articulate a struggle that transcends the boundaries between different national contexts. On one level, creeping state control of the media in, for example, Hungary over recent years looks a world apart from the UK context. But at root in both cases is a strengthening alliance between media and political elites. Whether that alliance is founded on favored media outlets or compliant political actors merely reflects two faces of the same problem: the reassertion of control over public conversations.

Ultimately, the split that Carroll and Hackett (2006b) identified between media activist groups that focus on engagement with hegemonic institutions and those that adopt an oppositional resistance toward them need not reflect a choice that has to—or indeed should be—made. Though it may take some time, media activist groups that are embedded in broad-based and grassroots movements for social change stand a better chance of mounting a meaningful challenge to the dominant neoliberal frameworks that overarch media policy debates. Those frameworks co-opt notions of resistance and autonomy in ways that obscure the real and fundamental threat to press freedom in the United Kingdom and other Western democracies: concentrated media power.

NOTES

1. Note that both authors have had an active role in the MRC from the outset and continue to do so.

2. BBC Radio 4 *Media Show*, May 8, 2013.

3. See, for instance, the commitment of Climate Justice Now at https://lists .riseup.net/www/info/cjn.

REFERENCES

Atton, C. 2003. "Reshaping social movement media for a new millennium." *Social Movement Studies* 2, no. 1: 3–15.

Bennett, D., and J. Townend. 2012. "The Scandal of Selective Reporting." *British Journalism Review* 23, no. 2: 60–66.

Calabrese, A. 2004. "The Promise of Civil Society: A Global Movement for Communication Rights." *Continuum: Journal of Media and Cultural Studies* 18, no. 3: 317–329.

Chomsky, N. 1967. "The Responsibility of Intellectuals." *New York Review of Books.* www.nybooks.com/articles/archives/1967/feb/23/a-special-supplement-the -responsibility-of-intelle/.

Colebatch, H. K. 2002. *Policy.* Buckingham, UK: Open University Press.

Curran, J. 2002. *Media and Power.* London: Routledge.

Curran, J., and J. Seaton. 2009. *Power without Responsibility: The Press and Broadcasting in Britain.* London: Routledge.

Dye, T. 1976. *Policy Analysis: What Governments Do, Why They Do It, and What Difference It Makes.* Tuscaloosa: University of Alabama Press.

Fairclough, N. 1989. *Language and Power.* London: Longman.

Foucault, M. 1980. *Power/Knowledge: Selected Interviews and Other Writings, 1972–1977.* Brighton, UK: Harvester Press.

———. 1981 [1970]. "The Order of Discourse. Inaugural Lecture at the Collège de France, 2 December 1970." In *Untying the Text: A Post-Structuralist Reader*, edited by R. Young, 49–78. Boston, MA: Routledge and Kegan Paul.

————. 1991. "Governmentality." In *The Foucault Effect: Studies in Governmentality*, edited by G. Burchell, C. Gordon, and P. Miller, 87–104. Hemel Hempstead, UK: Harvester Wheatsheaf.

Freedman, D. 2014. "Truth over Justice: The Leveson Inquiry and the implications for democracy." In *How We Are Governed: Investigations of Communication, Media and Democracy*, edited by P. Dearman and C. Greenfield, 53–74. Cambridge: Cambridge Scholar.

Galperin, H. 2004. "Beyond Interests, Ideas, and Technology: An Institutional Approach to Communication and Information Policy." *Information Society* 20, no. 3: 159–168.

Garnham, N. 1986. "The Media and the Public Sphere." In *Communicating Politics: Mass communications and the political process*, edited by P. Golding, G. Murdock, and P. Schlesinger, 37–54. Leicester: Leicester University Press.

Gilligan, A. 2013. "The Truth About Hacked Off's Media Coup." *Telegraph.co.uk*, March 30. www.telegraph.co.uk/news/uknews/leveson-inquiry/9963263/The-truth-aboutHacked-Offs-media-coup.html.

Gramsci, A. 1996. *Quaderni Dal Carcere*. Torino: Einaudi.

Hackett, R. A., and W. Carroll. 2004. "Critical Social Movements and Media Reform." *Media Development* 51, no. 1: 14–19.

————. 2006a. *Remaking Media: The Struggle to Democratize Public Communication*. New York: Routledge.

————. 2006b. "Democratic Media Activism Through the Lens of Social Movement Theory." *Media, Culture and Society* 28, no. 1: 83–104.

Ham, C., and M. Hill. 1993. *The Policy Process in the Modern Capitalist State*. Brighton, UK: Harvester Wheatsheaf.

Horwitz. R. B. 1997. "Broadcast Reform Revisited: Reverend Everett C. Parker and the 'Standing' Case." *Communication Review* 2, no. 3: 311–348.

Howlett, M., and M. Ramesh. 2003. *Studying Public Policy: Policy Cycles and Policy Subsystems*. Oxford: Oxford University Press.

Jansen, S. C. 2011. "Media, Democracy, Human Rights, and Social Justice." In *Media and Social Justice*, edited by S. Jansen, J. Pooley, and L. Taub-Pervizpour, 1–23. New York: Palgrave Macmillan.

Juris, J., and M. Razsa. 2012. "Occupy, Anthropology, and the 2011 Global Uprisings." In *Fieldsights—Hot Spots, Cultural Anthropology Online*. http://culanth.org/fieldsights/63–occupy-anthropology-and-the–2011–global-uprisings.

Keane, J. 1991. *The Media and Democracy*. London: Polity Press.

Leveson, Lord Justice. 2012. *An Inquiry into the Culture, Practices and Ethics of the Press: Report*. 4 vols. London: Stationery Office.

McChesney, R. W. 2007. *Communication Revolution: Critical Junctures and the Future of Media*. New York: New Press.

Mouffe, C. 1999. "Deliberative Democracy or Agonistic Pluralism?" *Social Research* 66, no. 3 (Fall): 745–758.

Napoli, P. M. 2007. "Public Interest Media Activism and Advocacy as a Social Movement: A Review of the Literature." McGannon Center Working

Paper Series. New York: Fordham University. http://fordham.bepress.com/
mcgannon_working_papers/21.

Pilger, J. 2012. "The Leveson Inquiry into the British Press—Oh, What a Lovely
Game." *Johnpilger.com.* http://johnpilger.com/articles/theleveson-inquiry
-into-the-british-press-oh-what-a-lovely-game.

Roberts, A. 2012. "Why the Occupy Movement Failed." *Public Administration
Review* 72, no. 5: 754–762.

Scannell, P. 1989. "Public Service Broadcasting and Modern Public Life." *Media,
Culture and Society* 11, no. 2: 135–166.

Sen, A. 2009. *The Idea of Justice.* Cambridge, MA: Harvard University Press.

Sparks, C. 2007. "Extending and Refining the Propaganda Model." *Westminster
Papers in Communication and Culture* 4, no. 2: 68–84.

CHAPTER TWELVE

Media Reform Movements in Taiwan

HSIN-YI SANDY TSAI, *National Chiao Tung University, Taiwan*
SHIH-HUNG LO, *National Chung Cheng University, Taiwan*

MEDIA REFORM STRATEGY

This chapter discusses how media reform groups in Taiwan and the Media Watch Foundation and the Campaign for Media Reform in particular worked together to engage in media reform movements ranging from the promotion of public media to the campaign against big media mergers. The strategies they utilized are context-sensitive and mostly effective in terms of raising public awareness, setting public agendas and having an impact upon government decisions and policies concerning media freedom and media democratization. These strategies include, among others, engaging in public debates, lobbying the legislature, and, where necessary, mobilizing public support in both online and offline protests. However, these strategies have limitations too. Most of all, the mainstream media were usually hostile to media reform movements and the government was passive in its response to the cause of media reform. To overcome this extremely difficult situation, media activists in Taiwan have taken advantage of new media and social media to pursue their media reform goals and have increasingly engaged in creating more diverse information alternatives to the mainstream media.

In recent years, Taiwan has been ranked as one of the top countries in terms of press freedom in Asia (Freedom House 2014). This is remarkable given that the country was still ruled by an authoritarian, one-party regime less than three decades ago. Until 1987, the martial law, which aimed to suppress the Communists in mainland China, was finally lifted and the press liberalized (AP

1987). Taiwanese citizens were allowed to form different political parties. The political transition of democratization in the late 1980s put an end to the world's longest period of rule under martial law (thirty-eight consecutive years) and the longstanding bans on organizing political parties and publishing new newspapers. From then onward, three waves of media reform movements aimed at media democratization started apace. Each wave of media reform pursued the same goals but utilized somewhat different strategies.

Over the last few decades, the interests of the ruling party usually commanded civil society; the party opinions and interests usurped the public ones. Therefore, the first wave of media reform movements in Taiwan called for media privatization and liberalization because the government maintained its monopoly over terrestrial television right up to 1997, when the first privately owned terrestrial television station was launched in Taiwan. Yet, another eight years would pass before the government was forced to finally give up its ownership of the other three terrestrial television stations.

However, the media privatization and liberalization that ran riot in the 1990s also caused new problems. Private media in pursuit of commercial interests could not manage the responsibility of handling the public service and the public sphere. The big media consolidation, thanks to a series of mergers, had resulted in "market-driven journalism" (McManus 1994) or even "audience ratings journalism" (Lin 2009). The second wave of movements for media reform in Taiwan thus strived to campaign against big media and irresponsible journalism prevailed in the market. Since the late 1990s, established by nonpartisan media activists, the Media Watch Foundation has done a great deal to promote media literacy education and fight media commercialism.

The third wave of media reform movements has entered new terrain. Since 2003, media activists and media scholars have formed the Campaign for Media Reform with the intent to pursue new goals. This new movement seeks to expand public media that correspond more closely to the ideal of public service broadcasting in terms of quality, universality, independence, and diversity and strengthen alternative, independent citizen media.

In this chapter, we focus on how media reform groups in Taiwan engaged in the aforementioned media reform movements and the strategies utilized by these groups. We begin by providing a brief account of media history and the changing media landscape in Taiwan. Next, we discuss the media reform movements in Taiwan and the strategies media reform activists utilized respectively in the campaign for the expansion of public media and the campaign against big media mergers. We also examine the increasing relevance of the alternative, independent citizen media in Taiwan's media reform movements. Our discussion concludes with the future prospects for media reform movements in Taiwan.

Taiwan is a small island country with a population of twenty-three million. Before 1987, because of the martial law, the broadcast media in Taiwan were mainly owned and controlled by the ruling Nationalist Party, the government, and the military. Most of the newspapers were in the hands of those who were close and loyal to the authoritarian regime.

With the abolition of the martial law, the ban on publishing new newspapers was lifted and newspapers were liberalized. In 1993, cable systems were legalized. The media industry started to grow rapidly. By 2015, there were 4 main daily newspapers, 171 radio stations, 5 terrestrial broadcasters, 57 cable systems, 5 multiple-system operators (MSOs), and 294 satellite TV channels. About 59.6 percent of households subscribed to cable TV (NCC 2015).

However, the huge number of media outlets did not guarantee a free marketplace of ideas. The fierce competitive market did not bring quality news to the Taiwanese people. Since 2003, tabloid journalism and sensational news has become influential in the media industry. Sensational news was commonly seen on TV and in newspapers. With the influences of media commercialism, professional journalism came to be at risk. The threat of big media consolidation and the influence of political power still remained severe to media freedom and the freedom of speech in Taiwan. Therefore, the need for media reform became more and more urgent. Since 1987, there have been three main media reform waves in Taiwan.

MEDIA REFORM MOVEMENTS IN TAIWAN

These three waves of media reform movements can be categorized as: (1) media privatization and liberalization starting in 1987; (2) the campaign for expanding public media starting in 2003; and (3) the fight over big media mergers in recent years.

MEDIA PRIVATIZATION AND LIBERALIZATION

Similar to the United States, the 1980s can also be characterized by many major political and societal changes (Mueller 2004). In 1987, the government abashed the martial law and Taiwan started its first step out from authorized government toward a more democratized country (Ministry of Culture 2011). However, despite the lifting of the martial law, Taiwanese society and media still had a long way to go before it could be considered fully democratic.

The fight for freedom speech happened at the same time as many social activists were still fighting for basic democratic rights in Taiwan. For instance, at that time, advocating for an independent Taiwan was still prohibited by the government. Right after the lifting of martial law, many social activists fighting for freedom of speech were organizing protests and strikes. A notable dissident, Nylon Cheng, even sacrificed his life in 1989 to fight the erstwhile authoritarian state in an attempt to awaken Taiwanese society.

In line with fighting for a democratic society and media, the push for media privatization and liberalization led to the first wave of media reform in Taiwan. An association formed by scholars and social groups argued for "no political parties, government, or military ownership of the three broadcasters." A big protest for media privatization and liberalization took place in 1995 in the capital city Taipei despite concerns about the power of big financial consolidation (Kuang 2004). The protest successfully led to the establishment of a new law regarding the sale of government-owned stocks of the broadcasters in order to privatize and liberalize them.

A series of media privatizations and liberalizations occurred in the 1990s. In 1993, cable systems were liberalized, and the government-owned telephone company was privatized in 1996. Between 1993 and 1998, there were limited foreign investments on the media in Taiwan. After the Cable Radio and Television Act was modified in 1999, foreign investments increased substantially. Three major MSOs were owned by big global media players after 2007 (Huang 2009). These major changes did not come out of nowhere: Comments made by scholars and social movements on the streets all contributed to the development of media privatization and liberalization. The experiences learned from other regulators in the world also contributed to the changes. In hindsight, the seemingly inevitable trend of media privatization and liberalization since the 1990s has catered for the interests of foreign private equity while foreign owners of the MSOs did not in return invest significantly in the production of local content. Instead, the foreign owners of the MSOs were seeking arbitrage profits. As a result, the trend of privatization and liberalization since the 1990s exacerbated the already serious lack of quality and diversity in terms of media content available to local audiences.

Strategies Utilized

According to Mueller (2004), there are three kinds of advocacy modes in information and communication policies: content, economic, and rights. Content is related to the problems of media messages; economic advocacy is about the supply of communication products and services; rights advocacy concerns individual rights, such as freedom of expression or property rights. At this stage, advocates asserted the rights of communication, including communication rights for public and media workers. Freedom of expression and anticensorship rights were the central arguments advanced by social movement activists in Taiwan during the first wave of media reforms.

Hackett and Carroll (2006) indicate five action frames for remaking media: Free press, cultural environment, media democratization, the right to communicate, and media justice. At the time, Taiwanese society was still in the process of fighting for a free press, the right to communicate, and media democratization. Political and social activists conducted protests and established new alternative magazines.

THE CAMPAIGN FOR EXPANDING PUBLIC MEDIA

Media liberalization and privatization in the 1990s did not bear fruit in terms of media democratization in the public interest. The mainstream commercial media in pursuit of ratings and advertising revenues at the expense of news credibility gradually lost the public's trust. It is no surprise that, according to an Edelman Asia-Pacific survey, Taiwanese broadcasters are ranked by viewers as being the least trusted in Asia (Lo 2012a). Many foreign commentators have also observed that the irresponsible journalism prevailing in Taiwan has caused widespread public mistrust (Fuchs 2014; Harding 2010). For example, a writer for Foreign Policy recently criticized that "over the last decade, Taiwanese media have come to be known for in-your-face, no-holds-barred reporting that manages to be simultaneously sensationalist and mundane" (Fuchs 2014).

The worsening situation of media performance as a whole in the early 2000s gave impetus to a new movement for media reform in Taiwan. Two major media reform organizations were established: the Campaign for Media Reform and the Media Watch Foundation. The Campaign for Media Reform was established in 2003 by a group of scholars, news workers and social movement activists from the Democratization of Broadcasters Coalition. The campaign goal is to improve media performance and environment, educate the public about media literacy, and improve media policies in Taiwan. Since 2008, it has hosted three annual media reform meetings to invite the public to talk about media reform in Taiwan (Lo 2012b). The campaign has initiated many important media reform movements and press conferences to make the public aware of pressing social issues. The Media Watch Foundation was established in 2002. The organization's goal is to protect the labor rights of media workers and to improve media literacy education for the public. The foundation also helped to form a public media mechanism, initiate many media reform movements, and monitor media performance and self-regulation in Taiwan. The foundation provided many training workshops at community colleges to help the public learn about media literacy and citizen journalism and has also supported youth programs in Taiwan. Both organizations have had a great influence on the major media reform movements in Taiwan over the last decade.

The establishment of public media—Public Television Station (PTS) and Taiwan Broadcasting Service (TBS)—is an important achievement for media reform movements in Taiwan. Aimed at serving the public interest and balancing the inadequacy of commercial televisions, PTS started broadcasting in July 1998. Back in 2003, when media reformers and social movement activists fought for separating political power, government, and military influences from broadcasters, the urge for a public broadcasting system was widely discussed. As a result, TBS was established in July 2006.

Since the 1970s, communication scholars had argued the importance of constructing a public television service in Taiwan (Feng 2006). However, the public television conceived in the 1970s and the 1980s was more like a television that did not have commercials, dealt with programs for social education, and played the role of fulfilling government's need for policies and education. In 1992, eighteen communication scholars at the National Cheng-Chi University made a public comment criticizing the draft of the Public Television Act (PTA), focusing in particular on the formation of the board for PTS. In 1993, the Student Group for Observing the Public Television was established. Communication scholars collaborated with other scholars (e g , economists, lawyers, and sociologists) and citizen groups to form a preparation group for the PTS. The movements for establishing a public broadcasting system were initiated by the Coalition for Establishing Public Media in 1997. However, on April 16, 1997, the major political party (KMT) announced that Taiwan would be giving up the establishment of public television. Scholars and others who cared about public media made their discontent with the decision known via public newspapers. The Association for the Establishment of Public Media, scholars, and students protested in front of the Legislative Yuan. Because of these comments and protests, the KMT decided not to give up on the act. The draft of the Public Television Act was finally approved by the Legislative Yuan.

In 1997, the Democratization of Broadcasters Coalition, which was the predecessor of the Campaign for Media Reform (Kuo 2010), was established to plan the formation of a public television system in Taiwan. By 2014, four stations were included in TBS. Moreover, public interest is the main principle for the maintenance of TBS which is owned by the public, instead of the government. However, the management of TBS is not straightforward. There have been countless arguments and debates concerning the nomination of the board members and the CEO of TBS. The need to have proper legislation rule for TBS is still an ongoing debate in Taiwan and has led t o many protests urging legislative reform.

Strategies Utilized

Similar to the claims made by advocacy movements in the first wave of media reform, this movement also asserted the rights of freedom of speech and communication: The right to public television, without interference by the government and the military during the second wave of media reform. In addition, they also were advocates for better content (Mueller 2004), a need that was not always covered by commercial media.

The main strategy used in establishing PTS was to form an agenda against government and military intervention, later forming positive arguments as to how to establish PTS (Feng 2006). Media reform activists tried to emphasize the importance of free broadcasters and a real *public* television station (not a government-controlled station). The initial argument was proposed by scholars in communication departments and then extended to scholars in other fields. Trying to persuade the public and the government was critical. However, during these movements, the intellectuals initiated the discussions. Still, they could not persuade the government to accept the values they proposed (Feng 2006).

THE FIGHT OVER BIG MEDIA MERGERS

Following the stream of media privatization, the legislative Yuan passed the National Communications Commission Organization Act in 2005. The act calls for the establishment of the National Communications Commission (NCC) to separate the influence of political parties, government, and military on the mass media.

In February 2006, the NCC was established to take charge of media regulation and related policy-making on telecommunications and broadcasting industries. Originally, the telecommunications industry was regulated by the Directorate General of Telecommunications and the broadcasting industries were regulated by the Department of Broadcasting Affairs of the Government Information office. However, with the increase in media convergence, there was a need to have a central regulator to take charge of both telecommunications and broadcasting industries.

In the age of media convergence, the issues of media concentration and media monopolies are more complicated and consolidation can be bigger and more influential. A media company can own multiple newspapers, online media, TV, radio, and cable systems. This makes companies even more powerful and influential than if they only owned one media outlet (such as a newspaper), making antimedia monopoly movements even more important.

Since 2006, several big mergers have been reviewed by the NCC, including the merger of BCC (radio) and CTV (TV) in 2006; the *China Times* (one of the four major newspapers) with two major TV news stations (CTV and CtiTV) in 2007; Taiwan Mobile with several cable systems in 2008; as well as Want Want China Broadcast's proposed buyout of China Network Systems (Lo 2013). Among these cases, the establishment of Want Want China Media Corp. and the approval of its subsequent mergers have been the most debated topics in Taiwan.

The Want Want China Times Incidents

Want Want Corp., originally a food product company, merged with one of the main newspapers (the *China Times*) and became a multiplatform media corporation that owned two TV news stations and a newspaper company (Want Want China Times Media Group) in 2009. This incident was called "Want Want merges China Times, CTV and CtiTV," also know as "Three Chung Incident." The approval of the merger was accompanied by significant debates. Nevertheless, the Want Want merger was approved by the NCC in 2008 and established Want Want China Times Corp in 2009.

Owning two news stations and one newspaper did not satisfy the corporation. In 2011, Want Want China Times Media Corp (Want Want) proposed to buy one of the main MSOs (CNS) in Taiwan. This merger ("Wang Chung Incident") would allow the corporation to own twelve TV channels, eleven cable systems, one publisher, two magazines, three online media companies, four newspaper companies (including one major newspaper), and at least nine other investment companies (such as travel agency and production companies). This would lead to a great monopoly in the media industry in Taiwan. The International Federation of Journalist (IFJ) also pointed out that the merger would allow Want Want to own about one-third of audiences and readers in the whole media industry. This caused great concern (*Liberty Times* 2012). Therefore, a series of media reform movements began to fight this media monopoly. In February 2013, after about two and a half years' investigation and debates, the NCC did not approve the merger proposal due to the public concern over the formation of a media monopoly (and the concentration of a marketplace of ideas).

Strategies Utilized

The media reform movement used a variety of strategies to achieve its goal including public comments, exposure to media, protests, and allies.

The media reform movements against the Want Want merger started with a series of public remarks in traditional newspapers made by communication and journalism scholars in September, 2011. These scholars were also active members of the Campaign for Media Reform in Taiwan which also issued a public statement opposing the merger. After September 2011, a series of letters to newspapers, antimonopoly workshops, and public statements were made by scholars in different fields and other media reform and social groups. For example, in September 2011, a workshop on cross-media concentration was held by the Institute of Journalism at National Taiwan University (NTU).

On October 24, 2011, the NCC held the first public hearing on the merger of Want Want. At the same time, scholars and media reform groups (including Taiwan Media Watch, Citizen Congress Watch, the Association of Taiwan Journalists, and the Campaign for Media Reform) started an online protest via a campaign platform created by social movement groups (see http://campaign .tw-npo.org/index2.php). By using blogs and other social media, including Facebook and Twitter, more than three thousand groups and individuals signed up to oppose the merger.

Another critical strategy to raise public awareness is to make full use of the media (Ryan, Jeffreys, and Blozei 2012). In the same month, the NTU and media campaign groups held a press conference to urge the NCC to fight the media monopoly. Later on, while the NCC was hosting the first public hearing, a group of students and media reform groups rallied in front of the NCC building. These events attracted the attention of the government. As a result, the Legislative Yuan invited one of the scholars to attend their meeting. After that, several workshops were held by these journalism and communication scholars in order to attract more media coverage and the attention of the public and the governments. The public pressure urged the NCC to postpone the approval.

On February 6, 2012, sixty-six people signed up with the "No China Times Movement" against the interference of *China Times* on media freedom. There was also a concern on how Want Want might have unbalanced news reports that favored the Chinese government. On May 7, 2012, the NCC held a second public hearing on the merger. One thing worth mentioning is that there were different points of view at the time. Some scholars opposed the merger, while some argued that it was difficult to have a real media monopoly in Taiwan because there were plenty of news sites. Other scholars argued that a high standard of the investigation on what stakeholders said was similar to the time when the freedom of speech was highly restricted in Taiwan.

In July 2012, the NCC announced a conditional approval of the merger. These conditions focused on maintaining the independence of news stations owned by the corporation, which was one of the emphases of those series of anti–media monopoly movements. After the announcement of conditional approval in September 2012, a protest involving more than nine thousand individuals was organized by the Association of Taiwan Journalists and other social groups. This was one of the biggest protests in Taiwan in the past twenty-two years.

In November 2012, Taiwan Law Society, Taipei Society, and Taiwan Democracy Watch filed an *Amicus curiae* brief to urge the NCC to ensure that those three conditions were satisfied before approving the merger. In 2013, the corporation did not fulfill those conditions; hence the NCC announced its disapproval of the deal.

ASSESSMENT

For the first wave of media reform movements, activists' strategies were made based on their previous decisions and the constraints of the environments (Meyer and Staggenborg 2012). In this phase, the majority of the intellectuals were people who fought for the freedom of speech and the freedom of the Taiwanese people. The mainstream media were controlled and influenced by the government (the main political party). Hence, having protests and publishing and commenting on alternative news magazines and newspapers were the main strategies used.

For the second wave of media reform movements, scholars and media reform activists had tried to gain the awareness of the government and the public for the importance of "no political power, no government influence on media." Writing critiques/letters in the mainstream media and conducting protests and strikes were the main strategies they used.

For the third wave of media reform movements, the approach to information and communication policies was more mixed and involved a combination of content, economic, and rights issues. Compared with two other wave movements, media activists had more resources (more possible media outlets) and networks (more institutions to form allies). As Fetner and Sanders (2012) point out, building up institutions for social movements is critical to the success of movements.

In fighting against media monopolies, all five actions mentioned in Hackett and Carroll (2006) were utilized. Specifically, these movements emphasized the specific cultural environment (the relationship between China and Taiwan) for freedom of speech in Taiwan (to avoid influence by the political power of China). Fighting for media justice (for people who were sued by the *China Times* for criticizing the newspaper) was another action taken. Participating in multiple actions and using multiple networks were critical strategies in the movement's success.

THE RELEVANCE OF ALTERNATIVE AND CITIZEN MEDIA

Media reformers in Taiwan also engaged in the expansion of independent and citizen media. For example, they launched several websites with the purpose of empowering citizen journalists. Among these websites, civilmedia.tw and we-report.org are the most active in producing citizen videos and crowdfunded independent investigative journalism. Both websites were established and managed by media reformers who are also the key figures of the Campaign for Media Reform in Taiwan as well as the Media Watch Foundation.

Civilmedia.tw is an alternative, independent media website in Taiwan (see www.civilmedia.tw/about for details). The site originated from a research project funded by the National Science Foundation, Taiwan, in 2007–2012, and was officially established in 2012. The site aims to record social movements that are ignored in mainstream media. By using interviews and video recordings, these movements and events can be shared with the public as another resource to understand social events. There are more than 1500 records in the database in the past five years. These records are not limited to media-related events. In fact, the database includes details on human rights, cultural issues, immigration, tax reform, community news, education, social welfare, and other events and topics. Civilmedia.tw also works with a PTS program to broadcast important social movements every week. In addition, civilmedia.tw is connected to social networking sites—including Facebook, Twitter, and Google Plus—to reach more people and have more interaction with the public.

WeReport.org was established by the Quality News Development Association at the media reform annual meeting in May 2011. On December 3, 2013, wereport.org was officially launched online. It is the first nonprofit investigative report platform and is supported by donations from the public. By posting one's proposal for a news reporting project, donors can choose projects they would like to support. Since 2011, there have been more than five hundred donors and more than thirty-seven projects proposed on the platform. Similar to civilmedia.tw, weReport.org also has outlets on social networking sites, including Facebook, Google Plus, Twitter, Plurk, and RSS.

MEDIA STRATEGIES IN TAIWAN

Overall, these important media reform organizations and websites were originally proposed by media scholars and social movement activists. They started by forming a small association and then looked for alliances and, most importantly, the participation of the public. Websites and social networking sites (such as Facebook, Twitter, and Plurk) are essential platforms to reach more people. Open platform and open access are the principles shared by these organizations. They also cooperate with each other with regard to improving media performance and citizen journalism in Taiwan.

Similar to the strategies used by Free Press and other media reform activists in the United States, social persuasion and grassroots movements have also been utilized by media reform activists in Taiwan (Lo 2012). Media watch and media literacy education, alternative media, media labor protests, and media policy advocacy are all important strategies used by media reform activists in Taiwan (Lo 2012).

Besides, many different institutions and organizations have been involved in these media-reform–related movements. With regard to media reform, collaboration is more helpful and more useful than letting every tub stand on its own (Ryan and Gamson 2011).

CONCLUSION

Media reform movements in Taiwan have fought for freedom of speech and a more democratic media environment. These movements started from fighting for media privatization and liberalization to remove the influence of political parties, government, and military over the mass media, to fighting media monopolies in order to protect the diversity of media and a wide range of ideas.

During the past ten years, media reform movements have been growing, leading to the establishment of civil society in Taiwan.

The fight against government control is not easy. Still, it is no less difficult for the public to see the importance of separating political parties, government, and military from broadcasters. To achieve this goal, protests and public critiques concerning mass media were made, leading to public discussions and eventually to legal reform.

The path for fighting against a media monopoly in Taiwan mentioned in this chapter can be concluded as COMEPA: comments, media, exposure, protest, allies. This chapter argues that a coalition that consisted of a number of scholars, media reform groups (such as Taiwan Media Watch and the Campaign of Media Reform in Taiwan), various kinds of social groups, and the public can be a powerful force to express the opinions of the public. First of all, scholars and media reform groups need to create an agenda to draw the public's attention. Second, it is essential to form alliances with various social groups to generate more comments, host workshops, and create protests (online and offline). Both the traditional media and online social media can be used to express their views. Furthermore, public participation is the key to success. These are strategies used in the anti–media monopoly movements in Taiwan to influence the government to prevent the formation of media monopoly that might impede the freedom of speech. However, these strategies have their limitations and challenges.

As mentioned above, with great media exposure through traditional mass media and social media, these movements reached a considerable amount of people to participate in the movements. However, there remains an "unreachable" public; reaching *every* person continues to be a big challenge for media reform movements.

In addition, the existence of a "cold" public is another challenge. To improve the awareness of the importance of media reform, similar to the NCMR (National Conference for Media Reform) hosted by the Free Press in the United States, the Campaign for Media Reform and Media Watch Taiwan have hosted media reform meetings annually. Media Watch also provides courses on media education, including media literacy and the production of citizen journalism. However, only limited people can sign up for these courses. It is essential to cooperate with more local communities and schools to provide media literacy education and to persuade the public on the importance of media reform.

Last but not least, the ongoing media reform movements that are fighting media monopoly and pressing for a more independent and well-functioning public media system have presented a real challenge. Still, as McChesney (2007, 1) has pointed out, "we are in the midst of a communication and information revolution." Fighting for a completely independent public broadcasting system remains difficult. When the media system is partially funded by the government, maintaining its accountability while ensuring its sustainability has been a dilemma in many countries around the world. Media reform still has a long way to go.

REFERENCES

AP. 1987. "Taiwan Ends 4 Decades of Martial Law." *New York Times*. www.nytimes.com/1987/07/15/world/taiwan-ends-4-decades-of-martial-law.html.

Feng, C. 2006. "Roles of Academic Intellectuals, Social Campaign and Political Authority." [In Chinese.] *Communication and Society*, no. 1: 47–67.

Fetner, T., and C. Sanders. 2012. "Similar Strategies, Different Outcomes: Institutional Histories of the Christian Right of Canada and of the United States." In *Strategies for Social Change*, edited by G. Maney, R. Kutz-Flamenbaum, D. Rohlinger, and J. Goodwin, 245–262. Minneapolis: University of Minnesota Press.

Freedom House. 2014. "Freedom in the World." www.freedomhouse.org/report/freedom-world/2014/china-0#.UuEqZGQo7Zs.

Fuchs, C. 2014. "Why Taiwanese Are Getting Fed Up with the Island's Salacious, In-Your-Face Media." *Foreign Policy*, February 20. http://foreignpolicy.com/2014/02/21/freedom-fried/.

Hackett, R. A., and W. Carroll. 2006. *Remaking media: the struggle to democratize public communication*. New York, NY: Routledge.

Harding, P. 2010. "PeoPo Helps Taiwanese Public Broadcaster to Restore Trust." *Guardian*, February 15.

Huang, X. 2009. "The Entry Mode to Taiwan Cable Television Market for Foreign Investment and Regulations Analysis." Paper presented at the 2009 annual conference of the Chinese Communication Society. http://ccs.nccu.edu.tw/UPLOAD_FILES/HISTORY_PAPER_FILES/1156_1.pdf.

Kuang, C. 2004. "Broadcasting Media Reform in Taiwan." [In Chinese.] *Media Digest*. http://rthk.hk/mediadigest/20040915_76_120091.html.

Kuo, L.-H. 2010. "The Roles of the Intellectuals in the Society." [In Chinese.] www.feja.org.tw/modules/news007/article.php?storyid=597.

Liberty Times. 2012. "International Media Worried About the Diversity of News in Taiwan." [In Chinese.] July 28. www.libertytimes.com.tw/2012/new/jul/28/today-fo10.htm?Slots=All.

Lin, Z. 2009. *TV Ratings-Driven Journalism: The Commodification of TV News in Taiwan*. [In Chinese.] Taipei: Lianzin.

Lo, S. H. 2012a. "It Is Time for Media Reform: The Observation at the NCMR." [In Chinese.] *Mass Communication Research*, no. 110: 257–278.

———. 2012b. "Public Television Service and Its Citizen Journalism Initiative in Taiwan." In *Citizen Journalism: Valuable, Useless or Dangerous*, edited by M. Wall, 71–81. New York: IDEBATE Press.

———. 2013. "How to Prevent Media Monopoly? a Composite Regulatory Approach." [In Chinese.] *Journal of Communication Research and Practice* 3, no. 2: 1–25.

McChesney, R. W. 2007. *Communication Revolution: Critical Junctures and the Future of Media.* New York: New Press.

McManus, J. H. 1994. *Market-Driven Journalism: Let the Citizen Beware?* Thousand Oaks, CA: Sage Publications.

Meyer, D. S., and S. Staggenborg. 2012. "Thinking About Strategy." In *Strategies for Social Change*, edited by G. Maney, R. Kutz-Flamenbaum, D. Rohlinger, and J. Goodwin, 3–22. Minneapolis: University of Minnesota Press.

Ministry of Culture. 2011. *Introduction. Community: Explore the Real Taiwan.* http://sixstar.moc.gov.tw/eng-2011/engIndexAction.do?method=go IntroductionDetail.

Mueller, M. 2004. "Reinventing Media Activism: Public Interest Advocacy in the Making of U.S. Communication-Information Policy, 1960–2002." SSRN eLibrary. http://ssrn.com/paper=586625.

National Communications Commission (NCC). 2015. www.ncc.gov.tw/chinese/show_file.aspx?table_name=news&file_sn=44103.

Ryan, C., and W. Gamson. 2011. "Sustaining Collaboration: Lessons from the Media Research and Action Project." In *Media and Social* Justice, edited by S. C. Jansen, J. Pooley, and L. Taub-Pervizpour, 71–82. New York: Palgrave Macmillan.

Ryan, C., K. Jeffreys, and L. Blozie. 2012. "Raising Public Awareness of Domestic Violence: Strategic Communication and Movement Building." In *Strategies for Social Change*, edited by G. Maney, R. Kutz-Flamenbaum, D. Rohlinger, and J. Goodwin, 61–92. Minneapolis: University of Minnesota Press.

CHAPTER THIRTEEN

Organizing for Media Reform in Canada

Media Democracy Day, OpenMedia, and Reimagine CBC

KATHLEEN CROSS, *Simon Fraser University, Canada*

DAVID SKINNER, *York University, Canada*

MEDIA REFORM STRATEGY

A review of three contemporary media reform initiatives in Anglophone Canada offers four insights on organizing media reform initiatives: (1) Understand your regional and national context. *To frame campaigns and create a broad base of support, organizations need to know the history of media development in their jurisdictions, the current competing power structures and discourses about the media, and the unique barriers to media reform. (2)* Seek collaborations and coalitions. *Networking and collaborating with other organizations, campaigns, and events can help enhance organizational reach and resources, as well as increase awareness of the need for media reform in other social-justice fields and organizations. (3)* Seek support from academic institutions. *Support from universities and colleges can help legitimate reform organizations; add expertise to campaigns, regulatory proceedings, and public debates; and generate volunteers through student placements and internships. (4)* Use multiple campaign approaches and modes of engagement. *Online campaign tactics (such as e-petitions or social media organizing) combined with offline traditional communications tactics (such as town hall discussions, street teams, or presentations to regulatory committees) engage the largest number of people. Similarly, campaign goals should be framed in simple, straightforward terms that can be easily grasped by people of different ages and diverse backgrounds.*

This chapter examines three contemporary media reform initiatives in Anglophone Canada: (1) Media Democracy Day (MDD), an annual public event in

Vancouver British Columbia that brings together community groups, scholars, activists and independent media practitioners to develop strategies for democratizing media content and structure; (2) OpenMedia (OM), an organization that has had substantial success bringing pressure on the media regulatory arena using social media and online organizing tools; and (3) Reimagine CBC, a national campaign conducted by a coalition of three youth-oriented activist groups in response to political and financial challenges mounted against the Canadian public broadcaster.

Set against the larger backdrop of media reform in Canada, it briefly highlights the historical conditions and circumstances that frame contemporary media reform efforts in Canada and considers the successes and challenges these campaigns have had in animating broader public interest and achieving particular media reform—related goals. It considers their communication strategies, use of social media, and involvement in coalition building, and it examines their lessons learned in developing campaigns, engaging with those new to media reform goals, and approaches to developing relationships with wider movements for social justice. Of particular interest are the ways they have managed to overcome traditional barriers to reform, as well as the ways in which the larger political economic conditions have worked to circumscribe their growth and development. Finally, drawing from the experiences of these cases we suggest four broad organizational strategies for media reform initiatives.

DEVELOPING ACTIVISM

Canada has a long history of citizen's groups, labor organizations, and academics agitating for comprehensive media and telecommunications systems that serve the public interest (Babe 1990; Girard 1992; Goldberg 1990; Hackett and Carroll 2006). But such activism has been spotty and much of it has been confined to simply testifying at the many federal government enquiries and commissions over the years (Canada 1970, 1981, 2005, 2006). Once these formal inquiries were over, there was very little pressure on the government of the day to follow up on recommendations and, generally, they have not done so (Skinner and Gasher 2005). In some quarters, this has led to the conclusion that perhaps the only way to get reform on legislative and regulatory agendas is through broad and sustained public pressure (Creery 1984). However, with a few exceptions, public pressure has not been developed in a sustained way.[1]

There are at least three reasons for this difficulty. First, there is the structure of the regulatory process itself. Participating in that process is expensive and requires considerable resources in terms of time, money, and knowledge. Take the Canadian Radio-Television and Telecommunications Commission (CRTC), Canada's broadcast and telecommunications regulator, for example. Hearings are generally conducted in Ottawa, making attendance in person difficult. At the same time, keeping up with the CRTC's regulatory agenda and undertaking the research necessary to understand the possible social, political, and economic impacts of the issues it addresses is very difficult without considerable time for research, or the benefit of paid staff, something very few besides major media corporations can afford. Participating in other regulatory processes requires similar resources.

A second, related problem is funding for nongovernment and noncorporate organizations in Canada. In the United States, media activist organizations such as Free Press and the Media and Democracy Coalition receive significant funding from foundations, philanthropic organizations, and institutional members. In Canada, with its small population and history of branch plant industry, access to such resources is severely limited (Skinner 2012).

Third, prior to the Internet, developing a broad base of support for media reform without the resources to mount high-profile publicity campaigns, complete with direct mail or telephone contacts, was an almost impossible task. This is further complicated by the fact that media reform is often viewed by activists in fields such as labor, race, gender, or environment as secondary to their primary concerns (Hackett and Carroll 2006, 199–209).

All three examples of media reform activism discussed in this section address these obstacles: Media Democracy Day through utilizing a range of techniques to build public awareness of, and participation in, media reform issues and concerns; and OpenMedia.ca and Reimagine CBC through using both Internet-based social networks and broadly focused campaign strategies to develop and expedite media reform campaigns.

MEDIA DEMOCRACY DAY

The first Media Democracy Day (MDD) was organized in Vancouver and Toronto in 2001 in response to escalating concentration of media ownership and its attendant threats to the diversity of media content.[2] Since that time, similar events have been organized across Canada and the United States, as well as in Argentina, Brazil, Germany, Indonesia, Spain, and the United Kingdom. But it has enjoyed its greatest success in Vancouver where over the last four years it has evolved into one of the premiere media reform events in Canada where—with two days of panel discussions, film screenings, keynote talks, workshops, and a media fair—it attracts over three thousand attendees.

MDD's action plan is reflected in the tag line, "Know the media. Be the media. Change the media." To *know the media,* refers to the need for evidence-based research on ownership structures, production, and distribution patterns, labor practices, and patterns of media content. This forms the basis for effecting change. To *be the media* signifies the commitment to citizen input and alternative media content. In this regard, MDD activities include a trade show–style media fair of independent media groups, as well as hands-on training workshops. To *change the media* refers to questions of policy, regulation, and laws, including copyright legislation, Internet throttling, concentration of ownership, and support for culturally important areas such as independent filmmaking and public broadcasting.

MDD seeks to build the media reform movement through emphasizing the importance of "communicative democracy" (Hackett 2000) to new audiences and organizations every year, consciously pitching MDD as a gateway event to communities and organizations whose interests overlap with media reform goals, such as social groups and activist organizations whose issues are misrepresented, underreported, or ignored by the corporate media. In this way, MDD organizers follow the advice of US media activist Robert McChesney (2005, 11) who—when speaking to other social justice activists—argued that "whatever your first issue of concern, media had better be your second, because without change in the media, progress in your primary area is far less likely." To develop program content, MDD also seeks input from this social justice community using both web-based targeted communication and small community meetings, and then approaches specific organizations directly to take part. For example, MDD 2012 included a panel discussion called "Muzzling the Scientists," about the increasing media restrictions placed on government scientists by the Conservative federal government, and contacted numerous environmental organizations, scientist associations and science faculties at universities and colleges to advertise the event. Another panel, "Decolonizing the Media," included representatives from the South Asian and aboriginal leadership and was targeted at their respective communities. Further, all the events are free of charge and held at the easily accessible main branch of the Vancouver Public Library in downtown Vancouver.

MDD has been a successful strategy for increasing information sharing with both the general public and citizen engagement groups (demonstrated by the fact that over 50 percent of attendees in 2012 were new to MDD), and one that has also been a catalyst for new forms of media activism and independent journalism. For example, the award-winning online magazine *The Tyee* (www .thetyee.ca) was launched at MDD and similar collaborations have occurred through the media fair.

The structure of MDD is based on collaborations between scholars, media practitioners, and social justice activists. The organizing committee is presently housed in the School of Communication at Simon Fraser University, a critical communication program for the last forty years, and both faculty members and representatives of community organizations sit on the steering committee. As a joint university-community organization MDD seeks to provide a link between research and practice and works to ensure a combination of scholars, practitioners, and activists on every panel. The 2012 event program listed over thirty speakers, as well as additional partnerships with fifteen groups and associations from seven postsecondary institutions, seventeen community and media sponsors, four labor unions, and five other social justice organizations and media practitioners. In addition, over thirty independent media-making groups were highlighted at the media fair.

Some partnerships have been particularly beneficial over the last twelve years. The UBC School of Journalism has been a supporter of MDD and sees the event as an avenue for its students to learn about and conduct journalism in the public interest. Library and archivist associations have been mainstays at the event, as have civil liberties organizations and groups concerned about freedom of information and privacy rights.

The collaborative nature of MDD appears to have helped sustain the organization and its longevity can be attributed to a number of factors including its affiliation with Simon Fraser University, attention to broad-based community collaborations, and its triple foci of media analysis, policy change, and creating alternative media. Reaching out to new groups and organizations each year with program content that speaks to the importance of democratic media helps to recruit new supporters, volunteers, and donors. However, as with many projects, it is limited by resources and has yet to develop an ongoing and reliable source of funding. MDD has periodic part-time staff and relies on volunteer efforts primarily from the School of Communication and other academic institutions and social justice groups in Vancouver. Thus, an inordinate amount of time is spent fundraising and seeking small donations from other social justice groups.

OPENMEDIA

In May 2007, University of Windsor's Communication Studies department convened a conference to celebrate the twentieth anniversary of *Manufacturing Consent* by Edward S. Herman and Noam Chomsky (1989). Riding a wave of enthusiasm for media reform generated at that event, Steve Anderson, a graduate student at Simon Fraser University's School of Communication, set up Canadians for Democratic Media (CDM) in Vancouver, British Columbia. Aiming to establish an umbrella group of organizations drawn from both those involved in media reform as well as newcomers to the enterprise, CDM's initial mandate was to "create a common front among groups promoting reform of print, broadcast, and web-based media" (CDM 2008, 2). Funding was cobbled together from small grants and in-kind donations from labor unions, other activist organizations, and individuals. A handful of academics and activists worked with Anderson to craft an organizational mandate and policy positions.

For the first two years CDM was involved with a range of campaigns broadly focused on issues surrounding concentration of ownership, community media, the Internet, and journalism. Their website provided campaign information, a place to sign people to their mailing list, and a page where people could make a direct submission to government hearings on these issues. CDM delegations made several trips to Ottawa to appear before hearings. The organization also published several reports related to media issues (Hackett and Anderson 2010). Based upon preliminary findings of one of these reports, CDM was rebranded in 2010 to reflect the growing strength of the open movement. The new name, OpenMedia (OM), reflected the organization's commitment to pursuing and creating an open, accessible media system in Canada in both the public and private sectors. Money was very tight, however, and Anderson spent much of his time fundraising.

About the same time, some large Canadian media companies began pursuing vertical integration between content and carriage (Krashinsky 2010; Ladurantaye 2010). The Internet and the wireless markets were key to these corporate plays, and the mergers sparked a series of struggles over the dimensions of public policy in this emerging field. Net neutrality became a key site of struggle and, along with a number of other activist organizations, OM participated in several campaigns in this area.

A major turning point for the organization and online organizing in Canada came in 2009 when OM launched the Stop the Meter campaign. This campaign sought to intervene in a CRTC decision to allow wholesale Internet providers the power to impose "usage-based billing" (pay per byte) on independent Internet service providers and, therefore, many Canadian Internet users. Deploying social media, and particularly Facebook, to accelerate the message, over the course of several months the Stop the Meter campaign garnered over three hundred thousand signatures to its online petition and flooded the CRTC offices with over one hundred thousand comments—an unprecedented level of public engagement with telecommunications policy. Interestingly, and without prompting from OM, many supporters created videos, comics, posters, T-shirts, websites, and other campaign materials, held protests in their local communities, and donated money for ads in major newspapers. The campaign pressured the CRTC to change their decision from one that favored the large telecommunications companies to one that has been characterized as more friendly to other stakeholders. The petition remains active to keep pressure on the CRTC and now has over five hundred thousand signatures. In the face of this apparent broad public concern and commitment to Internet access and affordability, OM's focus turned from media issues in general to those specifically related to the Internet.

Recent OM campaigns have not matched the level of support of Stop the Meter; but some have drawn well over one hundred thousand signatures. In this regard, OM has set a new benchmark for online organizing in Canada, working through social media, campaign videos, and graphics. At the same time, OM has not abandoned traditional media strategies such as newspaper ads and letters to the editor. Campaign websites include a "letter to the editor" tool that allows supporters to easily draft and send letters, usually aimed at newspapers located in the target politicians' home constituencies (cf. openmedia.ca/

letter). These tactics are combined with direct lobbying of both elected and nonelected officials and other traditional methods, such as presenting petitions at politicians' offices, to help build and sustain pressure for policy changes. While it is difficult to gauge the impact of any one group or individual on public policy processes, OM does appear to have had influence on the outcome of a number of Internet-related issues (Payton 2013). Given its successes, OM's work has also been recognized with a national award and employees are now regularly consulted by regulators, policy-makers, and the mainstream media.

Over the last five years OM has grown from a loose association of like-minded groups and individuals to become what the organization itself refers to as a citizen-driven, pro-Internet community or movement (openmedia.ca/operate). Social media, and particularly Facebook, have been important in constructing this larger community. OM operates with a full-time staff of more than ten, and their raised public profile has enhanced fundraising opportunities. Indeed, while other activist organizations continue to play an important role in providing resources and funding, almost 75 percent of donations in 2011 came from individuals (OM 2011).

One of the keys to OM's success in developing and maintaining their supporters is that they frame complex issues in simple, populist language that relates to everyday experience, particularly for young people. For instance, Stop the Meter (openmedia.ca/meter) took a rather arcane regulatory process dealing with the technicalities of Internet traffic management and presented it as putting a parking meter on individual Internet accounts. No Internet Lockdown (openmedia.ca/lockdown) presented legislation dealing with complex copyright and intellectual property law as containing provisions to lock down or restrict use of the Internet. And the highly secretive Trans-Pacific Partnership trade negotiations were framed as an Internet trap (stopthetrap.net) which may result in a number of highly repressive laws and regulations negatively affecting Internet activity. Campaign narratives also utilize a recognizable set of villains—Big Media or Big Telecom—and goals are framed in generalized populist terms such as the need for more choice, affordability, openness, transparency, or accountability. Both campaign framing and language are also carefully scrutinized to avoid alienating supporters who identify as coming from across the political spectrum. To promote political inclusivity, OM identifies itself as a postpartisan organization (openmedia.ca/operate).

The large constituency constructed through social media using these rhetorical strategies, coupled with the fact that OM has been able to motivate people to sign petitions and directly contact politicians and regulators, legitimates the claim that they are a citizens' movement. They have successfully overcome some of the financial and geographical barriers to participation in regulatory processes in Ottawa and provided people a voice in that venue. But while OM has enjoyed considerable success, the organization's growth and development has been circumscribed by its financial circumstances. Revenue streams require constant tending and current reliance on supporter-based revenue has gone hand in hand with an increasing focus on Internet-related issues.

REIMAGINE CBC

While OpenMedia's major focus shifted to the Internet, it has at times continued to collaborate with other organizations on broader issues of media and public service. One of the major projects that emerged in this regard was Reimagine CBC which focused on protecting and strengthening Canada's public broadcaster, the Canadian Broadcasting Corporation (CBC).

While public broadcasting in Canada, as in many other jurisdictions, has been under attack for over a decade, the election of Stephen Harper's Conservative government in 2006 saw the acceleration of budget cuts and calls for the complete defunding of the CBC. In the wake of these attacks, and an impending CBC license renewal hearing, OpenMedia joined forces with Leadnow and Gen Why Media in January 2012 to launch a campaign designed to engage Canadians in a discussion about the value of the CBC and to encourage suggestions for making it more responsive to the interests of Canadians (Reimagine CBC n.d.). Importantly, all three partners were known as youth-oriented and new media-savvy groups that emphasized civic engagement, participatory processes and public dialogue. Leadnow.ca was launched in March 2011, just before a federal election, with the intention of increasing democratic participation and voter turnout among youth Canada. Gen Why Media began in 2010 as a film project (the name refers to generational theorists' use of "generation Y" to categorize those born in the 1980s) and expanded to work toward changing social structures through events to encourage civic engagement. While Gen Why Media was primarily local, both OM and Leadnow.ca had the reach to make Reimagine CBC a national project.

With these organizations' youth-oriented supporters in mind, the campaign sought to influence the CBC's strategic planning by developing a "crowd sourced plan to enable the CBC to take on a leadership position in the digital era" (Reimagine CBC 2012). The campaign began with a website launch in January 2012. They explicitly stated their starting principles as supporting the role of public media as "good for democracy" and the importance of the public broadcaster to become nimble in terms of new media and technology. Based on questions of how the CBC could be more interactive, community based, participatory, encourage citizen-powered culture, and foster creativity and collaboration, they asked Canadians to submit their ideas about the kind of public broadcaster they wanted the CBC to be. In this way, they built an online community and an initial list of ideas and opinions. The website also encouraged people to post their own ideas and comments. The site also included testimonials from artists, writers, educators, and others to advertise the campaign.

Another section of the website allowed people to celebrate their favorite CBC moments, collecting stories and memories of the role of the public broadcaster in people's lives. Running from January to April 2012, almost 1,500 Canadians took part in this first phase of the campaign.

At the same time, Reimagine CBC created an online and paper-based petition that called on the government to stop the cuts to the CBC in the upcoming federal budget. The petition garnered more than twenty-five thousand signatures, and volunteers hand-delivered these petitions to thirteen Member of Parliament (MP) constituency offices. Nonetheless, the spring budget announced a 10 percent cut to the CBC, and the Reimagine CBC campaign switched its focus to the CBC license renewal hearings upcoming in the fall.

Importantly, the campaign was not limited to web-based tactics, but used multifaceted strategies to engage Canadians in this broad-based consultation. They conducted person-to-person engagement events, including street teams, visits to senior homes, and a youth-oriented public event. Four months into the campaign, Reimagine CBC was ready to report to the public about what they had heard so far. They did so with a large public event held on May 7, 2012, in Vancouver. Over one thousand people attended and some five hundred people watched via webcam. They reported on the first phase of the campaign and included performances from spoken-word poets, musicians, writers, and a panel discussion about how to make the CBC stronger and more relevant to Canadians.

Over the next few months, organizers developed a modular survey from the crowdsourcing feedback that reflected a list of six major priorities highlighted by the participants. During this second phase of the consultation process, people were encouraged to rank these priorities in order of importance. In all, almost eleven thousand people participated in this online survey. At the same time the campaign held twenty-eight gatherings across the county to seek feedback from those who may not yet have had a chance to comment. By the time the campaign held its second large public event in Toronto in October 2012, it was clear that Reimagine CBC had conducted one of the most creative and comprehensive public consultation processes ever seen about the future of the CBC or the role of media in Canada. In all, over thirty-five thousand people participated in the consultation process in one way or another. Its final report, *Make It Yours*, was submitted to the CRTC at the CBC's licensing hearing in November 2012. It was clear from the consultation that Canadians were deeply supportive of the public broadcaster. The survey put the highest priority on "courageous" and "in-depth" reporting as the key to CBC purpose. The other four recommendations included prioritizing radio, enriching the digital archive, collaborating with communities more to develop authentic Canadian stories, and using open and participatory processes to help shape CBC's vision. Upon presentation of the report to the CRTC the campaign earned praise from the CRTC chair as a model of citizen participation.

Why did this campaign have such buy-in from the public? Collaboration, public engagement, a combination of web-based and traditional communication methods, creative approaches, cross-generational focus—all of these elements combined to help build public participation.

ORGANIZATIONAL STRATEGIES FOR MEDIA REFORM

Each of these Canadian examples takes a different approach to media reform. MDD's focus is to hold public events, create a platform for media reform collaborations, and increase the capacity and opportunity for alternative media content. One of the advantages of MDD pitching its events to new communities of interest, and facilitating collaborations, is that it has succeeded in creating a sustainable event that attracts thousands of people a year, many of whom are new to media reform. However, while building public awareness of media issues, there is no way to gauge the organization's impact on public policy or the media environment in general.

Working from another position, OM was forged in the peculiar political economy of the Canadian media system to become effective Internet-oriented campaigners for policy change around primarily Internet-oriented issues. Utilizing social media and careful rhetorical framing of issues, it has been able to develop a very large set of supporters and thus overcome some of the traditional financial barriers to organizing media reform efforts in Canada. But in its efforts to establish a firm base of support, it appears to have somewhat limited the range of its campaigning work.

Finally, Reimagine CBC was a collaborative project that had the express purpose of conducting a short intensive campaign in support of the public broadcaster and focused on its license renewal hearing as the final goal. It was unprecedented in its scope, use of multifaceted tactics, and public engagement, but was disbanded soon after the CRTC hearing and had no follow-up capacity.

These three examples offer some interesting ideas and insights on how to engage people in media reform, notwithstanding their different goals and unique positioning. Four of these insights are discussed in the following sections.

UNDERSTAND YOUR REGIONAL AND NATIONAL CONTEXT

While it may seem obvious, it is worth emphasizing the importance of understanding the issues, debates, and struggles that give rise to the policies and structures media reform groups are trying to change. National media systems arise out of different political and economic pressures, and thus social and historical context offers insight to tailoring strategies for media reform. For instance, in Canada, regional differences in media access and coverage sometimes call for different campaign approaches in different locales. What works on the West Coast where media ownership concentration is particularly high, does not always play the same way in the center of the country where there is more media choice. Similarly, drawing upon the country's peculiar mythology about media development and the role of the media in democracy can offer opportunities for framing campaigns. For example, Canada's historical debates about cultural sovereignty vis-à-vis the United States have been evoked in a number of campaigns (cf. OM's Fair Deal Campaign). At another level, learning from and strategically positioning oneself in relation to the work of other domestic reform groups can be important. The Reimagine CBC campaign did not merely repeat the work of others who have championed the CBC; rather it drew upon traditional narrative and rationales for public broadcasting while using knowledge of youth and web cultures and an understanding of crowd-sourcing methods to fashion a unique dialogue about a renewed public service model. Thus, it created new tactics for championing an old institution. Finally, understanding the national context also means acknowledging barriers to media reform. For example, MDD has recognized the lack of public knowledge and opportunities for engagement about media reform issues, and responded by pitching its work at an entry level, offering discussions and events that help educate people about media reform issues and concerns. Similarly, recognizing the difficulty in accomplishing fundamental political and policy change, OM's campaigns generally focus upon incremental policy change. And, because Canada lacks many philanthropic foundations from which to raise core funding, organizers have had to be particularly creative in fundraising.

SEEK COLLABORATIONS AND COALITIONS

All three examples networked with a range of social justice groups and organizations who were not strictly involved in the media reform movement, thereby building their reach and resources. OM works to create a large constituency by making a claim to postpartisanship which can, at times, limit the range and depth of their progressive actions but, at the same time, has been successful at creating coalitions with groups that might not have traditionally supported its work. For its Stop the Meter campaign, it joined with consumer interest groups whose political orientation was different from OM's, but whose goals in this campaign were similar: affordable Internet access. Depending on the situation, it has also forged alliances with unions as well as small telecommunications companies and other private sector organizations whose goals and ambitions meet with the objectives of particular campaigns. While such partnerships can draw criticism for being opportunist, they can also provide much needed resources and can make the difference between success and failure. Further, both MDD and Reimagine CBC are examples of coalition-building approaches that generated interest with groups and citizens outside of the media reform sector.

Short-term campaign-specific collaborations with other social justice groups can also offer the benefit of doubling up on memberships and mail-lists, without the long-term headache of requiring agreement on long-term goals. Seeking collaborations, however, may have drawbacks. For instance, media reform groups need to be careful about overidentifying with particular political parties rather than specific policy proposals.

SEEK SUPPORT FROM ACADEMIC INSTITUTIONS

While support from institutions such as foundations and unions can be invaluable, we cannot emphasize enough the important role of scholars and their academic institutions in providing resources and support for the work of media reform. While, as a scholar, taking on the role of reform advocate may have consequences, such academic alliances create a level of legitimacy for the organization; add expertise to campaign outcomes, regulatory proceedings, and public debates; and can serve as vehicles of critical pedagogy for postsecondary students. All three examples reviewed here emerged or were assisted by academic involvement. The first Media Democracy Day in 2001 originated in part through the work of graduates and faculty in the School of Communication at Simon Fraser University. This affiliation with SFU continues today, as other faculty members have become leaders in the organization and new graduate and undergraduate students have become involved. The university is a great source of student volunteers and annual funding for MDD which, while extremely limited, still provides a basis of continuity. Similarly, the academy played an important role in animating the development of OM and provided both technical and material support in the early days of its development. Several of the founders were university professors and Steve Anderson was an alumnus of both graduate and undergraduate communication programs. Because OM is based in Vancouver, the School of Communication at Simon Fraser is an important source of student interns and staff, and scholars affiliated with other academic institutions across the country have also been involved as board members and collaborators, while student clubs have formed at other universities, such as McGill University in Montreal. As the organization grew and became self-sustaining, the role of the academy became less important, but scholars continue to legitimate OM campaigns by delivering expert testimony at regulatory hearings and assisting with policy and research documents.

USE MULTIPLE CAMPAIGN APPROACHES AND MODES OF ENGAGEMENT

Finally, all three of these examples show the importance of using both traditional and new media campaign techniques to reach both the public and policy makers. OM has created impressive online campaigns and Reimagine CBC utilized a simple yet effective online consultation and collaboration approach. Yet, all three examples also highlight the importance of traditional campaign tools, such as town hall discussions, street teams, large public events, circulating hard copy petitions, and directly engaging politicians and regulators. For instance, Reimagine CBC visited community centers and senior citizens' homes. OM conducts street interventions, delivers hard copy petitions to political offices, sends out traditional press releases and encourages letters to the editors of newspapers. MDD seeks out radio interviews, submits editorial columns to newspapers and makes classroom visits to post-secondary classes.

CONCLUSION

All three examples emphasize participation in their campaigns, providing people with a range of possible ways to become involved—from organizing and creating campaign materials, to directly voicing specific concerns, to attending public events or simply signing a petition. At the same time, all three illustrate that campaign goals need to be framed in simple, straightforward terms that can be easily grasped by people of different ages and with diverse backgrounds and education. All three have also illustrated that public meetings can be an important means of developing and galvanizing public interest in reform. And OpenMedia and Reimagine CBC have shown that online activism can help break down traditional barriers to the regulatory process and enable large numbers of people to make their voices heard. While some media reform groups may be more comfortable with either traditional *or* new media organizing, it appears that it is the combination of these activities that fueled the success of these three examples.

NOTES

David Skinner is cofounder of both Media Democracy Day and OpenMedia. He also sat on the board of OpenMedia until May 2012. Kathleen Cross chaired the Media Democracy Project, the organization that runs Media Democracy Day, for four years.

 1. Notable exceptions include Friends of Canadian Broadcasting and the National Campus and Community Radio Association (NCRA).

 2. See www.mediademocracydays.ca and www.mediademocracyproject.ca.

REFERENCES

Babe, R. 1990. *Telecommunications in Canada.* Toronto: University of Toronto Press.
Campaign for Democratic Media (CDM). 2008. *Annual Report.* Vancouver, BC.

Canada. 1970. *Special Senate Committee on the Mass Media.* Davey Report. Vol. 1: *The Uncertain Mirror.* Ottawa: Queen's Printer.

———. 1981. *Royal Commission on Newspapers.* Kent Report. Ottawa: Supply and Services Canada.

———. 2005. *Reinforcing Our Cultural Sovereignty. Second response to the report of the Standing Committee on Canadian Heritage.* www.parl.gc.ca/Content/ HOC/Committee/381/CHPC/GovResponse/RP1726418/CHPC_Rpt02 _GvtRsp/GvtRsp_Part2-e.pdf.

———. 2006. *Senate Standing Committee on Transport and Communications. Final Report on Canadian News Media.* Vol. 1. www.parl.gc.ca/Content/ SEN/Committee/391/TRAN/rep/repfinjun06vol1-e.htm.

Creery, T. 1984. "Out of Commission." *Ryerson Review of Journalism.* http://rrj.ca/ out-of-commission/.

Girard, B. 1992. *A Passion for Radio: Radio Waves and Community.* Montreal: Black Rose Books.

Goldberg, K. 1990. *The Barefoot Channel: Community Television as a Tool for Social Change.* Vancouver: New Star Books.

Hackett, R. A. 2000. "Taking Back the Media: Notes on the Potential for a Communicative Democracy Movement." *Studies in Political Economy*, no. 63: 61–86.

Hackett, R. A., and S. Anderson. 2010. *Revitalizing a Media Reform Movement in Canada.* http://openmedia.ca/revitalize.

Hackett, R. A., and W. Carroll. 2006. *Remaking Media: The Struggle to Democratize Public Communication.* Routledge: New York.

Herman, E., and N. Chomsky. 1989. *Manufacturing Consent: The Political Economy of the Mass Media.* New York: Pantheon.

Krashinsky, S. 2012. "Shaw to Buy Control of CanWest." *Globe and Mail*, February 12. www.theglobeandmail.com/globe-investor/shaw-to-buy-control-of -canwest/article4305910/.

Ladurantaye, S. 2010. "Bell Ushers in New Era with CTV Deal." *Globe and Mail*, September 11. www.theglobeandmail.com/globe-investor/bell-ushers-in -new-era-with-ctv-deal/article1320795/.

McChesney, R. 2005. "The Emerging Struggle for a Free Press." In *The Future of Media*, edited by R. McChesny, R. Newman, and B. Scott, 9–20. New York: Seven Stories Press.

OpenMedia. 2009. *Annual Report.* http://openmedia.ca/blog/openmediaca-annual -report-2009.

———. 2011. *Annual Report.* https://openmedia.ca/annualreport

Payton, L. 2013. "Government Killing Online Surveillance Bill." *CBC News*, February 11. www.cbc.ca/news/politics/story/2013/02/11/pol-rob-nicholson -criminal-code-changes.html.

Reimagine CBC. N.d. *Reimagine CBC.* http://reimaginecbc.ca/share.

———. 2012. *Make It Yours: Reimaging a Brave and Nimble CBC for the Age of Participation. A submission to the CRTC.*

Skinner, D. 2012. "Sustaining Independent and Alternative Media." In *Alternative Media in Canada*, edited by K. Kozolanka, P. Mazepa and D. Skinner, 25–45. Vancouver: UBC Press.

Skinner, D., and Gasher, M. 2005. "So Much by So Few: Media Policy and Ownership in Canada." In *Converging Media, Diverging Politics*, edited by D. Skinner, M. Gasher, and J. Compton, 51–76. Lanham, MD: Lexington Books.

CHAPTER FOURTEEN

The Battle Over Low-Power FM in the United States

HANNAH SASSAMAN AND PETE TRIDISH, *Prometheus Radio Project, United States*

MEDIA REFORM STRATEGY

Five lessons learned from the fight for low-power FM (LPFM) radio in the United States: (1) Nothing beats a passionate grassroots effort. *Even though Prometheus was a tiny outfit working out of a two-room church basement, we were able to harness the power of people who felt robbed of the community radio they needed. (2)* Local voices with meaningful stories will be heard. *We encouraged people to share detailed, personal stories with legislators about how their community could use a radio station. In crowded Washington, DC, with big corporate media lobbyists shaking hands and sharing meals with our elected representatives, we found that stories about local issues spoke louder than the richest broadcasters in the world. (3)* Change takes time. *It took our grassroots efforts ten years to get a propublic interest correction through, so choose your issues wisely. (4)* Focus on the long term. *The industry always has its insider advantages in the short term. Choose campaigns that will keep on making a difference in the long term, and stick with them long enough to win them. (5)* Community-building events strengthen movements for the long haul. *If we can come together outside of campaign work—outside of calling Congress, signing petitions, taking action— we will build a more permanent understanding of the lines that divide us and keep big media strong.*

In the mid-1990s, the movement for media reform in the United States was at a low ebb. The Telecommunications Act of 1996 moved through Congress with the fawning support of the corporate deregulatory agenda of both parties. Americans started to feel the effect within months, with hundreds of mergers and buyouts by big media corporations leading to waves of firings and loss of vital local programming.[1]

In response, hundreds of small bands of activists started unlicensed, or pirate, FM radio stations, in hopes of recapturing a sliver of the public airwaves for their communities. Their goal? To give voice to those being silenced by the commercial media. We saw that whether we were fighting for housing reform, an AIDS cure, peace, the environment, or anything else, a small number of corporations had a chokehold on the public's information and were using their media properties to narrow the contours of debate and the imagination about possible solutions. We took our inspiration from Mbanna Kantako, a blind DJ in a housing project in Illinois who started a tenants' rights radio station; and Stephen Dunifer, a peace activist and engineer who ran an antiwar radio station out of a backpack from the hills in Berkeley, California. In Dunifer's case, the Federal Communications Commission (FCC) tried and failed for four years to get the courts to issue an injunction against his broadcasts. Eventually, his station was shut down in 1998, but not before close to one thousand pirate radio stations had started up in the shadow of the legal whirlpool caused by his (almost successful) case against the system of broadcast laws.

The FCC's response to the unlicensed stations was twofold. First, the enforcement branch pursued fines, equipment confiscations, and injunctions against radio pirates. The legal confusion created by the Dunifer case created a situation much like the historic (and mythic) Wild West in the United States: the rule of law was temporarily in question as the courts reevaluated the FCC's performance in distributing radio licenses fairly. In the vacuum of authority that followed, freewheeling radio pirates, from immigrants to anarchists to artists, broadcast voices that had never been heard in the mainstream. Second, FCC Chairman William Kennard, who expressed concern about the trend toward radio industry consolidation and loss of diversity of ownership, engaged the pirates to create a new, licensed, hyper-local community radio service for the country: low-power FM radio.

Former radio pirates formed a national group, the Prometheus Radio Project, to help steer the political debate around the new licenses, and to assist groups technically and organizationally that wanted to give voice to their communities using radio. With Kennard's leadership, the FCC designed the LPFM service and opened the door for thousands of community groups to apply for licenses.

The battle with big broadcasting was far from over, however, as incumbents pushed Congress to pass an anti-LPFM bill in late 2000 that limited the FCC's authority to grant the licenses, adding onerous restrictions that kept LPFM out of most of America's big cities. Prometheus led a coalition in a ten-year effort to restore FCC authority over community radio, and to allow community radio stations into the cities. This effort was finally successful when Congress passed the Local Community Radio Act, signed by President Obama in January 2011.[2]

THE PROMETHEUS RADIO PROJECT

The campaign to expand low-power FM was a transformative one for the Prometheans who took it on. Prometheus's founders had experience organizing demonstrations and direct actions against the City of Philadelphia, against national Republicans setting up shop in town for the Republican National Convention, as well as for many other local struggles. But we'd never had the impact to change federal regulations before, or pass laws through Congress. We weren't really big league enough to take on the Federal government, and the Feds weren't quite ready for us either.

When thousands of LPFM applicants were denied their chance to get a community radio license, we saw a constituency spread across the country. But this constituency was made up of people very unlike each other: people who wanted to have radio stations for very different reasons. From two rooms in a West Philadelphia church basement, and from the back of an old car that had seen better days, we met the thousands who had lost their opportunity for their dream of building a community radio station including evangelical Christians in Merritt Island, Florida; Latino labor coalitions in Indianapolis; anarchists in Philadelphia; and the black Chamber of Commerce in Sacramento.

We developed a practice of deep conversation with all sorts of people about the unmet communications needs of their communities, and let their passion lead us. We listened to each group as they described why they wanted to be on the air, and helped them to respect the reasons provided by thousands of others. We needed to meet people where they were and to amplify to decision-makers the individual reasons why people wanted stations. One thing that helped us cross ideological lines was our emphasis on localism. In the wake of the wave of corporate mergers, many people had lost the traditional connection that they had with local radio. More and more of the program content was nationally syndicated, and radio hosts—from the Left or the Right—became more distant and unapproachable and less connected to the particular town that the signal was covering. Our belief was that just like you don't just have a president—you also have a governor and a mayor and a school board member—you need a media that has a mix of scales too. There is nothing wrong with having national media outlets, but some room on the dial needs to be reserved for smaller, local voices. This resonated for everyone, including Republicans who have often argued against the over-extension of federal powers. Libertarian Congress members, like Ron Paul of Texas, wanted to hear from local people who wanted the freedom to have their voice on the air, without government or corporate intervention. Left-leaning Congress members, like Dennis Kucinich of Ohio, wanted to hear about locals missing the diversity of local radio and the viewpoints of marginalized groups. Members of the Congressional Black Caucus and Congress members supporting immigrant voices wanted to hear about chances for minority nonprofits and businesses to share powerful stories, and to put power in the hands of marginalized groups to organize for themselves. Coming from a history where the FCC only considered comments from lawyers in the rulemaking process, and tossed comments from the public in the wastepaper basket, this strategy seemed unorthodox. Treating individual voices like unique flowers and delivering those flowers to Congress members one, two, three at a time? No way.

But we saw our methods start to work and change minds, one at a time, in Congress. In 2007, after a number of false starts in the House led by Democratic legislators without Republican backing and after a dogged bipartisan Senate team failed to push through big broadcaster opposition to LPFM expansion, Congressman Mike Doyle and his passionate legislative aide Kenneth Degraff approached us. They said that they had worked before with conservative Omaha, Nebraska, Congressman Lee Terry, and if we could demonstrate that there was powerful interest in LPFM in Omaha, he would have the wherewithal to stand up to the big broadcasters and cosponsor a bill to expand LPFM radio in the committee where it mattered the most.

We went deep in Omaha. We found applicants from African-American community organizations, local churches, and the YMCA who had wanted a radio station but who had lost out when Congress limited the LPFM service. In Omaha, the voices of these leaders meant a lot to Congressman Terry. He agreed to cosponsor the bill, and we had a bill to expand LPFM that was viable.

Once we saw this formula work, we just kept the turntable spinning with local variations on the same theme. We refreshed and maintained connections with new and former LPFM applicants at community radio station "barnraisings"—an event taken from the Amish tradition of communities coming together to build a barn in a day or two—where we build an entire LPFM radio station over the weekend. We moved through legislative cycles, getting a little further each time. Each breakthrough with a legislator was a small, nourishing victory that allowed us to keep going. Our allies in the Senate patiently moved the bill forward each session until they reached a ceiling and then we would try again. As the years went by, our "DC drag" got threadbare and sometimes our spirits did too. We got tired and angry when allies described our campaign as "low-hanging fruit" in the fight for media reform. We got exhausted explaining why National Public Radio and the big, conservative, commercial broadcasters at the National Association of Broadcasters were working together to kill our little stations.

We hit something of a turning point in 2005. When Hurricanes Katrina and Rita hit the Louisiana and Mississippi coasts and other parts of the Gulf, communications systems broke down for millions. Of the forty-two radio stations lining the Gulf coast, only four mostly stayed on the air during the storm—and two of those were LPFM radio stations. WQRZ-LP in Bay St. Louis, Mississippi, was so important for helping the Hancock County residents find shelter, family members, and resources that the local emergency services agencies invited them to set up shop with them and broadcast hyper-local emergency information along with the Zydeco and other local music that comforted survivors' frayed nerves.

Telling that story expanded our coalition further, bringing in emergency services leaders together with civil rights, social justice, libertarian, and right-and-left religious organizations all pushing for LPFM radio. After the wave of new Congress members in 2008, we were able to move bills through the House and Senate much more quickly. But we were working against time, as well. While 91.5 percent of Americans continued to rely on radio at least once every week (Radio Advertising Bureau 2014) LPFM advocates increasingly had to contend with statements like "Doesn't everyone have an iPhone now?" and "Why does radio matter?" We continued to get hundreds of calls a week from radio hopefuls including speakers of Hmong and Haitian Creole, students and seniors, pastors and parents. We knew however that our time was running out with Congress increasingly interested in shiny new things.

By December 2010, the campaign for LPFM had reached a make-or-break junction. We were burning out as a result of years of hitting our heads against intractable opponents and a do-nothing Congress. We were just days from the end of the legislative session, and if we did not pass our bill through the Senate, we would have to start from scratch in the new year. Our bill had passed the House, and passed the appropriate communications-focused subcommittee and committee in the Senate. It was up for "unanimous consent" in the Senate.

Our opponents at the National Association of Broadcasters (NAB), led by former Republican Senator Gordon Smith, were stopping our bill using a series of anonymous holds by Senators in red states. In order to break a hold like this, we had to discover who was behind it, then mobilize constituents, create public pressure and local press stories in the state until the Senators' staff buckled and released the hold.

We had little time to act—maybe two weeks before legislators went home for Christmas and closed all pending bills, which would have to be rebooted from scratch the next session. Experienced lobbyists advised us that we were better off hanging it up and trying again next session. We needed to turn up the pressure. This was very divisive in our group. Some of us felt we would lose all the legitimacy we had built up if we made fools of ourselves for nothing. But what was all that legitimacy for anyway, if we didn't win?

Danielle Chynoweth, the leader who had built WRFN-LP in Urbana, Illinois, and a seasoned campaigner who joined our team for the last years of the campaign, came up with the idea of a hula hoop demonstration. We called our friends who could juggle, do flips and splits, and were willing to "clown around" to help create a makeshift circus in front of the NAB's national headquarters. Our goal was to bring attention to the circus the NAB was making out of American democracy with their dirty Senate procedural tricks and insider string-pulling. Our slogan was "Stop making low-power FM radio jump through so many hoops!" (see Figure 14–1).

The demonstration was a success. The NAB locked their front doors, letting no one in or out of the building for an hour, clearly concerned that we might take over their lobby with our hooping. We got a ton of press, and thousands of people watched a three-minute video we made of the event.[3] The decisive story was an article in *Politico* (Martinez 2010). The reporter not only interviewed us and the NAB, but also other organizations that were having similar issues with the NAB at the same time, such as the cable industry and the music licensing industry. They echoed our claim that the NAB was using dirty political tricks—a move that definitely helped our cause.

The NAB surrendered two days later, and on January 6, 2011, President Obama signed the bill that would expand community radio to thousands of communities nationwide. The bill lifted the restrictions that made it impossible to apply for licenses in cities, making it possible that most moderate-sized cities would have at least four or five channels available, and even the densest markets (like here in Philadelphia) would have one or two. It was one of the biggest victories for media reform in a generation, passing new regulations and a bill through Congress that will lead to thousands of new media outlets in the hands of ordinary people around the country.

Keep it tuned to your radio dial. Looking forward to another generation of struggle.

FIGURE 14-1.   LPFM activists demonstrating outside of the National Association of Broadcasters.

LESSONS LEARNED

Five lessons we learned from the fight for LPFM radio in the United States:

1. There's no replacement for a passionate grassroots effort that wants to build something. Even though Prometheus was a tiny outfit working out of a two-room church basement, we were able to harness the power of people who felt robbed of the community radio they needed.
2. Local voices with meaningful stories will be heard. We encouraged people to share detailed, specific, personal stories with legislators about how their community could use a radio station. In a crowded Washington DC, with big corporate media lobbyists shaking hands and sharing meals with our elected representatives, we found that stories about local issues spoke louder than the richest broadcasters in the world.
3. Change takes time. It took our grassroots efforts ten years to get a pro-public interest correction through. Ten years is like forever in the current environment of technological change, so choose your issues wisely.
4. If you flit from hot topic to hot topic, you will never win. Industry always has its insider advantages in the short term. Choose campaigns well, campaigns that will keep on making a difference in long term, and stick with them long enough to win them.
5. Community-building events strengthen movements for the long haul. If we can come together outside of campaign work—outside of calling Congress, signing petitions, taking action—we will build a more permanent understanding of each other across communities, race, and class, that is, the lines that divide us and keep big media strong.

Prometheus kept the LPFM radio movement vibrant by bringing together practitioners, hopefuls, and friends for "radio barnraisings"—events that brought together communities to build a radio station over the course of a weekend—hand pulling a donated radio tower up to its full height in rural Tennessee; at the top of water towers in Woodburn, Oregon, and in the fields with strawberry pickers. Radio barnraisings included the construction of a studio, over forty workshops on radio technology, politics, content and operations, and erecting the antenna. In addition to the physical construction of the station, radio news stories and announcements were edited, organizing meetings were held, and live radio dramas were performed during the first broadcast of the new station on Sunday night before everyone went home. We built LPFM radio stations, and a movement that fought and won media reform.

NOTES

1. In the United States, radio companies used to be severely constrained from owning what, from the government's perspective, was too many stations.

Government policy enforced the notion that radio was broadcast on the public airwaves and had an accompanying public trust. Local stations were supposed to be assets to local communities. Early ownership rules were designed to keep ownership as diverse as possible and keep the stations' focus as local as possible. All that changed with the Telecommunications Act of 1996, which essentially did away with ownership restrictions on radio. Soon after the 1996 act, just a handful of companies controlled radio in the 100 largest American markets. For example, at its peak, Clear Channel owned 1,200 stations, generating more than three billion dollars annually in revenues (Boehlert 2001). For more information about the radio broadcast ownership landscape in the United States, see Turner (2007).

2. For more information at the Local Communication Radio Act, see http://www.prometheusradio.org/LCRA; for more information about the battle over LPFM in the United States, see Opel (2004) and Connolly-Ahern, Schejter, and Obar (2012).

3. You can view the videos at on YouTube at www.youtube.com/watch?v=gPh68vh_Hc8 and www.youtube.com/watch?v=29jIFXQRVMo.

REFERENCES

Boehlert, E. 2001. "Radio's Big Bully: Dirty Tricks and Crappy Programming: Welcome to the World of Clear Channel, the Biggest Station Owner in America." *Salon.com*, April 30. www.salon.com/2001/04/30/clear_channel/.

Connolly-Ahern, C., A. M. Schejter, and J. A. Obar. 2012. "The Poor Man's Lamb Revisited? Assessing the State of LPFM at Its 10th Anniversary." *Communication Review* 15, no. 1: 21–44.

Martinez, J. 2010. "National Association of Broadcasters Gives Congress Static." *Politico*, December, 16. www.politico.com//news/stories/1210/46433.html.

Opel, A. 2004. *Micro Radio and the FCC: Media Activism and the Struggle Over Broadcast Policy.* Westport, CT: Praeger.

Radio Advertising Bureau. 2014. "Why Radio Fact Sheet: Average Weekly Reach." www.rab.com/public/marketingGuide/DataSheet.cfm?id=1.

Tridish, P. 2014. "The Radio, the Internet and the Jetsons." Media Alliance, February 23. www.media-alliance.org/article.php?id=2372.

Turner, S. D. 2007. "Off the Dial: Female and Minority Radio Station Ownership in the United States: How FCC Policy and Media Consolidation Diminished Diversity on the Public Airwaves." Free Press. www.freepress.net/sites/default/files/fp-legacy/off_the_dial.pdf.

CHAPTER FIFTEEN

Ninety Percent Community, 10 Percent Radio

SANJAY JOLLY, *Prometheus Radio Project, United States*

MEDIA REFORM STRATEGY

For the first decade of the 2000s, the Prometheus Radio Project led the fight for the Local Community Radio Act, a federal law. With its passage in 2011, local community organizations had the opportunity to apply for low-power FM (LPFM) radio licenses in what had the potential to be the largest expansion of community radio in the country's history. In preparation for the LPFM application window in 2013, Prometheus maintained a robust system of outreach and support to facilitate the applications of a diverse cohort of social justice organizations nationwide. We informed and educated potential applicant groups through intensive networking, national tours, and press campaigns. We published detailed guides for how to complete the legal and technical portions of the LPFM application, hosted online tools that enabled applicants to find available frequencies and complete the application's engineering studies, provided legal support via our networks of discount and pro bono attorneys, and organized a team of staff and volunteers that fielded thousands of inquiries from applicant groups. When the LPFM application window was held in fall 2013, more than 2,800 organizations submitted applications in fifty states and Puerto Rico, and of those, approximately 1,000 received support from Prometheus.

In fall 2013, thousands of nonprofit organizations from around the United States, many with the support and guidance of the Prometheus Radio Project, submitted an obscure government form to the Federal Communications Commission (FCC). The form was an application to construct a low-power

FM (LPFM) radio station, and the organizations were taking part in the largest expansion of community radio in the country's history. Prompted by the passage of the Local Community Radio Act of 2010,[1] LPFM's expansion is a rare victory on a national scale for media justice activists. Since the mid-1990s, the Prometheus Radio Project has been on the forefront of activism for independent media in general and community radio in particular. This article provides relevant social and policy context for understanding community radio in the United States, describes the organization's strategy around the 2013 LPFM application window, and briefly discusses what lies ahead for LPFM stations.

THE US CONTEXT FOR COMMUNITY RADIO

In general, the descriptor *community* differentiates a radio station from its commercial and public (i.e., state) counterparts. The defining, though perhaps vague, property that enjoins all community radio stations is a local, participatory approach to programming and governance (Doyle 2012). The conceptual importance of community radio is that it is a mode for expressing local perspectives that for various reasons are unrepresented in other media. As Zane Ibrahim of Bush South Africa puts it, "community radio is ninety percent community and ten percent radio" (quoted in Tridish and Coyer 2005, 304).

In the United States, the media landscape features highly consolidated ownership of the commercial radio sector, consequently excluding a diversity of voices from the airwaves. The present-day superstructures of traditional American media follow directly from the 1996 Telecommunications Act, which accelerated the neoliberal movement toward deregulation of the telecom industry. For radio, the Telecom Act raised the maximum number of stations a single entity could own in a single market and eliminated the national cap altogether (Goldfarb 2007, 27). The resulting onslaught of mergers and acquisitions dramatically changed the makeup of American radio. By 2002, four companies controlled more than 70 percent of the national radio market (Dicola and Thompson 2002, 3). The Clear Channel network alone grew from 40 AM and FM radio stations in 1996 to more than 850 today (CCC 2014). Furthermore, in dominating a greater share of the radio dial, corporations pushed out other noncorporate actors, disproportionate among them women and people of color. In 2007, women owned just 6 percent of all full-power radio stations in the United States; African Americans and Latinos owned 3.4 percent and 2.9 percent, respectively, while non-Hispanic whites owned 87 percent (Turner 2007, 16).

This exclusion of independent and politically marginalized voices has profound consequences for democratic expression and civic participation. As if talking into a vacuum, the inability of communities to participate in their local media is in effect the silencing of speech. The contrast between the corporate megaphone and local communities' absence from the media actualizes in disparities of cultural and political representation, often along racial, sexual, and socioeconomic lines. Consider that in Philadelphia, one in three African Americans lives below the poverty line, and despite making up almost half the city's population, there is not a single black-owned station on the local FM dial (Shaw 2014).

This national policy landscape, shaped by an unrelentingly procorporate federal government, makes an innocuous intervention like community radio into something of an insurgency. Quoting Ted Coopman, Stavitsky, Avery, and Vanhala (2001, 347) argue that the unlicensed radio broadcasts that activists turned to in response to their own media exclusion were "on the cutting edge of a social and political movement based on 'reclaiming' the airwaves for the people." Within the so-called free radio movement, radio was seen as a critical tool for democratic expression. Radio's strength in this respect is its technological simplicity; it is easy to produce, free to consume, and accessible to more people across the world than any other mass medium (Fraser and Estrada 2001).

The free radio movement and later the campaign for the Local Community Radio Act were struggles between corporate and grassroots actors for control over the nation's communications infrastructure. Brandy Doyle (2012, 43) illustrates what happens when grassroots actors lose that fight: "Without control over communications infrastructure, communities are denied access to high speed Internet by service providers with local monopolies. They lose local news because national corporations close their local newspapers or cut the news team from local television stations. They lose access to public television because cable companies negotiate to stop paying franchise fees for public television." The tension in the US media landscape that Doyle expresses is one between civic democracy and hardline corporatism, wherein community radio is a political movement for the former.

PROMETHEUS: THE MEDIA'S MALCONTENTS

The Prometheus Radio Project descended from the unlicensed microradio outfit WPPR (West Philadelphia Pirate Radio), also called Radio Mutiny. The amateur DJs that made up Radio Mutiny were brought together not by a common interest in radio, but by grassroots activism across various social justice movements, including ACT UP, Food Not Bombs, and the American Indian Movement, among others (Shaffer 1998). In 1998, the FCC raided Radio Mutiny's illicit studio and confiscated its equipment, permanently shutting down the station (Klinenberg 2007, 251). That same year, four former Radio Mutiny DJs founded the Prometheus Radio Project with the following mission: "The Prometheus Radio Project builds participatory radio as a tool for social justice organizing and a voice for community expression. To that end, we demystify media policy and technology, advocate for a more just media system, and help grassroots organizations build communications infrastructure to strengthen their communities and movements" (Prometheus 2014). Alongside other radio activists, Prometheus's first efforts resulted in the legal normalization of community radio. In 2000, the FCC announced the creation of the LPFM service "to serve very localized communities or underrepresented groups within communities" and "to afford small, community-based organizations an opportunity to communicate over the airwaves and thus expand diversity of ownership" (FCC 2000). Local community organizations could apply for LPFM stations at a maximum power of 100 Watts, which produces a broadcast radius of about three and a half miles (ibid.).

Shortly thereafter, incumbent broadcasters lobbied aggressively for the so-called Radio Broadcast Preservation Act, which imposed restrictions on LPFM stations that barred them from urban areas (Klinenberg 2007, 257). The law forced the FCC to block approximately 80 percent of potential LPFM stations from getting on the air (Walker 2001, 263). Faced with a further weakened position, Prometheus's strategy in the first decade of the 2000s was two-fold: first, build stations where permitted; and second, fight to repeal LPFM's punitive statutory restrictions through the Local Community Radio Act.

The passage of the act in 2011, as described in the previous chapter, was a defining victory for Prometheus, and with it our strategic imperatives shifted considerably. For the first time since the creation of the LPFM service ten years earlier, government policies favored community radio's expansion. Within two or three years of the law's passage, the FCC was to hold a filing window in which local organizations could apply for LPFM licenses. Because of spectrum scarcity in urban markets, for many communities it was to be the last opportunity to include community radio stations on the FM dial.

Although a considerable triumph, the new law was no guarantee for the nationwide proliferation of authentic community radio. Of the eight hundred or so LPFM stations already on the air, many were smaller, local versions of their commercial counterparts and not necessarily bulwarks of community expression (Doyle 2012; Connolly-Ahern, Schejter, and Obar 2012). Our challenge, then, was not only to encourage and support groups to apply, but to target our efforts toward those who shared our vision of radio as a tool for community participation.

In practice, the mission of Prometheus relies on the mutually dependent values of localism and social justice. We seek to substantively shift control of the radio dial from corporate to community actors, and enable those community actors to use radio as an instrument for activism.

Perhaps the first outfit in the United States to use microradio primarily for social justice organizing was WTRA in Springfield, Illinois. In the 1980s, no commercial radio station was directed at Springfield's fifteen thousand African Americans (Foerstel 2001, 241). Named for the tenants' rights association of a local housing project, WTRA broadcast programming designed to organize

Springfield's African American community for housing improvements and against police brutality. Now broadcasting under the moniker Human Rights Radio (Gazi, Starkey, and Jedrzejewski 2011, 74), WTRA was a pioneer in what is essentially *subaltern radio*, the use of radio by communities that are structurally excluded from a society's media infrastructure.

The very existence of such broadcasters is in itself an act of anti-oppression, and their practical manifestations range from outright organizing to subtler forms of activism. Some play local music, others focus on community health or local history, and many stations serve their communities by broadcasting in languages other than English. While the missions and programming of these broadcasters may be diverse, their activism coheres by counteracting the dominant forces of the media landscape today—consolidation, homogeneity, the stifling of civic democracy—and collectively reasserting local control of the airwaves.

THE 2013 LPFM APPLICATION WINDOW

The Local Community Radio Act mandated that the FCC carry out an application window in which nonprofits, public safety groups, and Tribal organizations could apply for LPFM licenses (FCC 2012). Held in October and November of 2013, the LPFM application window was at once the culmination of and starting point for Prometheus's work, as it represented the most significant opportunity in a generation to redefine the FM dial. Our goal, then, was to mobilize a critical mass of groups to apply for LPFM licenses and to target our efforts toward organizations with social justice missions. From 2011 to 2013, we conducted a nationwide outreach campaign, produced materials to demystify the bureaucratic process, provided intensive legal and technical support to applicants, and maintained channels of advocacy at the FCC.

The Local Community Radio Act passed in a flurry of legislative activity in the waning days of the 111th Congress and received little news coverage (Stelter 2011). With the FCC making no effort to inform the public, few were aware of the opportunity to start an LPFM station. Thus, Prometheus and some ally organizations took on the role of outreach, a central feature of which was to build relationships with organizations of women, people of color, and urban membership. We toured the country to present at community meetings and conduct workshops, and our public communication via press, web, and telephone targeted progressive outlets and utilized the membership lists of allies. The purpose of this outreach was to inform groups about the LPFM application window and to persuade them to apply. For many social justice organizations, LPFM was an esoteric concept, so our challenge was to inspire them to see radio as a powerful tool for advancing their work.

Take, for example, OneAmerica, an immigrants' rights organization in Seattle, Washington. OneAmerica's organizing and advocacy work spans a diverse range of immigrant and refugee communities in the Seattle area that includes Latino, African, and Asian neighborhoods largely ignored by the commercial media (OneAmerica 2014). Rahwa Habte, a former organizer for OneAmerica, described the challenge of disseminating information and mobilizing action among such diverse groups: "We're going out to small businesses, apartment complexes, knocking on people's doors and trying to make a one-to-one connection. When you're working on a campaign, it's difficult to do and time consuming" (Sullivan 2013). The importance of mass communication to OneAmerica's organizing and advocacy work led them to apply for an LPFM license with the guidance of Prometheus. A key aspect of the proposed station was multilingualism. OneAmerica planned to broadcast in English, Somali, and Spanish to disseminate news, connect communities with city services, coordinate campaigns, and conduct media trainings so that members could produce and broadcast programming in their own languages (Minard 2013).

For OneAmerica and others, the legal and technical documentation required by the FCC bureaucracy was expensive and difficult to navigate. Most organizations we worked with had no previous media operations and never before imagined running a radio station. Rahwa Habte captures a sentiment that parallels the experiences of many of the hundreds of groups that Prometheus supported to apply for LPFM licenses: "We're not a media organization, we're a nonprofit whose mission is to build power in immigrant communities, and we can see how we can do that with radio. Beyond that, we have absolutely no idea. It's been a huge learning curve, and it's great to have someone to hold our hand and guide us through the process" (Sullivan 2013).

For over three hundred LPFM applicants, Prometheus provided in-depth and bilingual support. Our staff included engineers, educators, and organizers who worked with groups to navigate the FCC process and exhaustively prepare the legal and technical materials for the application. Furthermore, we worked with the National Lawyers' Guild and the Institute for Public Representation to answer legal questions and connect applicant groups with discount and pro bono attorneys. Prometheus also published a series of do-it-yourself guides and online seminars on how to complete all parts of the LPFM application.[2] To that end, we hosted open-source online software that enabled applicants to search for available frequencies and produce the LPFM application's required documentation.[3] Lastly, in the months leading up to the application window, our team of staff and volunteers fielded thousands of emails and phone calls from applicant groups asking for assistance.

When the LPFM application window was held in October and November of 2013, 2,830 organizations applied (FCC 2013), and of those, around 1,000 received support from Prometheus. Among them was the Gullah People's Movement, who planned to use their LPFM station to preserve the oral history of the Gullah community in the coastal Southeast (Clyburn 2012). The Chicago Independent Radio Project (CHIRP), a volunteer-run collective devoted to Chicago's underground music scenes, also applied (Lyon 2013). So too did the Fort Hood Support Network, a veterans-led antiwar group planning to use their LPFM station to "create a safe space for opposition to war and discourse about the issues that affect military communities" (Muncy 2013). Indeed, hundreds of unique organizations applied for space on their local FM dial, with a sizeable minority dedicated to prison reform, local art, environmental action, and a variety of other social justice causes.

RADIO NOT, HERE WE COME!

In 2014 and 2015, the FCC reviewed and processed virtually all all the applications filed in the 2013 LPFM application window (FCC 2013). Over the coming years, one thousand to two thousand new radio stations will receive licenses to get on the air in fifty states and Puerto Rico. Among them will be churches, unions, schools, civil rights groups, and local artists. While the average listener may not consider it, the inclusion of local and diverse voices on the radio dial is an extraordinary victory for democratic media given the corporate domination of traditional and, increasingly, digital communications.

Despite the important progress made under the Local Community Radio Act, however, the LPFM service faces significant obstacles as a media reform strategy. Stavitsky, Avery, and Vanhala (2001) argue that LPFM could meet a fate similar to FM's Class D, which was discontinued in 1978 at the hands of commercial and public radio trade lobbies and poorly run Class D stations. The story of Class D demonstrates the need for LPFM stations to be engaging and relevant to the communities they serve, a difficult task given LPFM's limited broadcast range and doubts over the future viability of terrestrial radio.

The 2013 LPFM application window provided the requisite policy conditions to carry out Prometheus's mission. The addition of thousands of community broadcasters to the ten thousand or so stations already on the FM dial represents a substantive shift toward localism in radio (FCC 2011). With an increase in corporate mergers, ongoing threats to net neutrality principles, and restrictions on municipal broadband, community radio remains one of the more promising developments in American media activism. But even with policy and physical infrastructures in place, the use of LPFM for the purposes of social transformation is still an open question. LPFM could follow a parallel course to Class D, or the low-power and public access television services, all of which are locally controlled media infrastructures that have done little to engage the public. Likewise, LPFM could become what the scholar Raymond Williams (1976, 186) called "a marginal culture; even, at worst, a tolerated play area." For community radio to become an authentic participatory medium in the United States on a large scale, LPFM stations and media reform activists (including the Prometheus Radio Project) must exhibit political creativity and a sustained commitment in the years ahead.

1. H.R. 6533 of the 111th Congress.
2. See www.prometheusradio.org/archives.
3. See http://rfree.radiospark.org/.

REFERENCES

Clear Channel Communications (CCC). 2014. Investor FAQ. www.clearchannel
.com/Investors/ Pages/faq.aspx.
Clyburn, M. 2012. Statement of FCC Commissioner Mignon L. Clyburn re: cre-
ation of a low-power radio service. Federal Communications Commission.
Connolly-Ahern, C., A. M. Schejter, and J. A. Obar. 2012. "The Poor Man's
Lamb Revisited? Assessing the State of LPFM at Its 10th Anniversary."
Communication Review 15, no. 1: 21–44.
Dicola, P., and K. Thompson. 2002. *Radio Deregulation: Has It Served Citizens
and Musicians? Report on the Effects of Radio Ownership Consolidation Fol-
lowing the 1996 Telecommunications Act*. Washington, DC: Future of Music
Coalition.
Doyle, B. 2012. "Low-Power Community Radio in the US: The Beginnings, the
First 10 Years and Future Prospects." In *Community Radio in the Twenty-
First Century*, edited by J. Gordon. New York: Peter Lang Publishing.
Federal Communications Commission (FCC). 2000. Report and Order in the
Matter of Creation of Low Power Radio Service. Mm Docket No. 99–25.
Washington, DC: Federal Communications Commission.
———. 2011. Broadcast Station Totals as of March 31, 2011. Washington, DC:
Federal Communications Commission.
———. 2012. Fifth Order on Reconsideration and Sixth Report and Order.
Mm Docket No. 99–25. Washington, DC: Federal Communications
Commission.
———. 2013. Public Notice: Media Bureau Provides Further Guidance on the
Processing of Form 318 Applications Filed in the LPFM Window. Washing-
ton, DC: Federal Communications Commission.
Foerstel, H. N. 2001. *From Watergate to Monicagate: Ten Controversies in Modern
Journalism and Media*. Westport, CT: Greenwood Publishing.
Fraser, C., and S. Estrada. 2001. *Community Radio Handbook*. Paris: UNESCO.
Gazi, A., G. Starkey, and S. Jedrzejewski. 2011. *Radio Content in the Digital Age*.
Chicago: University of Chicago Press.
Goldfarb, C. B. 2007. *FCC Media Ownership Rules: Current Status and Issues for
Congress*. Washington, DC: Congressional Research Service.
Klinenberg, E. 2007. *Fighting for Air: The Battle to Control America's Media*. New
York: Metropolitan Books.
Lyon, G. 2013. "An Interview with Jenny Lizak." www.prometheusradio.org/
future-community-radiochirp.
Minard, A. 2013. "New Neighborhood Radio Stations to Apply for Licenses—After
the Shutdown Ends, of Course." *Stranger*, October 15. http://slog.thestranger

.com/slog/archives/2013/10/15/new-neighborhood-radio-stations-to-apply-for
-licensesafter-the-shutdown-ends-of-course.

Muncy, M. 2013. "Community Powered Radio." *Under the Hood Café*, September 25. http://underthehoodcafe.org/2013/09/community-powered-radio.

OneAmerica. 2014. "What We Do." http://weareoneamerica.org/what-we-do.

Prometheus Radio Project. 2014. *Mission, Vision, and Values.* http://prometheus radio.org/mission.

Shaffer, G. 1998. "Radio in the Raw." *Philadelphia City Paper*, May 21.

Shaw, Y., and A. Lewis. 2014. "Going Black: The Legacy of Philly Soul Radio." Radio program. Philadelphia: Mighty Radio.

Stavitsky, A. G., R. K. Avery, and H. Vanhala. 2001. "From Class D to LPFM: The High-Powered Politics of Low-Power Radio." *Journalism and Mass Communication Quarterly* 78, no. 2: 340–354.

Stelter, B. 2011. "Low-Power FM Radio to Gain Space on the Dial." *New York Times*, January 25, C1.

Sullivan, P. 2013. "FCC Opens FM Airwaves for Nonprofits." *Non-Profit Times*, August 30. www.thenonprofittimes.com/news-articles/fcc-opens-fm-airwaves -for-nonprofits.

Tridish, P., and K. Coyer. 2005. "A Radio Station in Your Hands Is Worth 500 Channels of Mush: The Role of Community Radio in the Struggle Against Corporate Domination of Media." In *News Incorporated: Corporate Media Ownership and Its Threat to Democracy*, edited by E. Cohen, 287–314. Amherst, NY: Prometheus Books.

Turner, S. D. 2007. *Off the Dial: Female and Minority Radio Station Ownership in the United States: How FCC Policy and Media Consolidation Diminished Diversity on the Public Airwaves.* Free Press. www.freepress.net/sites/default/ files/fp-legacy/off_the_dial.pdf.

Walker, J. 2001. *Rebels on the Air: An Alternative History of Radio in America.* New York: New York University.

Williams, R. 1976. *Communications.* Third edition. London: Penguin.

CHAPTER SIXTEEN

Media Reform Initiatives in West Africa

KWAME KARIKARI, *Media Foundation for West Africa, Ghana*

MEDIA REFORM STRATEGY

As a leading media reform promotion organization in West Africa, the Media Foundation for West Africa (MFWA) initiated its advocacy work on major issues of reform through the following key strategies: Research and documentation of the issues at stake; the definition of objectives; the identification of primary, secondary, and vicarious beneficiaries of the projected reforms; the definition and identification of target authorities, institutions, or agencies to address; consultation with key stakeholders (including publishers, journalists, media professional associations, and human rights groups); the mobilization of stakeholders into activist committees or coalitions; the collective definition of key activities to engage in and the drafting of a plan or program of active intervention; detailed action plans for specific interventions or activities (including press statements, briefings, conferences, position papers, situation analysis, memoranda, petitions, public hearings and presentations, court litigation, protests and demonstrations); constant monitoring and evaluation of programs, activities, and interventions; and the continual sharing and communication of outcomes and challenges with the public and members of the advocacy constituency. The result of the implementation of these strategies has been massive public support for the objectives of the advocacy, namely the right and freedom of citizens and groups (who can afford it), to establish radio and TV broadcasting in Guinea without arbitrary state obstructions.

Advocacy work by civil society for media reform is central to democratic governance reforms in general. In West Africa, media reform activities demonstrate some of the most visible and vibrant civil society advocacy campaigns since the mid-to-late 1980s. It would not be an exaggeration to assert that the Media Foundation for West Africa (MFWA)—a nonprofit, nongovernmental organization set up in 1998 to promote media pluralism, media rights, and media development—has been the most influential NGO in the advocacy work for those liberal democratic reforms in West Africa over the past decade and a half. It is a nonmembership civil agency, set up by a small group of media academics and professionals, lawyers, labor, and human rights activists with long, active involvement in popular struggles for democratization in the region.

THE MEDIA FOUNDATION FOR WEST AFRICA

The MFWA is but one of many civil society rights advocacy groups that emerged as agencies of the region-wide popular movement for democratization and human rights across the region in the late 1980s–1990s. Many of these groups addressed issues at the community level, others at the national level. A few worked across borders on a regional scale, using the ECOWAS charter as reference point. Although regional in focus, the MFWA operates from a single bureau in Accra. Unlike other regional organizations, it has no membership base, branches, or affiliate structures in other countries.

MISSION

The founding mission of the MFWA is aimed at promoting and protecting media rights and freedom of expression, advocating for media law and policy reform, providing legal defense for victims of media rights violations, strengthening media to strengthen democracy, and supporting media in conflict and for peace-building.

GUIDING STRATEGIC APPROACHES

Throughout its existence, the organization's work has been guided by the following key strategies:

a. Clear and unambiguous definition of objectives and methods,
b. Action-oriented research and investigation,
c. Identification of and encouragement to local/national stakeholders and or potential beneficiaries of positive outcomes,

d. Broad and in-depth consultations with, and involvement of, national civil society agencies active especially in rights advocacy,

e. Identification of "favorable" links in government for possible access points (because no rights reform is possible without government participation)

f. Consultations—and cooperation where possible and realizable—with government officials and institutions,

g. Promotion of local and national stakeholders and actors as the primary and principal actors or agencies for advocacy and related programs,

h. Support to strengthening the capacity of local and national stakeholders and actors to play leading roles, and

i. Involvement, as much as possible, of UN sector agencies in the processes.

HISTORICAL AND POLITICAL CONTEXT

When the MFWA was established in May 1997 in Accra, Ghana, the West Africa region was throbbing with widespread popular demands for democratization on the one hand and ablaze with brutal civil wars or threats of violent conflicts in many countries on the other. One of the key and concrete expressions of the dawning atmosphere of democracy and freedom was the explosion of newspapers in the streets of the capitals and big cities throughout the region. This was soon followed by an even higher impact explosion of the airwaves: Radio and TV broadcast pluralism. Liberia and Sierra Leone, then caught in the vicious throes of civil war, were not left out of the media revolution. The authoritarian state's monopoly control of the mass media and the public space for public participation in public affairs discourse was broken almost simultaneously across the fifteen countries of the Economic Community of West African States (ECOWAS).

Popular agitations, complemented by international donor pressure, compelled the extant, mostly authoritarian, governments to adopt new constitutions promising generally liberal democratic governance systems. But it was only in Benin where the very first elections, based on the new constitutions, led to the replacement of the old, repressive military-based one-party Marxist-Leninist regime by a popularly elected government. All the authoritarian regimes, mostly of military origin, transformed themselves into civilian political parties and retained state power through the elections sanctioned by the new constitutions.

This partly, but significantly, explains why the new wave of media freedom was met with such widespread violent acts of intimidation, repression, and persecution as never before. The vicious reaction of the states to the unmuzzled voices of the new independent and irreverent press and radio, threatened the expansion and consolidation of the emerging freedom of expression and, of course, of other rights and freedoms.

Moreover, there was little, or insignificant, reform in the myriad legislation—dating back to colonial and early postcolonial regimes—constraining freedom of expression, journalists' freedom to function, and censoring media content. Until this day, only Ghana has abolished, by parliamentary repeal, criminal defamation and seditious libel. Cape Verde, Ghana, and Nigeria aside, all other countries maintain the old criminal legislation on media and journalism practice from colonial times. The Republic of Guinea was also the last to open up the airwaves to end state monopoly and control of broadcasting—in 2005.

Elsewhere in the region, media reform was acutely limited to structural questions concerning ownership liberalization, and the establishment of statutory regulatory mechanisms—institutions that, in most cases, came under governmental control.

The general situation, even after the explosion of media production and the end of state monopoly and control, was characterized by three broad tendencies: Popular demand for media freedom and freedom of expression; growing threats to the emerging freedoms; and, reluctance (often even hostility) of governments (even in regions where the parties in power have been democratically elected) to initiate or encourage broad reforms in media law and policy.

SOME KEY MFWA ADVOCACY STRATEGIES

The organization's Governing Council, the policy formulating and final decision-making body, includes lawyers, media professionals, academics, labor leaders, and other rights advocates. The MFWA initiated or joined coalitions and networks of media rights–related and human rights advocacy organizations at national, regional, or continent-wide levels to mobilize as many forces around thematic issues and to promote specific campaigns for justice or for reforms.

The MFWA was the first to initiate a successful region-wide system of regular, uninterrupted daily monitoring and publication of attacks on and violations of journalists and media rights—from the perspectives of Article 19 of the Universal Declaration of Human Rights. Started in 1999, the program continues to this day, issuing alerts whenever and wherever in the West Africa region there is a case of violation. The association was also the first to provide legal defense for journalists and media prosecuted by the state under laws that make journalistic infractions and media outputs criminal offenses.

The MFWA provided heightened publicity and gave considerable relevance to the regional judicial mechanism, the ECOWAS Court of Justice, when it sued the government of The Gambia to seek justice in a case of habeas corpus for a disappeared journalist and in another case of torture of a journalist by state security operatives.

The MFWA has provided safe haven for journalists fleeing threats of death, suffering arbitrary arrests and detention, or experiencing other criminal violence perpetrated by the state or dangerous nonstate actors. The MFWA has championed media law and policy reform advocacy activities and processes in nearly all of the fifteen ECOWAS countries. An initiative for an ECOWAS statute or protocol for a standard media legislation is pending before the regional economic bloc's Council of Ministers. In addition to these efforts, the regional rights advocacy organization has initiated African and international partnerships to promote media development program in postconflict transitional countries.

The MFWA also set up a network of correspondents in each of the sixteen West African countries (including Mauretania), for monitoring and publicizing attacks on and violations of media rights.

A final noteworthy strategy is that the Legal Defense program was executed through a network of volunteer lawyers who provided pro bono services for needy and persecuted journalists and media outlets. The same team of lawyers, from different countries, also provided advice on legal and policy reform activities, such as drafting proposals for alternative legislation.

OUTCOMES OF THE MFWA'S ADVOCACY WORK

The contribution of the organization's work to media rights promotion in West Africa includes:

a. Substantial reduction in the incidence of violent state attacks on and abuse of journalists and media rights in West Africa;
b. Increased cooperation and collaboration among media professional associations and NGOs working on media rights in Africa;
c. Improved, increased, and regular promotion of media rights issues;
d. Growth in activities and national groups addressing media rights promotion;
e. Heightened publicity, and possibly increased awareness, of the work and relevance of the ECOWAS Court in rights promotion among civil society groups in the region; and
f. Increasing use of ECOWAS and other continental and regional instruments and mechanisms in to promote liberalization of the airwaves, promoting rights issues.

SUCCESSFUL MFWA REFORM STRATEGIES

The MFWA's successes include the liberalization of the airwaves in the Republic of Guinea and the postconflict media development strategic plan in Liberia.

By the year 2000, state monopoly control of broadcasting in West Africa existed only in the Republic of Guinea (Conakry). General Lansana Konte, who had come to power in 1984 and "legitimized" himself by winning elections in 1993, had embarked on exercises of constitutional reform. The reforms, however, remained on paper. The regime ensured that opposition political parties and independent civil society organizations could not function freely to exercise their rights of association and of expression. The regime hung on to power by brute force. Headed by a sick and dying old soldier, the regime began to fall apart the more violently it resisted popular demands for political reforms and democratization.

With mounting mass protests and widespread agitation, led by the trade unions, bureaucrats and top officials abandoned the state and their offices. The crisis created a favorable political opening for promoting rights issues, including broadcast pluralism.

Through the monitoring of media rights violations and developments on the media scene in general, the MFWA monitored the political developments in the country. The organization initiated the liberalization of the airwaves project by using the following strategies:

a. Researching and critiquing Guinea's media laws;
b. Drafting a concept note on the project;
c. Identifying stakeholders and potential beneficiaries;
d. Setting up a national committee made up of newspaper publishers, representatives of some rights advocacy organizations, officials of the state broadcasting organization, officials of the Ministry of Information, and a few independent lawyers to drive the project; and
e. Organizing a national conference—attended by broad sections of media, civil society, and government—to promote a policy reform to open up the airwaves to citizens and their organizations.

All this was undertaken between late 2003 and June 2004. The result was massive public support for the objectives of the advocacy, namely the right and freedom of citizens and groups (who can afford it), to establish radio and TV broadcasting in Guinea without arbitrary state obstructions. The minister of information, representing the collapsing government, openly endorsed the resolutions of the conference. Today, the country has more than thirty privately owned FM radio stations.

LIBERIA

As the civil war in Liberia reached the phase of ceasefire and peace talks, there was an urgent need for media reforms so as to enhance media capacity to contribute meaningfully to restore peace, build democracy, and reconstruct the social and economic fabric of the society. The MFWA thus initiated an international committee to develop strategic plans for postwar media development. The Partnership for Media and Peace-building in West Africa included Liberian media associations and rights advocacy groups; West African, European, and other international organizations working on media rights and media development program. It also included UNESCO and UNDP and attracted the interest of bilateral aid agencies and private philanthropic donor organizations, which offered assistance to the process.

The group undertook several consultative meetings with the Liberian media, civil society groups, and officials of the transitional government. Out of the consultations and studies conducted, a comprehensive media development strategic plan was developed. The outcomes of the activities included:

a. The establishment of a national Liberian media rights and development advocacy agencies to promote implementation of the strategic plan;
b. Contribution to constitutional reforms to ensure the protection of media rights; and
c. Policy reforms including legislation of a new Right to Information law, making Liberia the second country in the region to achieve this.

CONCLUSION

Advocacy for media reforms in Africa during the last two decades and more has generally been boosted by the emergence in that same period of several media rights advocacy organizations and media professional associations. The need for media reforms arose from the fact that, while governments in most countries lost control of the states' monopoly of the ownership and operation of the media as a result of the popular agitations for democratic governance in the 1980s and early 1990s, little was done by the governments to write new laws and regulations to protect and strengthen the new freedoms won by the media. Instead, in response to the robust work of the new independent private media, governments unleashed a new, violent wave of repression of press freedom. The emergence of media rights advocacy organizations was a reaction to the threats government repression posed to the fragile freedoms of the new and young independent media. The international environment provided a favorable atmosphere for rights advocacy work. The governments had signed on to charters, protocols, and statutes of regional intergovernmental organizations (such as the African Union, formerly known as the Organization of African Unity, and the subregional Economic Community of West African States), which provided progressive legal instruments promoting these rights. These strategies provide useful reference points for engaging those governments that accommodated democratic engagement, or the basis for legal litigation against more recalcitrant regimes. Finally, these factors and developments have defined and determined the advocacy strategy for the MFWA and its local, regional, and international allies and collaborators.

PART FOUR

Media Reform as Democratic Reform

CHAPTER SEVENTEEN

Waves of Struggle

The History and Future of American Media Reform

VICTOR PICKARD, *Annenberg School for Communication, University of Pennsylvania, United States*

MEDIA REFORM STRATEGY

The 1940s was a contentious decade for US media policy. Activists, policy-makers, and media industries grappled over defining the normative foundations that governed major communication and regulatory institutions. At this time, a reform agenda took shape at both the grassroots social movement level and within elite policy circles. An analysis of the rise and fall of this postwar media reform movement holds at least three key lessons for contemporary media activists. First, it reminds us of the imperative to maintain a strong inside/outside strategy that keeps regulators connected to the grassroots. Second, we learn that media activists retreat on structural reform objectives at their own peril. Finally, we must remember that media reform rises and falls with other political struggles and radical social movements. With these lessons in mind, media reformers should seek to build liberal-left coalitions and, perhaps, a new popular front.

Most Americans learn in school that an independent press is necessary for democratic self-governance, but rarely do we stop to reflect on what this means.[1] How did we as a society determine media's primary democratic role? How did we decide upon media institutions' obligations to the public? How was the relationship between the state, the polity, and media institutions constructed, and how has this arrangement changed over time? Such inquiries require historical analyses that retrace policy discourses and trajectories to moments of conflict when normative foundations were fought over and assumptions about media's democratic role crystallized. This approach highlights contingency; it reveals that outcomes were neither preordained nor natural. At key junctures in this media system's development, amid multiple sites of struggle, certain claims won out over others. What quickly becomes clear is that the contours of our media system have resulted more from contestation than any consensual notion of what media should look like. Thus, a history of media is, in fact, a history of media reform. It is also a history of failed attempts to de-commercialize the media system.

Although perhaps best characterized as a continuous struggle, American media history is punctuated with moments of upheaval and reform when politicized groups saw media as a crucial terrain of contestation—often with the very structures of media themselves at issue. An overview of this history benefits from case studies in which conflicting interests and their respective discourses are cast into sharp relief. These moments tend to occur during what previous scholars termed "critical junctures" (McChesney 2007) or "constitutive moments" (Starr 2004) marked by crisis and opportunity. Drawing from archival materials to reflect on past struggles to change the American media system, this chapter focuses on one particular historical episode in the 1940s, a contentious decade for US media policy. As activists, policy-makers, and media industries grappled over defining the normative foundations that governed major communication and regulatory institutions, a reform agenda was taking shape at both the grassroots social movement level and inside elite policy circles. With the aim to help inform future media reform efforts, this chapter examines the tensions within this nascent media reform movement, many of which are still negotiated among media advocacy groups today. My analysis pays particular attention to the retreat from a structural critique of the commercial media system that was crystallizing in the immediate postwar years.

Historicizing media reform efforts and policy debates allows us to address the "how did we get here" question, which restores contingency and deferred alternatives—alternatives that merit recovery not only to correct the historical record by reclaiming resistance, but also to inspire future reform efforts. This kind of critical historical analysis problematizes and *politicizes* current media policies. An examination of any media system at any given time will most likely discover, to varying degrees, struggle against it. During constitutive moments or critical junctures—or whatever metaphor we choose—this resistance peaks, opening up fleeting windows of opportunity for reform. It is also during these moments of crisis when a media system's normative foundations become concretized. Focusing on these episodes sheds light on larger paradigmatic shifts. It shows that despite tremendous activism against US media in the 1940s, by the end of the decade, the ideological consolidation that girded a commercial, self-regulated media system emerged largely intact and further inoculated against structural challenges. The rise and fall of this postwar media reform movement holds at least several key lessons for contemporary media activists. Before explicating these implications, I first discuss a historical and theoretical framework that brings these struggles into focus. Next, I examine some specific 1940s reform initiatives with an eye on how they parallel our current political moment. I conclude with some general lessons that can be drawn from this failed media reform movement.

A GRAMSCIAN APPROACH TO POWER AND HISTORY

Like all social phenomena, media policy does not spring fully formed from Zeus's head but rather emerges from a multiplicity of sociopolitical influences. In making sense of these messy processes, a historical analysis of media policy invites a particular theoretical model, one that underscores contingencies without obscuring the evolving contours of power relationships. At its best, this kind of theoretical approach encourages and guides action by underscoring what is at stake and by bringing into focus political arrangements and power relationships. In general, historicizing is valuable for allowing us to see present relationships, practices, and institutions as historical constructs contingent upon contemporaneous factors instead of simply natural phenomena. Historicizing current media debates allows us to reimagine the present and reclaim alternative trajectories.

The rubric of historical research encompasses a number of theoretical approaches. While high theory is unnecessary for understanding the history of American media reform, a particular framework may prove to be useful if we are to understand recurring patterns of struggle. Combining intellectual, social, and political histories, my theoretical approach to understanding how power operates and history unfolds vis-à-vis media processes and institutions can best be described as Gramscian. Emphasizing contingencies, contradictions, and ruptures, this analysis assumes that, rather than being independent and linear, historical processes transpire in complex dialectical interplays that are mutually constitutive. Rendered correctly, such a Marxist historical approach avoids over-determination. The critical media studies scholar Deepa Kumar (2006, 83) argues that "far from being reductionist, the Marxist method enables us to understand the world in all its complexity and opens up the possibility for change." Despite its concern with showing how power triumphs, this framework does not presuppose that policy outcomes always reflect the most powerful interests' intentions, but rather strives to encourage engagement and resistance.

Much Gramscian theory centers on the notion of *hegemony*, a contentious political process by which assumptions that serve elite interests becomes commonsensical (see Gramsci 1971, 323–334, 419–425). Seeing the formation of common sense as a crucial terrain for constant political struggle, this framework focuses on how hegemonic forces operate via a complex interplay between dominant interests and those they attempt to subjugate. These power relationships are messy and inherently unstable and must be recreated daily, constantly opening up new areas for resistance. Stuart Hall (1988, 7) observed that hegemony "should never be mistaken for a finished project." A Gramscian historical framework restores the promise of agency, recovers contingency, allows for unexpected outcomes, and assumes human events reflect not just societal consensus but also ongoing conflict. This conflict is greatest during realignments of what Gramsci termed "historic blocs" of the ruling elite. Such reconfigurations produce new political opportunities, which in turn allow for new policy formations.

Gramsci referred to these periods in which historic blocs are challenged as "conjunctural moments." A conjuncture marks the immediate terrain of conflict. Explicating this useful concept, Gramsci (1971, 178) wrote:

> A crisis occurs, sometimes lasting for decades. This exceptional duration means that incurable structural contradictions have revealed themselves . . . and that, despite this, the political forces which are struggling to conserve and defend the existing structure itself are making every effort to cure them, within certain limits, and to overcome them. These incessant and persistent efforts . . . form the terrain of the "conjunctural," and it is upon this terrain that the forces of opposition organize.

Conjunctural moments have taken place at different points in American history when industry control of major media institutions was challenged. For example, during the Depression and the New Deal, a counter-hegemonic regime was struggling to take hold. With historic blocs in flux, new alliances emerged to foment what Gramsci (1971, 210, 275) termed a "crisis of the ruling class's hegemony" and a "crisis of authority." "If the ruling class has lost its consensus," Gramsci argued, "this means precisely that the great masses have become detached from their traditional ideologies" (ibid., 275–276). According to this Gramscian analysis, "the crisis consists precisely in the fact that the old is dying and the new cannot be born."

That these counter-hegemonic attempts ultimately failed makes them no less significant, especially since even in failure they often have lasting material effects. The cultural historian Michael Denning (1996) makes a similar argument in his examination of the Popular Front, a coalition of left-wing radicals, New Deal liberals, and political progressives that drove reform during the 1930s and 1940s. Denning notes how during the Great Depression and World War II a prolonged "war of position" unfolded between conservative forces and a Popular Front social movement that attempted "to create a new historical bloc, a new balance of forces." Denning argues that the "post-war settlement" that eventually emerged—exemplified by the corporatist arrangement of big labor, big capital, and big government—resulted from "the defeat of the Popular

Front and the post-war purge of the left from the CIO [Congress of Industrial Organizations] and the cultural apparatus." He observes, "If the metaphor of the front suggests a place where contending forces meet, the complementary metaphor of the conjuncture suggests the time of the battle." Based on this view, Denning sees the history of the Popular Front as "a series of offensives and retreats on the 'terrain of the conjunctural'" (ibid., 22).

This focus on evolving historical patterns of power struggles and shifting institutional life cycles around conjunctural moments brings Gramscian analysis into conversation with concepts associated with historical institutionalism. At first glance, Gramscian theory and historical institutionalism may seem like an odd pairing. After all, the latter seeks to move beyond grand theorizing and basic material explanations for institutional change that do not adequately explain all variations of institutional behavior (Skocpol, Evans, and Rueschemeyer 1985). However, historical institutionalism assumes that larger macrolevel historical and political relationships—the focus of much Marxist analysis—are mediated through interconnected institutional discourses, habits, and imperatives to impact micro-level processes. Both Gramscian theory and historical institutionalism focus on how struggles over foundational assumptions during times of crisis can result in intellectual paradigm shifts. Furthermore, in seeking to shed light on sharp breaks from the past, combining these theoretical models can help account for how moments of equilibrium are ruptured, ushering in a new set of policies that may result in new institutional arrangements. Both theoretical approaches also assume that institutions and their relationships reflect historical experiences of conflict and compromise among organized constituencies. Historical institutionalism tries to make sense of how some trajectories are chosen over others (Skocpol 1985). Thus, discerning "paths not taken" is as significant as identifying the chosen trajectories and resulting path dependencies (Pierson and Skocpol 2002).

The most important overlap in the two frameworks is between Gramsci's notions of crisis and conjuncture and historical institutionalism's focus on "critical junctures." The concept of critical junctures brings into focus how institutional regimes and relationships see long periods of relative stability and path dependency punctuated by sudden ruptures in which the system is jolted and new opportunities for change arise. Much literature within policy studies shows how decisions made during such periods profoundly impact systemic development (see, for example, Collier and Collier 1991; Kingdon 2002; Stone 2001). Kathleen Thelen (1999, 385) notes that although politics always involves some degree of happenstance like agency or choice, once a path is established, it can become "locked in" because "all the relevant actors adjust their strategies to accommodate the prevailing pattern." Applying this theory to understanding media reform, Robert McChesney (2007, 9–12) sees increasing evidence that the US media system is undergoing a critical juncture in the early twenty-first century.

Critical junctures tend to invite more public engagement with and scrutiny of media systems than less contentious periods and typically emerge during times of technical, political, and social change. This theoretical and historical framework brings into focus recurring moments of contestation when a media system's normative foundations are challenged and, possibly, redefined.

WAVES OF MEDIA REFORM

Periods of media reform are often marked by an explosion of activist media. Reform-oriented media have been a crucial resource for American social movements and marginalized groups, who have often resorted to contesting representations in the mainstream press or creating their own media to advance activist causes. Revolutionary pamphleteers helped to mobilize the struggle for independence against the British. A vibrant abolitionist press galvanized reformers for decades preceding the Civil War. A popular working-class press was integral to the burgeoning labor movement in the first half of the twentieth century. In the early 1900s, the advertising-supported socialist newspaper *The Appeal to Reason* reached nearly a million subscribers and helped advance the socialist candidate Eugene Debs's presidential ambitions. During the nineteenth and twentieth centuries, an ethnic press provided support for various marginalized cultural groups (Gonzalez and Torres 2011). In the 1960s, an underground press helped sustain the civil rights movement and other activist groups. Today, digital social media are central to a myriad of activist efforts.

In the twentieth century, media reform struggles also increasingly centered on questions of policy to effect change in the structures of media themselves. The 1930s and 1940s witnessed widespread reform efforts, particularly toward broadcasting and newspapers, when both elites and grassroots activists considered a relatively wide range of policy options. In the 1930s, a spirited media reform coalition attempted to establish a more public-oriented broadcast system, while the Newspaper Guild challenged commercial publishers' control over print media production (Scott 2009). These efforts were ultimately crushed, but they established benchmarks for future reformers to carry on various struggles, particularly during the 1940s, which I expound on shortly. During the 1960s, media reform projects were carried forward by the civil rights, antiwar, and other social movements (Lloyd 2007). These years witnessed notable media reform victories: public broadcasting was established, the Fairness Doctrine enjoyed its golden age, and civil rights activists won the historic WLBT case in which a racist broadcaster was denied a license renewal (Horwitz 1997). Also at this time, advocacy groups began coordinating around a number of key policy issues (Mueller, Kuerbis, and Page 2004). In the 1970s, during what were called the New World Information and Communication Order (NWICO) debates, a global media reform movement, perhaps the first of its kind, took place around communication rights (Pickard 2007). Similar global media reform efforts would later reemerge during the World Summit on the Information Society (WSIS) debates (McLaughlin and Pickard 2005).

More recently, issues like media ownership and Internet policy—as well as ongoing struggles over representation and alternative media—have galvanized vibrant reform movements. Media democracy, media justice, and media reform movements coalesced in the late 1990s to take on excesses in the corporate media system. These movements often took advantage of new digital media, especially the Internet. The latter would help drive a radical indymedia movement that established independent media centers in communities across the globe (Pickard 2006). These activist efforts also helped establish the influential media reform organization Free Press, which would successfully engage and organize broad constituencies around what had previously been seen as obscure media policy issues. For example, in 2003, nearly three million people wrote letters to the FCC contesting proposed plans for loosening ownership restrictions (Klinenberg 2007, 238–244). In 2006, over one million people petitioned Congress to protect a then obscure policy called net neutrality to maintain a nondiscriminatory Internet (Kirchgaessner, Waldmeir, and Waters 2006). In 2012, four and a half million people signed a petition to roll back two proposed Internet piracy bills that would have given government and corporations tremendous power over web content (*Washington Post* 2012).

These activist energies are often put on full display during a series of large national media reform conferences that have attracted several thousand attendees. As reformist interventions against media institutions become increasingly common, it might be tempting to assume that these instances are unique in American history. However, as this brief historical sketch suggests, these waves of struggle are, in fact, continuities of ongoing media reform traditions. For its parallels and important antecedents to today's policy challenges, one moment of media reform—a largely forgotten episode within dominant historical narratives—holds particular contemporary relevance.

THE RISE AND FALL OF THE POSTWAR
MEDIA REFORM MOVEMENT

Building on earlier progressive and New Deal criticism, the 1940s saw public discontent toward media institutions rise once again.[2] Commercial radio came under vehement public condemnation, and the newspaper industry underwent a crisis that drew calls from policy-makers for structural reform (for newspapers, see Davies 2006; for radio, see Fones-Wolf 2006). The disintegration of the liberal New Deal consensus and a rightward shift in the political landscape led to sociopolitical turmoil and uncertainty. Technological change was afoot with the rise of FM radio and television just over the horizon. These factors combined to produce a fleeting opportunity for a fairly radical overhaul of an entrenched media system—reforms that would have been unthinkable during less contentious times.

At this time, policy elites and social movement groups fought over a number of crucial media debates that helped define the relationships between media institutions, various publics, and the state. These included the FCC forcing NBC to divest itself of one of its two major networks; the Supreme Court's 1945 antitrust ruling against the Associated Press, which called for government to encourage "diverse and antagonistic voices" in media; the 1946 FCC Blue Book, which outlined broadcasters' public service responsibilities; the 1947 Hutchins Commission on Freedom of the Press, which established journalistic ethics; and the 1949 Fairness Doctrine, which mandated public interest parameters for broadcasters. The logic driving these reform efforts can best be described as "social democratic," in the sense that it assumed a progressive role for the state in protecting media's public service responsibilities from excessive commercialism (Pickard 2011b).

By the mid-1940s, these media reform efforts were further bolstered by increasing dissent from below, manifesting in a media reform coalition composed of dissident intellectuals, civil libertarians, African American groups, religious organizations, educators, labor unions, and other progressive activists (Pickard 2013) who pressured broadcasters via petitions, call-ins, and letter-writing campaigns to radio stations and the FCC, urging them to democratize the public airwaves and to improve programming.[3] Within these debates, significant challenges arose to confront the reigning libertarian notion of media industry self-regulation.

A signature model of 1940s radio activism was the neighborhood listener council. Created by local communities—often by parents, educators, and minority groups—listener councils lobbied radio stations and the FCC to privilege localism and curb excessive commercialism. By the mid-1940s, listener councils were established in Cleveland, Columbus, and New York City, and in more rural areas like central Wisconsin and northern California. As these councils emerged across the country, reformers advocated for their broader deployment to countervail against commercial broadcasters' political and economic dominance. Some reformers hoped that the councils could extend to the state level and coordinate nationally, thereby harmonizing their standards and survey techniques with the FCC and perhaps via an overarching public national council. Reformers proposed that councils receive local government support or be volunteer-based to provide institutional support for community groups to conduct research on the state of broadcasting. Armed with hard data, listener councils could pressure radio stations with the threat of their representatives presenting critical evaluations to the FCC during broadcasters' renewal procedures (White 1947, 122–125, 222–234). Listener councils' potential generated much enthusiasm, even if they never became a major force within policy debates. The veteran media reformer and NYU communication professor Charles Siepmann (1946, 51–52), for example, saw listener councils as the "community's best safeguard against the exploitation of the people's wavelengths and the surest guarantee [that radio stations consider] . . . its needs." By devising blueprints for high-quality local programming, Siepmann opined, the councils served as "watchdogs for the listener, ready and able to protest the abuse of airtime and to promote its better use."

Many activist groups participating in the 1940s media reform coalition also created their own alternative media, contested negative imagery, and exploited opportunities within commercial media to disseminate their political messages. Labor unions succeeded in buying a number of stations and instructed their members on how to produce their own shows or insert labor-friendly scripts into commercial radio programming. The Congress of Industrial Organizations galvanized reformers with its *Radio Handbook* (1944), a pamphlet that contained instructions for getting on the air and promoting "freedom to listen" and "freedom of the air." It insisted that workers had not exercised "their *right* to use radio broadcasting." "Labor has a voice," the pamphlet stated, and "the people have a right to hear it." Although broadcasters owned radio stations' equipment, "the air over which the broadcasts are made does not belong to companies or corporations. *The air belongs to the people*" (CIO 1944, 6). The handbook instructed activists on how to gain radio time for a labor perspective via different techniques and formats, including the Straight Talk, the Round Table Discussion, the Spot Announcement, and the Dramatic Radio Play. It encouraged activists to generate good publicity and to coordinate with consumer groups, cooperatives, women's organizations, and religious organizations (ibid., 25). In addition to labor-related concerns, the groups constituting this media reform coalition shared major grievances, critiques, tactics, and strategies for engaging in media policy debates in Washington, DC. Media reformers continuously pressured the FCC and commercial broadcasters to include diverse perspectives in radio programming. Their objectives often would converge around specific policy interventions, especially with progressive allies at the FCC.

This media reform alliance of grassroots activists and progressive policymakers led to some of the most aggressive public interest interventions in American media history. As mentioned above, in the early 1940s the FCC broke up NBC—an action that is unthinkable today. In the mid-1940s the FCC attempted to establish meaningful public service requirements on broadcasters as a requirement for license renewal. But as the decade progressed, structural criticism and reform efforts petered out and policy activism both in- and outside Washington DC underwent a period of decline. To summarize a much longer story, the postwar media reform movement ultimately failed, both by its own shortcomings and by external pressures and other political events beyond its control. In particular, it was suppressed by a corporate backlash that used Cold War politics to effectively red-bait and silence reformers. Radicals were purged from social movement groups and New Deal progressives were chased out of the nation's capital. Once this reform movement was demobilized, its initiatives were variously ignored, contested and co-opted by industry-friendly arrangements. I call this the "postwar settlement for American media," which kept in place a self-regulating, commercial media system based on a "corporate libertarian" arrangement that continues to shape much of the media Americans interact with today (Pickard 2015a). The media reformers would, however, succeed in advancing some progressive policies, including what would become the Fairness Doctrine, and their discursive gains would help set the stage for future victories like the creation of a public broadcasting system in the late 1960s.

LESSONS FROM THE 1940s MEDIA REFORM MOVEMENT

The rise and fall of the 1940s media reform movement is relevant today for a number of reasons. The recent collapse of commercial journalism makes the 1940s reformers' inability to effect structural change and the passed-over alternatives especially timely for reexamination. The persistence of similar crises suggests that many of our media problems are structural in nature and therefore require structural alternatives. Specifically, given the failure of the market to provide viable journalism, structural alternatives like nonprofit or worker-owned newspapers, as well as more proactive regulatory interventions and resources devoted to public media, are worthy of reconsideration (Pickard 2011a). Another parallel is that similar issues complicating new media then are reoccurring today, like questions of gatekeeping (e.g., net neutrality), corporate capture of policy discourse and regulatory agencies (as well as new forms of red-baiting), and questions of spectrum allocation and management (Meinrath and Pickard 2008; Pickard and Meinrath 2009). As the FCC and other regulatory agencies take up questions regarding the future of journalism and broadband provision, they would do well to remember that the lack of clear public interest standards can be traced back to earlier policy battles. And finally, just as a media reform movement was coalescing then, a vibrant one is emerging now, albeit in fits and starts. History can help guide this new movement away from past mistakes.

Indeed, the purpose of this research is not to mourn a lost golden age or to lament what could have been. Rather, it aims to draw linkages between previous struggles and alternative futures, to learn lessons from past failures, and to see contemporary media reform movements as part of a long historical tradition. This kind of research reminds us that our media system could have developed differently. If policy initiatives like the Blue Book had been given fair consideration, we would likely have a different media system today, one based more on public service and less on commercialism. But beyond denaturalizing the status quo, there are at least three key lessons for today's media reformers that can be gleaned from the decline of the 1940s media reform movement.

The first lesson is fairly intuitive for contemporary reformers: this history reminds us that it is imperative to maintain a strong inside/outside strategy that keeps regulators connected to the grassroots (and not only to corporate lobbyists). Had progressive policy-makers in the 1940s coordinated more with community activists, they may have been able to better withstand the ensuing onslaught. In more recent times, this vulnerability arguably recurred in the United States around net neutrality and other Internet policy issues that ultimately focused too much on a Washington, DC, strategy and failed to connect—despite noble efforts and notable exceptions—with less technocratic circles. Moreover, when liberal regulators come under pressure from industry—as they inevitably will—they will require popular support, which is difficult to maintain when compromises are being hatched behind closed doors, often in the public's name but without public consent.

Second, we learn that media activists retreat on structural reform objectives at their own peril. Postwar media reformers faced many difficulties beyond their control, but their decline also came about in part due to their failure to maintain a structural critique of the commercial media system. In the early 1940s, reformers were attempting to break up media conglomerates, but by the end of that decade they were trying to shame media corporations into being good. A structural approach recognizes that, short of public ownership, the most effective safeguard against an undemocratic commercial media system is a combination of aggressive government regulation at the federal level and local control and oversight at the community level. In light of the current struggle to prevent an overly commercial and concentrated media system from becoming even more so, it is instructive to recall a time when the FCC fought to bolster public interest safeguards instead of throwing them out. For a reinvigorated media reform movement to rise up, however, also requires an intellectual project that maintains a clear structural critique, one that penetrates to the root of the problem with a commercial media system. This structural critique could potentially unite diverse constituencies and lead to not just reform, but transformation of the media system.

Third, we are reminded that media reform rises and falls with other political struggles and social movements. Most activists are well aware that coalition-building between diverse social movements is paramount. For example, we must convince activists associated with voting rights, the movement against the carceral state, anti–death penalty campaigns, the environmentalist movement, immigrants' rights, and so on that media reform should be a central piece of their platforms. However, we also have to seek out ideologically diverse coalitions. This does not just mean linking up arms with social conservatives and libertarians to create strange bedfellow coalitions—as has happened in the recent past—around issues like excessive commercialism, media concentration, indecency, and privacy; it also means liberals should be finding common cause with more radical social movements. An open letter to the left-leaning *Nation* magazine reminded liberal reformers that they needed radicals to advance their issues, that history shows how radicals often provided punch and coherence to liberal reform agendas (Sunkara 2013). Radicals tend not to make the strategic error of retreating from structural critique and activism, nor do they lose sight of the longer struggle and the over-arching normative vision.

CONCLUSION

To summarize, what American media policy advocates should consider aiming for is a new popular front, one that unites inside-the-beltway liberals with more radical grassroots activists and intellectuals. While the DC policy liberals understand the political process and can remain focused on steering reform initiatives through legislative and regulatory channels, radicals can maintain a "big picture" structural critique, remind liberals what is at stake, and keep their eyes on the long-term vision. Bold ideas for new media policies typically exist at the margins of political discourse. It is incumbent upon activists and intellectuals—including radical scholars—to bring those alternatives to light, to challenge dominant ideologies and relationships, and to assist reform movements working toward a more just and democratic media system.

NOTES

1. Much of this chapter is adapted from Pickard (2015b).
2. For more details on the postwar media reform movement, see my discussion in Pickard (2012, 2013).
3. Other groups involved in 1940s media reform campaigns include the ACLU, Jewish organizations, and women's groups. For an interesting case study of the latter that was often pro-industry and anti-media regulation, see Proffitt (2010).

REFERENCES

Collier, R. B., and D. Collier. 1991. *Shaping the Political Arena: Critical Junctures, the Labor Movement, and Regime Dynamics in Latin America*. Princeton, NJ: Princeton University Press.

Congress of Industrial Organizations (CIO). 1944. *Radio Handbook*. New York: CIO, Political Action Committee.

Davies, D. 2006. *The Postwar Decline of American Newspapers, 1945–1965*. Westport, CT: Praeger.

Denning, M. 1996. *The Cultural Front: The Laboring of American Culture in the Twentieth Century*. London: Verso.

Fones-Wolf, E. 2006. *Waves of Opposition: Labor, Business, and the Struggle for Democratic Radio.* Urbana: University of Illinois Press.

Gonzalez, J., and J. Torres. 2011. *News for All the People: The Epic Story of Race and the American Media.* London: Verso.

Gramsci, A. 1971. *Selections from the Prison Notebooks.* New York: International Publishers.

Hall, S. 1988. *The Hard Road to Renewal.* London: Verso.

Horwitz, R. 1997. "Broadcast Reform Revisited: Reverend Everett C. Parker and the 'Standing' Case." *Communication Review* 2, no. 3: 311–348.

Kingdon, J. W. 2002. *Agendas, Alternatives, and Public Policies.* New York: Longman.

Kirchgaessner, S., P. Waldmeir, and R. Waters. 2006. "Google Action Tests Power of Cash vs. Votes in Washington." *Financial Times*, July 18.

Klinenberg, E. 2007. *Fighting for Air: The Battle to Control America's Media.* New York: Metropolitan Books.

Kumar, D. 2006. "Media, Culture, and Society: The Relevance of Marx's Dialectical Method." In *The Point Is to Change It*, edited by L. Artz, S. Macek, and D. Cloud, 72–86. New York: Peter Lang.

Lloyd, M. 2007. *Prologue to a Farce: Democracy and Communication in America.* Urbana: University of Illinois Press.

McChesney, R. 2007. *Communication Revolution.* New York: New Press.

McLaughlin, L., and V. Pickard. 2005. "What Is Bottom Up About Global Internet Governance?" *Global Media and Communication* 1, no. 3: 357–373.

Meinrath, S., and V. Pickard. 2008. "The New Network Neutrality: Criteria for Internet Freedom." *International Journal of Communication Law and Policy*, no. 12: 225–243.

Mueller, M., B. Kuerbis, and C. Page. 2004. *Reinventing Media Activism: Public Interest Advocacy in the Making of U.S. Communication-Information Policy, 1960–2002.* Syracuse, NY: Convergence Center, School of Information Studies, Syracuse University. http://ccent.syr.edu/wp-content/uploads/2014/05/reinventing.pdf.

Pickard, V. 2006. "Assessing the Radical Democracy of Indymedia: Discursive, Technical, and Institutional Constructions." *Critical Studies in Media and Communication* 21, no. 1: 19–38.

———. 2007. "Neoliberal Visions and Revisions in Global Communications Policy from the New World Information and Communication Order (NWICO) to the World Summit on the Information Society." *Journal of Communication Inquiry* 31, no. 2: 118–139.

———. 2011a. "Can Government Support the Press? Historicizing and Internationalizing a Policy Approach to the Journalism Crisis." *Communication Review* 14, no. 2: 73–95.

———. 2011b. "Revisiting the Road Not Taken: A Social Democratic Vision of the Press." In *Will the Last Reporter Please Turn Out the Lights: The Collapse of Journalism and What Can Be Done to Fix It*, edited by R. McChesney and V. Pickard, 174–184. New York: New Press.

———. 2012. "The Postwar Media Insurgency: Radio Activism from Above and Below." In *Media Interventions*, edited by K. Howley, 247–266. New York: Peter Lang.

———. 2013. "'The Air Belongs to the People': The Rise and Fall of a Postwar Radio Reform Movement." *Critical Studies in Media Communication* 30, no. 4: 307–326.

———. 2015a. *America's Battle for Media Democracy: The Triumph of Corporate Libertarianism and the Future of Media Reform.* Cambridge: Cambridge University Press.

———. 2015b. "Media Activism from Above and Below: Lessons from the 1940s American Reform Movement." *Journal of Information Policy* 5, 109–128.

Pickard, V., and S. Meinrath. 2009. "Revitalizing the Public Airwaves: Opportunistic Unlicensed Reuse of Government Spectrum." *International Journal of Communications*, no. 3: 1052–1084.

Pierson, P., and T. Skocpol. 2002. "Historical Institutionalism in Contemporary Political Science." In *Political Science: State of the Discipline*, edited by I. Katznelson and H. V. Milner, 693–721. New York: W. W. Norton.

Proffitt, J. 2010. "War, Peace, and Free Radio: The Women's National Radio Committee's Efforts to Promote Democracy, 1939–1946." *Journal of Radio and Audio Media* 17, no. 1: 2–17.

Scott, D. B. 2009. "Labor's New Deal for Journalism: The Newspaper Guild in the 1930s." PhD diss. University of Illinois at Urbana-Champaign.

Siepmann, C. 1946. *Radio's Second Chance.* Boston: Little, Brown and Co.

Skocpol, T. 1995. "Why I Am a Historical-Institutionalist." *Polity*, no. 28: 103–106.

Skocpol, T., P. B. Evans, and D. Rueschemeyer. 1985. *Bringing the State Back In.* Cambridge: Cambridge University Press.

Starr, P. 2004. *The Creation of the Media.* New York: Basic Books.

Stone, D. 2001. *Policy Paradox: The Art of Political Decision Making.* New York: Norton.

Sunkara, B. 2013. "Letter to *The Nation* from a Young Radical." *Nation*, May 21. www.thenation.com/article/174476/letter-nation-young-radical#axzz2fuJjTC5K.

Thelen, K. 1999. "Historical Institutionalism in Comparative Politics." *Annual Review of Political Science* 2, no. 1: 369–404.

Washington Post. 2012. "SOPA Petition Gets Millions of Signatures as Internet Piracy Legislation Protests Continue." January 19.

White, L. 1947. *The American Radio.* Chicago: University of Chicago Press.

CHAPTER EIGHTEEN

Policy Hacking

Citizen-based Policy-Making and Media Reform

ARNE HINTZ, *Cardiff University, United Kingdom*

MEDIA REFORM STRATEGY

Policy hacking is a media reform strategy that focuses on the citizen-based and self-organized development of policy alternatives. Rather than merely advocating for policy change, it focuses on analyzing existing and writing new laws and regulations. Two different types of policy hacking are introduced here: (a) civil society-based policy reform initiatives that identify policy gaps and regulatory challenges, and assemble new legal or regulatory packages, and (b) policy hackathons that seek to understand and upgrade policies.

Bringing together policy amateurs from both activism and the wider public, policy hacking offers new avenues for citizen engagement, crowdsourcing expertise, and the opening-up of traditionally closed political processes. Successful instances of policy hacking are often connected with traditional action repertoires and conditions, including protest and public pressure, the appropriate framing of issues, and the existence of policy windows.

Civil society–based advocacy has been a key factor for media reform. From local citizen initiatives to national campaign organizations and transnational civil society networks, nonstate actors have exerted pressure on policy-makers to respect communication rights, curb media concentration, legalize citizen and community media, reduce copyright restrictions, maintain an open Internet, and implement many other demands and concerns. Scholars have analyzed strategies and practices of advocacy, and disciplines such as policy and social movement studies have developed a significant body of theory that can explain motivations, tactics, necessary conditions, successes, and failures. They have observed how advocates use information, norms, and public pressure to convince policy-makers of a certain course of action, or try to alter their interests and align them with reform agendas. The role of advocacy has thereby been to provide input for a policy process that is controlled by established policy actors, typically in government and international institutions.

Some civil society advocates have moved beyond normative interventions and the exercise of public pressure and have instead developed regulatory proposals, written legal text, and taken advantage of openings in policy debates that allow for open, collective, and crowd-sourced development of policies. Expanding, and potentially surpassing classic forms of advocacy, they have thus taken a more active role in the policy-making process and have tried to take law-making into their own hands.

In this chapter I explore these practices and new forms of interactions between advocates and the policy process. I will trace and investigate media reform initiatives that seek to change policy by developing new regulatory frameworks, rather than exerting normative and argumentative pressure, and that engage with media policy change through citizen- and grassroots-based do-it-yourself policy-making. I call this practice *policy hacking*, and discuss its backgrounds, characteristics, and possible implications for policy, advocacy, and, more broadly, democracy.

In the first section of this chapter, I present theoretical and conceptual foundations, and discuss the context of multiactor and multilevel global governance from which policy hacking emerges. In the second section, I explore several cases of local and international policy hacking. In the final section, I analyze key characteristics and implications of policy hacking. I investigate commonalities with technological development, review potential changes to how we understand policy change and civil society's role, explore the international dimension of policy hacking, and ask whether this form of media reform activism allows for more direct democratic interventions or whether it may actually have problematic implications for democracy.

Despite using the wording of hacking and code, this chapter does not primarily concern the use of technology for policy-making, and it is not exclusively about information and communications technology (ICT) policy. Rather, it describes and analyzes a particular set of practices that we find in contemporary policy activism and that shows similarities with typical tech development practices of assembling, upgrading, and tinkering.

COMMUNICATION POLICY, ADVOCACY, AND HACKING

The study of media and communication policy addresses the regulatory rules and norms that shape the media landscape, and explores how they are created, based on which values and interests, and how they shift. It encompasses the analysis of existing legislation, but also the process of policy-making as political negotiation between a variety of actors and interests. It highlights interactions between social forces, the conditions and environments of interaction, and prevalent societal norms and ideologies that underlie and advance specific policy trends (see Freedman 2008).

National policy increasingly intersects with developments taking place at other levels and is subject to both normative and material influences by a variety of actors. Both the local and the national have "become embedded within more expansive sets of interregional relations and networks of power" (Held and McGrew 2003, 3) and policy-making is thus located at "different and sometimes overlapping levels—from the local to the supra-national and global" (Raboy and Padovani 2010, 16). Policy debates such as the World Summit on the Information Society (WSIS) and the Internet Governance Forum (IGF) have experimented with new forms of multistakeholder processes that include nonstate actors such as civil society and the business sector. The vertical, centralized, and state-based modes of traditional regulation have thus been complemented by collaborative horizontal arrangements, leading to "a complex ecology of interdependent structures" with "a vast array of formal and informal mechanisms working across a multiplicity of sites" (Raboy 2002, 6–7).

Civil society advocates have been able to use this complex environment for interventions in the "consensus mobilization" dynamics of policy debate (Khagram, Riker, and Sikkink 2002, 11). They define problems, set agendas, exert public pressure, hold institutions accountable, sometimes participate in multi-stakeholder policy development, and maintain significant leverage by lending or withdrawing legitimacy to policy goals, decisions, and processes. The mix of strategies applied by advocates to influence policymaking includes, for example, the ability to create successful conceptual frames to articulate the characteristics of an issue to policy-makers, potential allies, and the wider public. Framing an issue makes it comprehensible to the audience, attracts attention, and makes it fit with predominant perspectives in an institutional venue (Keck and Sikkink 1998). Further, advocates have created networks and collaborations across movements, both domestically and transnationally, and alliances both within and outside the institutional arena that help shift predominant institutional ideologies as well as public opinion. Policy scholars have observed that interventions are greatly enhanced by the existence of a "policy window," that is, a favorable institutional, political, and sometimes ideological setting that provides a temporary opening for affecting policy change (Kingdon 1984). International conferences can provide such a window of opportunity, as can unforeseen revelations, catastrophes, economic crises, and political change. A crisis in the social, economic, or ideological system can cause disunity among political elites and create a dynamic in which established social orders become receptive to change. "Policy monopolies"—stable configurations of policy actors—may be weakened or broken up as political constellations change and the balance of power shifts (Meyer 2005).

Strategies and action repertoires like these have been applied and observed, and favorable conditions have been exploited, across different policy fields and disciplines. However, this perspective on policy interventions has also been criticized as too limited to encompass the full range of responses by activists and civil society groups to policy windows and policy challenges. Two (related) sets of limitations concern (a) the range actors involved in this form of advocacy, and (b) the range of tactics and approaches. As for the former, classic advocacy has typically centered around the activities of large, professionalized nongovernmental organizations (NGOs) which are better resourced than smaller activist groups, enjoy easier access to policy institutions that are often highly selective in inviting civil society actors to policy debates and are tightly constraining those actors' influence on policy results (Wilkinson 2002). Social movement scholarship has been criticized for its strong focus on these forms of "organised civil society" (de Jong, Shaw, and Stammers 2005), and networked governance arrangements that include such actors in multistakeholder policy processes have been described as "neo-corporatism" (Messner and Nuscheler 2003; McLaughlin and Pickard 2005). Even at the UN WSIS summit, which was celebrated as highly inclusive and innovative, opportunities of participation for less formalized sections of civil society were limited, thereby excluding many experimental, grassroots- and activist-oriented practices of infrastructure development that form important parts of "information society" (Hintz 2009). New informal connections, loose collaborations, and temporary alliances among engaged individuals—which have been termed "connective action" (Bennett and Segerberg 2012) and "organised networks" (Lovink and Rossiter 2005)—have had little opportunity to be involved in the process, even though they are increasingly recognized as complementing "organised civil society."

Secondly, classic advocacy neglects the many forms of prefigurative action that exist in media activism and technological development. Technical communities have long engaged in latent and invisible "policy-making" by practically setting technical standards and developing new protocols and infrastructure that allow some actions and disallow others (see Braman 2006; DeNardis 2009). Media activists, equally, have focused on the creation of an alternative infrastructure that bypasses regulatory obstacles, rather than lobbying against those obstacles. Instead of campaigning for privacy rights and against online surveillance, many of them develop communication platforms (e.g., e-mail services and social media) that respect user privacy, and instead of advocating for community media legalization, many of them broadcast their own unlicensed ("pirate") radio. This prefigurative approach challenges the classic division of repertoires of action, as suggested by social movement scholars, into "insider" and "outsider" strategies (see Banaszak 2005; Tarrow 2005), where insiders interact cooperatively with power-holders through advocacy and lobbying while outsiders question the legitimacy of power-holders and address them through protest and disruptive action. Prefigurative action, in contrast, interacts with the policy environment neither inside nor outside institutional or governmental processes, but *beyond* those processes by creating alternatives to hegemonic structures and procedures and by adopting a tactical repertoire of circumvention (Hintz and Milan 2010; Milan 2013).

Hacking as it is commonly understood—as changing, manipulating, revising, or upgrading a technological system, such as a computer or wider communication infrastructure—plays a role in the *beyond* approach as it means to bypass restrictions and enable new uses. However, the more established definitions of hacking allow for a broader perspective. According to the requests for comments (RFCs) that have accompanied the development of the Internet, a computer hacker is "a person who delights in having an intimate understanding of the internal workings of a system, computers, and computer networks in particular" (Network Working Group 1993). Based on curiosity, a "positive lust to know" (Sterling 1992) and a search for innovation, a hacker thus seeks to understand a complex system and to experiment with its components, often with the aim of changing or modifying its structure. This system can be broader than a narrowly confined computer system. The trend of civic hacking, for example, has applied the notion of hacking and its practice of making, creating, and tinkering as a method to change social, cultural, and political aspects of one's environment. Beyond the realm of technological applications, science- and bio-hacking have encompassed amateur scientific experiments, and life sciences have opened up to do-it-yourself (DIY) approaches that exhibit similarities with hacker ethics (Delfanti 2013). The development of alternative licenses such as the Creative Commons license has applied this DIY practice to the legal realm (Coleman 2009), and activist practices such as culture jamming and the manipulation of meanings (e.g., of advertising) have been discussed as a form of hacking of our cultural, social, and political environment (Harold 2004). Whether the main concern is the enclosure of scientific knowledge, restrictions through proprietary software, or copyright limitations to cultural practices, questions of control and participation are at the center of hacking. Hacking thus becomes a way of discussing questions of openness and enclosure, and of exploring new ways for enhancing participation, public intervention and grassroots-based development—not just in the realm of ICT systems but in the wider reaches of society. In the following section, I take a brief journey through several localities, countries, and regions to explore whether these characteristics of hacking can be applied to the policy field.

FROM LOCAL TO TRANSNATIONAL POLICY

The first stop of our journey is the city of Hamburg in Northern Germany. As activists, in 2010, tried to investigate a scandal over public spending for a new opera house, they realized that necessary data was unobtainable and was instead stored in a data room to which only selected local government members and parliamentarians had access. Unhappy with the result, they brought together a coalition of local and national NGOs, including Transparency International, Mehr Demokratie (an initiative for participatory and direct democracy) and the Chaos Computer Club (a group of hackers and technology experts). Together, they created the Hamburg Transparency Law Initiative with the goal to have all public data, such as the contracts and financial commitments regarding the opera house, publicized and made accessible to citizens without the need for specific requests and without the possibility to keep information secret—a radical open data law for the city.

Rather than starting an advocacy campaign to put pressure on the city council to create and pass a transparency law, the members of the initiative decided to write the law themselves. Without significant legal knowledge, they approached an experienced lawyer who suggested drafting the law for a fee. However, they envisaged the policy project as a participatory process that involved citizens. So they drafted the core of a law, published it on a wiki page, and invited other civil society organizations and members of the public to comment and develop it. The final product was handed over to the local government and all political parties. After a series of negotiations and revisions, the local parliament unanimously adopted the law in June 2012 (Lentfer 2013).

Thus the initiative generated significant policy change not by merely demanding it but by developing its core components themselves, and in an open and participatory process. A key part of this endeavor was a review of best practices in other German states (*Länder*) and other European countries. Components of information laws in, for example, Berlin and Slovakia were used for the Hamburg proposal. The new law did not emerge from scratch but was based on existing experiences and incorporated established legal language. Since the successful conclusion of their campaign, the group has advised organizations and coalitions in other states and in Austria. Several of these have managed to draft laws for their specific localities, often using components from the Hamburg law, and to have them adopted by government (Lentfer 2013).

From the local space of a city council, we move to the national level, stopping first in a country with, actually, far less inhabitants than the city of Hamburg—Iceland. The financial collapse of the Icelandic economy in late 2008 offered the backdrop for the Icelandic Modern Media Initiative (IMMI) to emerge. Set up by local social and media activists, including tech developers and Internet experts, and supported by international civil society organizations, its goal was to change no less than the development model of the country which until then, had thrived as a safe haven for banks and financial services. Instead of the secrecy and the suppression of information that accompanied the old model that had become disastrous for Iceland's economy, society, and democracy, IMMI aimed at transforming Iceland into a transparency haven and a favorable environment for media and investigative journalism.

As in Hamburg, the strategy was not merely to demand but to develop new policy. IMMI created a bundle of legal and regulatory proposals, with the aim to "protect and strengthen modern freedom of expression" (IMMI 2012). These included the development of a new Freedom of Information Act to enhance access for journalists and the public to government-held information and to end the previous culture of secrecy, and measures to limit libel tourism, prior restraint, and strategic law suits that serve to block legitimate information. The group also initiated a new law on source protection, making it illegal for media organizations to expose the identity of sources for articles, books, and other formats, if the source or the author requested anonymity. IMMI has developed policy proposals on whistleblower and intermediary protection, it has made proposals on safeguarding net neutrality, and IMMI activists have engaged with debates on the European Data Retention Directive and, more broadly, surveillance (Gudmundsson 2011). IMMI was adopted in principle by the Icelandic Parliament in 2010, and many of its components have been implemented since then. The IMMI package, if fully implemented, would provide a legal environment that protects not just national but potentially international publishers. All information originating from (or routed through) Iceland would be governed by the new set of laws and therefore be more difficult to suppress. Blogs, websites, and all kinds of online publications would thereby fall under Icelandic jurisdiction and be less vulnerable to censorship (Bollier 2010).

The practice of cherry-picking laws and regulations from other jurisdictions, which we have seen in Hamburg, was implemented more excessively in Iceland. IMMI created a puzzle of tried-and-tested components, including, for example, parts of the Belgian source protection law, the Norwegian Freedom of Information Act, Swedish laws on print regulation and electronic commerce, the EU Privacy Directive, the New York Libel Terrorism Act, or the Constitution of Georgia (IMMI 2012). It was supported by a wide-open window of opportunity, as the economic breakdown had affected large parts of the population and the secrecy of banks was widely debated and criticized. A significant section of the public was in favor of a radically new model, parts of the old political class were delegitimized, thus traditional policy monopolies were broken, new social actors were swept into politics and helped facilitate the adoption of the new laws in Parliament.

While the previous two examples were inspired by the possibilities of new digital media, similar practices exist in relation to a more traditional broadcast environment. The Argentinian Coalition for Democratic Broadcasting emerged in 2004 as a civil society coalition to campaign for a new national audiovisual media law and to replace the old law which stemmed from the times of the military dictatorship. Bringing together unions, universities, human rights groups, and community media, the coalition developed a set of key demands to serve as core pieces of a new policy framework, and a coalition member and university professor was charged by the government to draft a law based on this input. The first draft was discussed at twenty-eight open hearings, comments by civil society groups were included in the document, and a demonstration of twenty thousand people brought the final text to Parliament where it was adopted, making it a true "law of the people" (Loreti 2011).

The resulting Law 26.522 on Audio-Visual Communication Services, adopted in 2009, broke radically with established policy traditions. In particular, the law recognizes three sectors of broadcasting—state, commercial, and not-for-profit—and guarantees a 33 percent share of the radio frequency spectrum for each sector, making it "one of the best references of regulatory frameworks to curtail media concentration and promote and guarantee diversity and pluralism," according to World Association of Community Radio Broadcasters (AMARC 2010). Applying similar strategies as the initiatives in the previous two cases, the coalition did not write the new policy framework from scratch but drew heavily on existing legal frameworks and policy documents, both from international institutions (for example, the UNESCO Media Development Indicators and the UN Rapporteur on Freedom of Expression) and from civil society organizations (such as a model legal framework developed by AMARC called Principles on Democratic Regulation of Community Broadcasting [2008]). Just as the new law in Hamburg has inspired initiatives elsewhere, the Argentinian law has become a model law for the region. Groups and coalitions elsewhere have tried to replicate this model in their countries, thus influencing legislative development across the region (see Klinger 2011).

Neighboring Uruguay has hosted similar policy innovations, created through similar processes. A new law that legalized community broadcasting was based on a draft submitted by civil society groups. Its adoption was influenced by significant pressure from a broad civil society coalition of media, labor, educators, and human rights organizations, and by AMARC's advocacy work. As in Argentina, community media advocates crossed the lines between inside and outside the policy-making realm: AMARC expert Gustavo Gomez became (temporarily) the National Director of Telecommunication (Light 2011), while the new regulatory advisory body COFECOM in Argentina elected Nestor Busso, a representative of community radios, as its president (Mauersberger

2011). The policy development processes in both countries interacted with each other as Argentine activists integrated parts of the earlier Uruguayan reform into their proposal for a new comprehensive law and Uruguayan activists, in turn, used that law as a model for further reforms.

Regional coalitions have sometimes applied the same practices as national groups. When the issue of net neutrality came onto the European agenda, a network of advocacy groups picked up the idea of developing a model law. The net neutrality debate had a slow start in Europe, with the European Commission initially rejecting the need for policy interventions, but gained momentum in 2011 when the Council of Europe drafted a declaration in support of Internet freedoms, including net neutrality (Council of Europe 2011). In 2012 the Dutch parliament passed a law in favor of net neutrality, while telecom operators like the German Telekom announced they would start managing certain types of online traffic. At the Chaos Communication Congress in Germany in December 2012, members of digital rights groups, including organizations such as the Dutch Bits of Freedom and the French La Quadrature du Net, came together to discuss what to do. Several options were debated, including a Europe-wide coordinated public campaign, but what materialized in the aftermath of the congress was an attempt to draft a model law and promote its adoption by European governments. By June 2013, the loose network developed a comprehensive inventory of (actual and proposed) legislation and soft law on net neutrality in countries such as Norway, Chile, Netherlands, Slovenia, Korea, Peru, and Finland as well as a set of principles for a model net neutrality law through an open process and the use of wiki pages and etherpads.[1] Utilizing the legal precedents and normative principles adapted from other countries, the idea was to draft a new regulation and to advocate for its implementation in each EU member state.

In July 2013, the new Dynamic Coalition on Net Neutrality that emerged within the Internet Governance Forum (IGF) picked up the project and moved it from the European to the global level. By the end of the year, it developed a "policy blueprint" which served as a model legal framework and was adaptable to different jurisdictions (Belli and van Bergen 2013). The European activists meanwhile redirected their efforts from policy development back to classic advocacy as the EU Commission presented its own proposed net neutrality law (EDRI 2013; Baker 2013).

POLICY HACKATHONS

While the aspect of hacking in the examples presented is rather indirect and figurative (more on that below), more direct connections between technological practices and policy exist in the form of policy-related *hackathons*. As hackathons—meetings of computer programmers and others involved in technological development aimed at solving specific technological problems and developing software applications—have become ever more frequent and widespread, they have increasingly targeted social and political concerns. Events such as Hack the Government have developed open data and transparency tools for government, while the Education Hack Day served to develop solutions for education problems, and Random Hacks of Kindness is devoted to disaster management and crisis response.[2] Policy hackathons bring together interested people (often, but not always, with tech skills) to analyze and, ideally, improve policies. The open source development company Mozilla, for example, has organized several hackathons to analyze and improve the privacy policies of websites. The Brooklyn Law Incubator and Policy Legal Hackathon (BLIP) brought together advocates, lawyers, and technologists in April 2012 to "explore how technology can improve the law and society, and, conversely, how law can improve technology." In addition to presentations and debates, BLIP encouraged participants to tackle issues of IP policy and propose policy reform through collaborative online tools.[3] On the other side of the Atlantic, a broad network of civil society groups including the European Digital Rights Initiative (EDRI), the Electronic Frontier Foundation, and Transparency International have teamed up with leading Internet companies like Google and Facebook to hold an annual EU Hackathon in Brussels. The series has included events such as "Hack4YourRights" in 2013 which brought together thirty participants to work on "data sets from network analysis, corporate Transparency Reports and freedom of information (FOI) requests [and] create apps and visualizations that shed light on the state of government surveillance in their country."[4]

Policy hackathons are distinct from the case studies described earlier as they constitute temporary events rather than sustained campaigns, typically focus on policy analysis rather than policy change and the development of alternative legislative proposals, and are often hosted or cohosted by companies, rather than exclusively by civil society coalitions. However, they share an approach that focuses on policy texts and artefacts, and aims to manipulate policy. All the cases highlighted imply the claim that citizens, civil society activists, and technological developers can be policy experts and are willing and able to take policy development into their own hands.

CHARACTERISTICS AND IMPLICATIONS

The examples described in the previous sections share a specific approach to advocacy that focuses on analyzing and rewriting existing policies and developing new laws. In some of the cases, this practice is combined with traditional campaigning and even public protest (such as the mass demonstration in Argentina), but at its center lies an attempt at citizen-based policy development and law-making. This is significant as it surpasses civil society's traditional role as advocator and provider of information and values, and puts media reform initiatives instead in the driving seat of the policy process. It is based on the prefigurative modes of action which we find in media activism and technological development: the creation of alternative standards and infrastructures. Moreover, it is often based on temporary and informal alliances and thereby relates to newer organizational forms of connective action and organized networks.

The cases of citizen-based policy-making which were presented here display several connections with the practice of hacking. First, the focus on DIY self-production is wide-spread in tech activism. Second, the specific practice of assembling policy precedents from around the world, repackaging them, experimenting with legal "code," and manipulating, improving, and upgrading it bears strong similarities with technological development. Most of the initiatives highlighted here started from an inventory of existing laws and regulations in other countries, picked out the components that seemed useful for the particular local situation, and reassembled them to a new legislative or regulatory package. Third, while technical tools do not always play the central role that they do in policy hackathons, the use of wikis and other open platforms to crowd-source policy expertise and contributions has been consistent and wide-spread. Fourth, the initiatives and coalitions presented here connect legal and technical communities and combine legal and technical expertise. Those two sets of expertise are closely related as attempts at "tinkering" with technology and with the law require similar skills and forms of reasoning (Coleman 2009).

These forms of policy hacking may allow us to review key questions of policy studies, such as policy diffusion and transfer. Traditionally, the spread of policies from one jurisdiction to another, and the transfer of rules and regulations across boundaries, are analyzed in the context of governmental actions and institutional processes. However, in the cases presented here, we can see that civil society groups have advanced the movement and distribution of policies. They have cherry-picked regulations from other countries, and their own legal "products" have become models that came to influence legislative development elsewhere. In Hamburg, Iceland, Argentina, and elsewhere, these groups have followed up their domestic policy development by advising civil society campaigns and policy-makers in other countries (or, in the Hamburg case, other states) and have, in some cases, triggered policy change elsewhere. They have thus provided examples for policy transfer and diffusion that is citizen-led, not government-led.

Grassroots policy development of the kind described here offers an enhanced role for citizens and civil society in political processes. It may compromise established processes of representative democracy as it questions the role of elected officials and intervenes into their prescribed area of authority. However, it opens up new channels for participatory engagement by citizens and for direct forms of democracy. Involving loose networks, informal coalitions, and concerned individuals, it relates to more recent ideas of digital democracy and liquid democracy. Moving beyond the classic focus on political parties and representative democratic processes, these democratic forms enable broader participatory processes, allow for interactive and flexible decision-making, enhance active citizenship, and make the public sphere more inclusive (with varying success, see Hindman 2008).

However, even if policy hacking opens new avenues for participation and intervention, it contains several potentially problematic aspects that lie in its connection with the debates and action repertoires of civic and technological hacking and of digital democracy. Civic hacking has been hailed as welcome involvement of experts and creative youth to solve important problems in government, public service, community life, and disaster response, but it has also been criticized as cheap outsourcing of state functions and as questionable public relations exercises (Baraniuk 2013). Tools and practices of digital democracy have often empowered elites and small groups of experts rather than the broader citizenry (Hindman 2008). Perhaps most significantly, the goals of hacking (both technological and civic) are typically to improve, fix, and solve a discrete problem, rather than the recognition of power imbalances, battles between different interests, and the political dimension of policy debates. A hack is "an elegant solution to a technological problem" (Levy, quoted in Meikle 2002, 164). With critics like Evgeny Morozov, however, we may question the reductionist "solutionism" of an approach that focuses on solving discrete problems, rather than understanding (and tackling) the deeper roots and broader contexts of an issue (Morozov 2013). The cases of policy hacking presented here largely follow this route as they focus on improving policy and fixing specific issues within a given and largely unquestioned political context. As civil society advocacy in the field of media and communication policy adopts policy-hacking practices, it exploits promising new paths for media reform, but it prioritizes a problem-solving approach over a more fundamental political and social struggle.

CONCLUSION

Policy hacking is a media reform approach that focuses on citizen-based DIY creation of concrete policy alternatives. It involves identifying policy gaps and regulatory challenges, analyzing existing policies that may address those, and reassembling these components toward a new legal or regulatory package. Its practice displays similarities with technological development and hacking. The examples presented in this chapter have demonstrated that it can be a viable and successful strategy for media and communications reform.

However, policy hacking is not entirely distinct from other established advocacy strategies and experiences. Text work and proposals of policy language have long been part of advocates' repertoires of action. From national policy reform to lobbying at a UN summit such as WSIS, civil society activists have invested significant time and effort into changing the wording of legislative and regulatory documents, and have developed alternative declarations (see Hintz 2009). The initiatives presented in this chapter have moved significantly beyond these practices by developing complete legislative proposals, but they are based on a similar approach. Also, policy hacking has been embedded in a broader range of traditional action repertoires, including protest and public pressure (as in Argentina), placing champions inside the relevant institutional setting (as in Iceland), and sophisticated framing exercises that communicated the need for policy change to government members, regulators, and parliamentarians.

In all cases, the existence of policy windows was crucial to the success of policy hacking and the adoption and implementation of DIY policy. In Iceland and Argentina, policy change was possible in response to economic crises, the failure of established economic models, and a legitimacy crisis of the political class; and in Hamburg, a public spending scandal (as well as the steep rise of the Pirate Party in the polls) helped convince political leaders that change was necessary. However, both examples have also shown that windows of opportunity can close again and conditions for policy change can deteriorate quickly. Policy hacking is thus a useful and promising strategy in specific historical moments; but as all advocacy strategies, it is context specific.

If we understand reform as a democratic practice, policy hacking provides an interesting perspective as it offers new avenues for citizen engagement, mechanisms for crowd-sourcing expertise, and the opening up of traditionally closed political processes, particularly for the less organized and less institutionalized parts of civil society. Its grassroots orientation, however, does not necessarily challenge established configurations of political power, as its problem-solving approach and its focus on improving and fixing specific issues distinguishes it from wider social struggles.

Considering these different concerns, from the need for policy windows to democratic effects, policy hacking may be limited as a stand-alone strategy. Yet as part of a broader strategic and tactical mix, it can play a significant role in future efforts to affect both democratic change and media reform.

NOTES

1. On net neutrality in countries such as Norway, Chile, Netherlands, Slovenia, Korea, Peru and Finland, see Overview of Net Neutrality Regulations, https://wiki.laquadrature.net/Overview_of_Net_Neutrality_Regulations. For more details on a set of principles for a model net neutrality law, see https://quadpad.lqdn.fr/XKTMzQ5Yg2.

2. On Hack the Government, see http://nationalhackthegovernment.word press.com/; on Education Hack Day, see http://educationhackday.org/; and on Random Hacks of Kindness, see www.rhok.org/.

3. See BLIP Legal Hackathon, http://legalhackathon.blipclinic.org/.
4. See EUHackathon, http://2013.euhackathon.eu/.

REFERENCES

AMARC. 2008. *Principles on Democratic Regulation of Community Broadcasting.* www.amarc.org/documents/14principles_EN.pdf.
———. 2010. "AMARC Deplores Suspension if New Communication Law in Argentina." *AMARC link* 13, no. 1 (January–March): 10. http://podcast .amarc.org/amarc_link/amarc_link_AVRIL_2010_EN_final.pdf.
Baker, J. 2013. "Digital Rights Activist: EU's Proposed Net Neutrality Law 'as useful as an umbrella in a hurricane.'" *IDG News Service*, September 11. www.pcworld.com/article/2048563/digital-rights-activist-eus-proposed-net -neutrality-law-as-useful-as-an-umbrella-in-a-hurricane.html.
Banaszak, L. A. 2005. "Inside and Outside the State: Movement Insider Status, Tactics, and Public Policy Achievements." In *Routing the Opposition: Social Movements, Public Policy, and Democracy*, edited by D. S. Meyer, V. Jenness, H. Ingram, 149–176. Minneapolis: University of Minnesota Press.
Baraniuk, C. 2013. "Civic Hackers: Techies Volunteer to Rescue Government." *New Scientist*, June 29. www.newscientist.com/article/mg21829232-000 -civic-hackers-techies-volunteer-to-rescue-government/.
Belli, L., and M. van Bergen. 2013. "A Discourse-Principle Approach to Network Neutrality: A Model Framework and Its Application." Network Neutrality Dynamic Coalition. http://networkneutrality.info/sources.html.
Bennett, L., and A. Segerberg. 2012. "The Logic of Connective Action." *Information, Communication and Society* 15, no. 5: 739–768.
Bollier, D. 2010. "A New Global Landmark for Free Speech." June 16. www.bollier .org/new-global-landmark-free-speech.
Braman, S. 2006. *Change of State: Information, Policy, and Power.* Cambridge: MIT Press.
Coleman, G. 2009. "Code Is Speech: Legal Tinkering, Expertise, and Protest Among Free and Open Source Software Developers." *Cultural Anthropology* 24, no. 3: 420–454.
Council of Europe. 2011. Internet Governance Principles. Council of Europe. https://wcd.coe.int/ViewDoc.jsp?id=1835773.
de Jong, W., M. Shaw, and N. Stammers. 2005. Introduction. In *Global activism, global media*, edited by W. de Jong, M. Shaw, and N. Stammers, 1–14. London: Pluto Press.
Delfanti, A. 2013. *Biohackers: The Politics of Open Science.* London: Pluto Press.
DeNardis, L. 2009. *Protocol Politics: The Globalization of Internet Governance.* Cambridge: MIT Press.
European Digital Rights Initiative (EDRI). 2013. "Leaked Regulation: Schrödinger's Net Neutrality on Its Way in Europe." *EDRI-gram*, July 11. www.edri.org/ schroedinger-NN.

Freedman, D. 2008. *The Politics of Media Policy.* London: Polity Press.

Gudmundsson, G. R. 2011. Research interview by Arne Hintz, Reykjavik, August 20.

Harold, C. 2004. Pranking Rhetoric: Culture Jamming as Media Activism. *Critical Studies in Media Communication* 21, no. 3: 189–211.

Held, D., and A. McGrew. 2003. "The Great Globalization Debate." In *The Global Transformations Reader*, edited by D. Held and A. G. McGrew, 1–50. Cambridge: Polity Press.

Hindman, M. 2008. *The Myth of Digital Democracy.* Princeton, NJ: Princeton University Press.

Hintz, A. 2009. *Civil Society Media and Global Governance: Intervening into the World Summit on the Information Society.* Münster: Lit.

Hintz, A., and S. Milan. 2010. Media Activists and the Communication Policy Process. In *Encyclopedia of Social Movement Media*, edited by J. Downing, 317–319. London: Sage.

Icelandic Modern Media Initiative (IMMI). 2012. International Modern Media Institute. https://immi.is.

Keck, M. E., and K. Sikkink. 1998. *Activists Beyond Borders. Advocacy Networks in International Politics.* Ithaca: Cornell University Press.

Khagram, S., J. V. Riker, and K. Sikkink. 2002. "From Santiago to Seattle: Transnational Advocacy Groups Restructuring World Politics." In *Restructuring World Politics: Transnational Social Movements, Networks, and Norms*, edited by S. Khagram, J. V. Riker, K. Sikkink, 3–23. Minneapolis: University of Minnesota Press.

Kingdon, J. W. 1984. *Agendas, Alternatives, and Public Policy.* Boston, MA: Little Brown.

Klinger, U. 2011. "Democratizing Media Policy: Community Radios in Mexico and Latin America." *Journal of Latin American Communication Research* 1, no. 2: 4–22.

Lentfer, D. 2013. Research interview by Arne Hintz, Hamburg, May 16.

Light, E. 2011. "From Pirates to Partners: The Legalization of Community Radio in Uruguay." *Canadian Journal of Communication* 36, no. 1: 51–67.

Loreti, D. 2011. Research interview by Arne Hintz, Montreal, February 11.

Lovink, G., and N. Rossiter. 2005. "Dawn of the Organised Networks." *Fibreculture Journal*, no. 5. http://journal.fibreculture.org/issue5/lovink_rossiter.html.

Mauersberger, C. 2011. "Whose Voice Gets on Air? the Role of Community Radio and Recent Reforms to Democratize Media Markets in Uruguay, Chile and Argentina." *Journal of Latin American Communication Research* 1, no. 2: 23–47.

McLaughlin, L., and V. Pickard. 2005. "What Is Bottom-Up About Global Internet Governance?" *Global Media and Communication* 1, no. 3: 357–373.

Meikle, G. 2002. *Future Active: Media Activism and the Internet.* New York: Routledge.

Messner, D., and F. Nuscheler. 2003. *Das Konzept global governance: Stand und Perspektiven*. INEF-Report 67. Duisburg: Institut für Entwicklung und Frieden.

Meyer, D. S. 2005. "Social Movements and Public Policy: Eggs, Chicken, and Theory." In *Routing the opposition. Social Movements, Public Policy, and Democracy*, edited by D. S. Meyer, V. Jenness, and H. Ingram, 1–26. Minneapolis: University of Minnesota Press.

Milan. S. 2013. *Social Movements and Their Technologies: Wiring Social Change*. London: Palgrave Macmillan.

Morozov, E. 2013. *To Save Everything, Click Here: The Folly of Technological Solutionism*. New York: Public Affairs.

Network Working Group. 1993. *Request for Comments: 1392. Internet Users' Glossary*. http://tools.ietf.org/html/rfc1392.

Raboy, M. 2002. *Global Media Policy in the New Millennium*. Luton: University of Luton Press.

Raboy, M., and C. Padovani. 2010. *Mapping Global Media Policy: Concepts, Frameworks, Methods*. http://www.globalmediapolicy.net/sites/default/files/Raboy&Padovani%202010_long%20version_final.pdf.

Sterling, B. 1992. *The Hacker Crackdown: Law and Disorder on the Electronic Frontier*. New York: Bantam Books.

Tarrow, S. 2005. *The New Transnational Activism*. New York: Cambridge University Press.

Wilkinson, R. 2002. "The Contours of Courtship: The WTO and Civil Society." In *Global Governance: Critical Perspectives*, edited by R. Wilkinson and S. Hughes, 193–211. London: Routledge.

CHAPTER NINETEEN

Reforming or Conforming?

The Contribution of Communication Studies to Media Policy in Switzerland

MANUEL PUPPIS, *University Of Fribourg*
MATTHIAS KÜNZLER, *University of Applied Sciences HTW Chur*

MEDIA REFORM STRATEGY

For over forty years, communication studies in Switzerland has been involved in media policy-making. While these activities helped in institutionalizing communication as an academic discipline and provided policy-makers with much needed knowledge that is unaffected by vested interests, results show that most studies fell short of driving reforms that aim at more diverse and participatory media systems. Instead of suggesting progressive reforms, many scholars' recommendations conformed to interests of political and media elites. The current media crisis—which led to new research devoted to critical analyses of ownership concentration, diversity, and funding options for media as well as to the formation of a new federal media commission—should be seen as a chance to reform the media system. Direct subsidies to new independent online news organizations, strong support for public service broadcasting and restrictions of further consolidation of media ownership promise to conduce to a media system that is able to fulfill its function for democracy.

Changes and crises in the media sector—both past and present—increase policy-makers' demands for insights on how to reform media regulation. For scholars this opens up possibilities of bringing forward ideas for more diverse and participatory media systems. However, the areas of study in media policy research are often shaped by politicians' needs, making funding available only for certain topics (Verhulst and Price 2008). Moreover, policy-makers are less interested in analyses that are too far outside the range of what they deem politically feasible (Braman 2003; Haight 1983). This, in turn, may induce scholars to narrow their analysis to the politically feasible.

Apparently, the relationship between communication policy research and communication policy-making is a complex one (Just and Puppis 2012). The involvement of researchers in the political process or the consideration of research results do not necessarily advance media reform. This raises the questions of what role scholarly research plays in shaping media policy and whether it spurs reform or conforms to prevalent conceptions of political and media elites.

This chapter presents an empirical analysis of the contribution of communication studies to media policy in Switzerland since the 1970s. In line with more recent publications claiming that research plays a meaningful role in policy-making (see Just and Puppis 2012; Napoli and Gillis 2006), we show that input from academia was indeed sought by policy-makers. Scholars have been involved in federal commissions and were charged with performing research accompanying new media developments (Bonfadelli and Meier 2007; Padrutt 1977; Saxer 1989, 1993). Many of their recommendations have been implemented in revisions of the Swiss Radio and Television Act. Yet it is necessary to qualify this positive assessment. Many recommendations made by scholars were not progressive and failed to challenge the interests of politicians and media owners. Proposals that called for more radical change (for example, concerning ownership regulation or press subsidies) did not succeed in getting political support. Consequently, despite being influential in media policy-making, communication scholarship has, thus far, mostly fallen short of driving progressive media reform.

In this chapter, we start by briefly discussing the role of communication and media policy research for communication and media policy-making and then present our results. After proposing five different phases of scholarly involvement in media policy-making in Switzerland, we focus on some reasons for the apparent demand of the political system for academic input and on the impact of media policy research both on policy-making and on the discipline itself. The final section then critically assesses the contribution of communication studies so far and argues that the funding opportunities offered by administrative research should be used by scholars as a chance to advance a reform of the media system.

COMMUNICATION POLICY RESEARCH AND POLICY-MAKING

In the last few decades, Western media systems have experienced massive changes. Switzerland is no exception. In the late 1960s and early 1970s the number of newspapers declined sharply, raising fears of the consequences of this development for democracy. In the 1980s, private broadcasting was cautiously introduced, which led to more cost-effective production principles and changes in up- and downstream markets. And in the last ten years, traditional media organizations have had to come to grips with digitization and the rise of the Internet. These latest developments have not only posed questions for the future development of public service broadcasting but also triggered a crisis of funding in the newspaper sector. On the one hand, in response to media convergence, the public service broadcaster SRG SSR implemented a complete reorganization which brought about the merger of previously separated subsidiaries for radio and television as well as the integration of newsrooms with the aim of being able to produce media content for radio, television, and the Internet. On the other hand, newspaper publishers are struggling to find new business models and to implement so-called pay-walls for their online content (Jarren, Künzler, and Puppis 2012; Künzler 2009; Meier, Bonfadelli, and Trappel 2012; Puppis and Künzler 2011).

Given these developments, policy-makers regularly require fresh insights for reforming regulation. Today, it seems, the demand for research in communication policy-making is greater than ever. This is also "due to the changing nature of the questions being asked by policymakers . . . and a growing recognition within the policymaking community of the limitations of economic analysis in answering important communications policy questions" (Napoli and Gillis 2006, 672). For scholars this opens up possibilities of bringing forward ideas for more diverse and participatory media systems. Hence, as Just and Puppis (2012) put it, communication policy researchers should be more self-confident when it comes to their contribution to policy-making.

Obviously informing the policy process also involves pitfalls (see Just and Puppis [2012]). To begin with, it can be very tempting to be involved in questions of public policy (Katz 1979, 85; Pool 1974, 40). Second, because critical analyses are not always welcome and will not receive funding from policy-making institutions, scholars may be inclined to narrow their analysis to the politically feasible. They might feel the need to limit themselves to a more pragmatic selection of subjects and approaches in order to meet the perceived needs of policy-makers (Corner, Schlesinger, and Silverstone 1997, 7–8; Freedman 2008, 102; Haight 1983, 231–232; Melody and Mansell 1983, 111). Finally, scholars should be aware that policy-makers might use research to legitimize decisions that have already been made: "It is not hard to think of situations in which evidence is adduced *after* a policy decision has already been made; or in which decisions are made regardless of the evidence, and the evidence is 'spun' in order to support a predetermined course of action" (Buckingham 2009, 204).

But the research field has too much to offer to stay silent. First of all, it can help in understanding change and in deciding how to address a policy problem. Without research, policy decisions are not as informed as they should and could be (Braman 2003, 6). Moreover, given its greater independence from vested interests, academic research is in a unique position to focus on issues that go beyond the normal short-term horizons of policy-makers (Melody 1990, 33). True, policy-makers are less interested in analyses that are too far outside the range of what is deemed politically feasible (Haight 1983, 230–231). Yet ideas matter a great deal. Scholars are in a unique position to propose inconvenient ideas and to expand the range of possibilities contemplated by policy-makers (Braman 2003, 6).

The question then is whether communication policy research lives up to these noble expectations. Given that previous research in Switzerland and elsewhere scrutinizes ownership concentration, commercialization, and its implications for media performance, one could expect that scholars might try to present policy proposals that address these issues. After all, the crucible of communication policy research is the advocacy of policy in the direction indicated by research results (Melody and Mansell 1983, 133). Looking at the case of Switzerland, this paper thus investigates the role that scholarly research plays in shaping media policy and whether it spurs reform or conforms to prevalent conceptions of political and media elites.

FORTY YEARS OF CONTRIBUTING TO MEDIA POLICY-MAKING

In order to examine the role of communication and media policy research in shaping media policy, we undertook a qualitative analysis of documents and semi-structured interviews.

METHODS

The document analysis was based both on publicly available research reports, studies, and papers of scholars involved in advising media policy and on articles about the state of media policy research in Switzerland. A number of these documents were written for federal commissions and the Swiss broadcasting regulator (e.g., EJPD 1975, 1982; Saxer 1983, 1989) or are critical reflections of scholars who were involved in media policy research themselves (Bonfadelli and Meier 2007; Padrutt 1977; Saxer 1993). These documents mainly offer information about the subjects of communication and media policy research, the methods employed and policy proposals formulated by scholars. In addition to this document analysis, nine semi-structured interviews with scholars and policy-makers provide additional insights into the motives of scholars in advising policy-makers as well as their perception by politicians and the regulatory agency. In order to collect rich data and account for different perspectives, a number of individuals who were involved in media policy-making and media policy research were interviewed; the interviewees included six scholars (three of them retired) who provided expertise to policy-makers, a representative of

the regulatory authority for broadcasting and telecommunications (OFCOM), and two members of parliament (a representative of a left-wing and a right-wing-party each). Both the documents and the interviews were analyzed by combining a deductive and an inductive method of content categorization (Kvale and Brinkman 2009; Mason 2002; Mayring 2010).

PHASES OF SCHOLARLY INVOLVEMENT IN MEDIA POLICY-MAKING

Since the 1970s, when media policy-makers started to involve communication scholars, the research field of communication and media policy research in Switzerland has developed significantly. Over the years, not only the number of studies has been increasing but also the variety of the subjects, theories, methods, and scholars involved. Based on the document analysis, this development of the research field may be classified into five different phases.

Phase 1: Extraparliamentary Federal Commissions (1973–1982). In the first phase, the input of communication studies mainly took the form of researchers being appointed to and performing research for extraparliamentary federal commissions. In the 1970s, the Swiss government established two extraparliamentry commissions dealing with the development of the media sector: the expert commission for press subsidies and the expert commission for a holistic approach to the media.

The former was a response to the increasing death rate of newspapers in the late 1960s and early 1970s due to a transition from traditional party newspapers to a nonpartisan press (Padrutt 1977, 34). Politicians across parties were afraid that the shrinking number of paid-for dailies would have negative implications for a well-functioning democracy. Several parliamentary proposals called for a more active press policy including direct subsidies. Thus, the commission was asked to look into the economic situation of the press and to evaluate possible subsidy schemes. Aside from representatives of newspaper proprietors, journalists, public service broadcasting, and the administration, researchers from media law, constitutional law, and communication studies were asked to join the commission. The commission also gave research projects to scholars (EJPD 1975, 9–10; Schade 2005, 30–31). In its 1975 report, the commission offered a series of recommendations for introducing press subsidies. However, the subsequent public consultation demonstrated that the subject matter was highly controversial, leading the government to abandon any attempts at implementing a more active press policy (Schade 2006, 265).

This failure to reform press policy was only one reason for the establishment of the second extraparliamentary commission in 1978. Another reason was that the Swiss government lost a referendum on the passage of a new constitutional article which constituted a prerequisite for allowing private broadcasting. Yet at the same time, pirate radio stations were spreading, increasing the pressure for liberalization. With governmental media policy in shambles, the new commission was briefed to analyze the state of the Swiss media system and to develop ideas for future regulation. As with the previous commission, representatives of publishers, journalists, and the administration as well as law and communication studies were invited to join. In the final report published in 1982 the commission proposed a three-level model for broadcasting (private broadcasting only allowed on the regional and international level; national level reserved for the public service broadcaster) which was eventually adopted when the government provisionally allowed private radio and television in the same year for a trial period (Meier, 1993). In contrast to broadcasting, no reform of press regulation was implemented.

Phase 2: Evaluation of Private Broadcasting Trial Period (1983–1991). After the government decided to try out private broadcasting in order to gain more experience and prepare for definitive liberalization, it became also apparent that communication studies would be needed to evaluate the consequences of these new channels for the media system and Swiss society (Saxer 1983, iii). The multiyear trial period was accompanied by researchers analyzing the content, use, and organizational structure of the new private radio and television stations. In the final report, the researchers acknowledged that there was demand for private broadcasters but also made suggestions for making changes to the regulatory environment (Saxer 1989).

Phase 3: Research Upon Application without Purpose of Consultancy (1992–2006). After the first Radio and Television Act (RTVA) which liberalized broadcasting came into force in 1992, the new regulatory agency OFCOM also started to fund research. Every year OFCOM invited tenders for research projects. These projects mostly did not have any direct use for the regulator and can thus be classified as not serving the specific purpose of providing consultancy to the administration. On average six research projects were supported each year.

Phase 4: Evaluation of Program Quality (since 2007). With the revised RTVA, OFCOM was assigned the task of evaluating the program and the quality management of licensed regional private broadcasters with a program remit. Thus, on the one hand, the organizational structures and practices of these broadcasters are regularly evaluated by independent companies that are accredited by OFCOM. On the other hand, and more importantly for communication research, OFCOM commissioned communication scholars with performing a content analysis of these broadcasters' news programs. Additionally, OFCOM also regularly commissions a survey investigating the satisfaction of viewers and listeners with Swiss media as well as an evaluation of the public broadcaster's online activities (Dumermuth 2012).

Phase 5: Research Upon Application with Purpose of Consultancy (since 2010). Since 2010 OFCOM is again funding research upon application (Dumermuth 2012). Yet the usefulness of such research projects for strategic decisions and the preparation of new regulation has come to the fore. For instance, following parliamentary postulates that demanded answers from government regarding the current crisis of media, OFCOM put several research projects out to tender in 2010. These projects had the task of investigating press and media diversity from various perspectives in order to deliver the government a basis for its own report (Bundesrat 2011). The research reports revealed a nuanced picture of the changing media landscape in Switzerland and suggested political action for dealing with the financial problems of news media. While the government in its 2011 report clearly stated that it agrees with the analysis in the scientific reports, it nevertheless decided to take no action, wait for the industry to find a solution and have another look at the situation in four years' time. Members of parliament were less than amused and submitted yet another motion in summer 2012, demanding that the government reports on the option for future media subsidies. Since then, OFCOM has commissioned a comparative analyses of media subsidies and public service broadcasting in eighteen different media systems in which the authors of this chapter were heavily involved. Moreover, the government installed a new federal media commission to consult on the future of the Swiss media system. This commission is chaired by a communication scholar and two more scholars are members. It still remains to be seen whether these new initiatives are just about delaying action or actually about preparing meaningful reform.

REASONS FOR THE INCREASING DEMAND FOR ACADEMIC INVOLVEMENT

The document analysis and the interviews clearly show that policy-makers actively sought input from academia; scholars have been involved in federal commissions and have been charged with performing research accompanying new media developments. Several reasons for this involvement can be distinguished.

One of the main reasons was that media policy-makers were in need of expert knowledge and fresh ideas on how to deal with changes, be it press concentration in the 1960s and 1970s, the introduction of private broadcasting in the 1980s, or the media crisis of our days. As one of the scholars interviewed points out:

> Members of Parliament had only little knowledge about how media policy works. They simply lacked expertise. Yet the introduction of teletext, private radio and television, and later of the Internet represented innovations which caused uncertainty regarding future developments. Politicians were interested in academic expertise in order to make sure that their decisions were suitable for the problems at hand.

Another important reason for involving communication scholars was that members of parliament, government officials, and the Swiss broadcasting regulator appreciate what they consider independent advice. As one scholar puts it: "For politicians, scientific knowledge provides a great opportunity. Communication research offers them a source of knowledge to draw on which is more independent from vested interests of the media industries." This assessment is shared by one of the politicians we interviewed: "Even though science is obviously never able to provide objective or neutral information, it offers consolidated knowledge with necessary elements of evidence for decision-making processes. Such scientific knowledge is not discredited as controlled by vested interests but seen as factual and informative and thus as true in a certain way." Thus, scholarly contributions add another perspective to policy-making processes mainly occupied and influenced by powerful elites. At best, research results can bring in alternative interpretations and point to neglected issues or undesired solutions.

Whereas the reasons of expertise and independent advice also help to explain why policy-makers invite scholarly contributions, at least in the earlier phases other reasons played a role as well. Especially in the 1970s and 1980s, communication scholars in Switzerland showed a self-interest in being involved in policy-making. They wanted to be involved in federal commissions and to perform research for the administration in order to enhance the discipline's own legitimacy. This was important since communication studies at the time was a very small academic discipline with scarce resources. To be seen as useful for society and to be able to point to the discipline's important role helped to institutionalize communication studies at universities across the country.

Nevertheless, on the side of researchers, motives for getting involved in media policy-making and performing research for official bodies went beyond mere self-interest. Some of the older interviewed scholars unequivocally pointed out that particularly their younger (and then still young) colleagues had a different understanding of what science is and should do. Young scholars wanted to inform policy-makers in order to contribute to the solution of societal problems, as a retired professor remembers: "My PhD students and associates were young people who believed that politics follows science—which is completely absurd to believe."

Last but not least, personal acquaintances between scholars and politicians played an important role in being asked to contribute and conduct research. One of the retired scholars indicated that at least until the 1980s the system of the militia army brought scholars, journalists and politicians together in the military division for press and broadcasting. This division played an important role in creating a network between communication specialists in a variety of areas, as a scholar remembers: "Everybody involved in the field of the media participated in this division. Thus, it also was a network to exchange ideas."

Assessing the impact of scholarly contributions is difficult. When only looking at the number of recommendations that were actually implemented, the impact on policy-making seems to be impressive. An analysis of research reports published in phases three to five reveals that out of eighty-two suggestions thirty-four have been completely and fourteen at least partly implemented by the legislature (mostly in the broadcasting act and its revisions). However, it is necessary to put the role of academia in policy-making into perspective. Most of the implemented proposals were already part of the political debate before the publication of research reports or they were uncontroversial and consequently uncontested. For instance, scholars' suggestions to institutionalize a press council or to limit private broadcasting to the local level, while reserving the national level for the public service broadcaster, was already in line with the dominant political parties. No wonder, therefore, that this proposal was implemented.

Moreover, there is another important aspect which should not be forgotten: involvement in policy-making certainly is not the same as proposing reform that leads to a more democratic or participatory media system. In contrast to younger scholars, most established scholars were not interested in shaping the policy-making process directly. As one interviewee put it: "science can neither solve policy problems nor expect that its suggestions are implemented directly." In addition, scholars were aware of the potential risk of the instrumentalization of academia by politicians. They felt that this would impair their academic neutrality and objectivity. Established scholars were also closely connected to elites in both politics and the media. Thus, most of them did not support radical change anyway, as a retired scholar clarifies: "Media policy research should remain neutral. The public has the right to know that we do not perform research on behalf of a certain political position. I always tried not to bother the public by political propositions or opinions." Younger scholars however felt that maintaining the status quo was as political as demanding change.

Getting involved in policy-making not only has an impact on media policy but also on communication as a scientific discipline itself (e.g., Löblich 2008). The development of media and communication policy as a research field goes hand in hand with the institutionalization of academic communication research. While the contribution of media policy research in the beginning only added up to little more than the appointment of individual scholars to federal commissions, the formation of the broadcasting and telecommunications regulator OFCOM in the early 1990s led to a continuous commissioning of research at various Swiss universities. Over time, commissioned research got more diverse. Whereas in the first phases projects mainly made use of secondary analysis of statistical data, interviews, and historic analysis, later on researchers applied the entire spectrum of research methods common in the social science. Research for federal commissions and the regulator also accelerated a transformation of the discipline away from the humanities to the social sciences (Meyen and Löblich 2007; Schade 2005). The scholars involved were aware that media policy research was a great opportunity for the development of their discipline: "By participating in the federal commissions, scholars hoped to attract third-party funded research projects in the future," one of the interviewed researchers stated. Nevertheless, these projects did not only lead to more funding but also helped to gain knowledge about the research subject itself: "It gave us the opportunity to better understand policy processes and professional practices."

In addition to this coevolution of media policy research and communication as a discipline, some recommendations of the federal commissions were in the self-interest of academia. In its final report, the second extraparliamentary commission suggested forming a new regulatory agency for communications that should continuously commission research projects. Today, one of OFCOM's duties is to award research projects. Even though this measure benefits academia, politicians and the regulator are satisfied as well. As a representative of OFCOM stated in an interview with the authors:

> I am pleased with media policy research. It is of high quality. And there is a demand for academic research: On the one hand, the government and the administration require research in order to answer inquiries and motions from parliament. On the other hand, one of our tasks is to evaluate if licensed broadcasters comply with regulations. Thus, from the perspective of the regulator, applied research like program analysis is useful.

CONCLUSION

Our study about the contribution of communication research to media policy in Switzerland shows that there was, and remains indeed, a demand for research. Politicians and the regulator OFCOM appreciate the contribution of communication studies in order to gain more independent knowledge about developments in the field of the media but also to get new ideas about media regulation in other countries. Communication studies were able to influence the debates about media policy by suggesting new models and regulatory changes, often inspired by comparative research. At the same time, the development of media policy research and of communication as a scientific discipline were closely interlinked. Researching media policy and contributing to policy-making supported the institutionalization of the discipline in Switzerland. The involvement of scholars in federal commissions and performing administrative research seems to be a win-win situation for policy-makers and scholars alike.

However, when questioning whether communication and media policy research successfully challenged the interests of political and media elites, it becomes clear that most studies fell short of driving media reform for the following reasons: First, more established scholars did object to getting directly involved in policy-making; they saw their role more in providing research results and then let the political system decide what to do with them. Second, independent from this self-conception of most scholars, researchers did not deliver too many progressive proposals that would severely challenge the interests of politicians and media owners. And third, even when they did, there was no political support for implementing such proposals. For instance, research calling for tougher regulation of ownership concentration or for direct media subsidies for journalism both offline and online in order to maintain and promote diversity was noticed but the proposals were never implemented.

Nevertheless, the regulator OFCOM explicitly commissioned research—and continues to do so—that can be expected to deliver critical results and proposals possibly unwelcome for many politicians. This deliberately creates room and funding opportunities for critical research. Moreover, the current media crisis led to new research devoted to critical analyses of ownership concentration, diversity, and funding options for media as well as to the formation of a new federal commission. These developments should be seen as a chance to reform the media system. Direct subsidies to new independent online news organizations, strong support for public service broadcasting and prohibition of further consolidation of ownership promise to support a media system that is able to fulfill its function for democracy. It is up to scholars to take these chances for adding alternative perspectives to the policy-making process and to compel all actors to at least acknowledge and reject ideas for a different media system.

REFERENCES

Bonfadelli, H., and W. A. Meier. 2007. "Zum Verhältnis von Medienpolitik und Publizistikwissenschaft—am Beispiel Schweiz." In *Ordnung durch Medienpolitik*, edited by O. Jarren and P. Donges, 37–58. Konstanz: UVK.

Braman, S. 2003. Introduction. In *Communication Researchers and Policy-Making*, edited by S. Braman, 1–9. Cambridge/London: MIT Press.

Buckingham, D. 2009. "The Appliance of Science: The Role of Evidence in the Making of Regulatory Policy on Children and Food Advertising in the UK." *International Journal of Cultural Policy* 15, no. 2: 201–215.

Bundesrat. 2011. *Pressevielfalt sichern*. Bericht des Bundesrates in Erfüllung des Postulats Fehr 09.3629 und des Postulats der Staatspolitischen Kommission des Nationalrates (SPK-NR) 09.3980. Bern: Schweizerische Eidgenossenschaft.

Corner, J., P. Schlesinger, and R. Silverstone. 1997. Editor's Introduction. In *International Media Research. A Critical Survey*, edited by J. Corner, P. Schlesinger and R. Silverstone, 1–17. London/New York: Routledge.

Dumermuth, M. 2012. "Das BAKOM und seine Forschungsaufträge." In *Im Auftrag des BAKOM. Aktuelle Studien zur Leistungsfähigkeit von Presse, Radio und Fernsehen in der Schweiz*, edited by M. Leonarz, 9–16. Zürich: SwissGIS.

Eidgenössisches Justiz- und Polizeidepartement (EJPD). 1975. *Presserecht— Presseförderung. Bericht der Expertenkommission für die Revision von Artikel 55 der Bundesverfassung vom 1. Mai 1975*. Bern: EDMZ.

————. 1982. *Mediengesamtkonzeption. Bericht der Expertenkommission für eine Medien-Gesamtkonzeption*. Bern: EDMZ.

Freedman, D. 2008. *The Politics of Media Policy*. Cambridge: Polity Press.

Haight, T. R. 1983. "The Critical Researcher's Dilemma." *Journal of Communication*, 33, no. 3: 226–236.

Jarren, O., M. Künzler, and M. Puppis, eds. 2012. *Medienwandel oder Medienkrise? Folgen für Medienstrukturen und ihre Erforschung*. Baden-Baden: Nomos.

Just, N., and M. Puppis. 2012. "Communication Policy Research: Looking Back, Moving Forward." In *Trends in Communication Policy Research. New Theories, Methods and Subjects*, edited by N. Just and M. Puppis, 9–29. Bristol/ Chicago: intellect.

Katz, E. 1979. "Get Out of the Car. A Case Study of the Organization of Policy Research." *International Communication Gazette* 25, no. 2: 75–86.

Künzler, M. 2009. "Switzerland: Desire for Diversity without Regulation: A Paradoxical Case?" *International Communication Gazette* 71, no. 1–2: 67–76.

Kvale, S., and S. Brinkman. 2009. *InterViews. Learning the Craft of Qualitative Research Interviewing*. London: Sage.

Löblich, M. 2008. "Das kommunikationspolitische Forschungsprogramm der Bundesregierung und der Wandel der Publizistikwissenschaft zu einer empirischen Sozialwissenschaft." In *Medien und Kommunikation in der Wissensgesellschaft*, edited by J. Raabe, R. Stöber, A. M. Theis-Berglmair, and K. Wied, 297–314. Konstanz: UVK.

Mason, J. 2002. *Qualitative Researching*. London: Sage.

Mayring, P. 2010. *Qualitative Inhaltsanalyse. Grundlagen und Techniken*. Weinheim/ Basel: Beltz.

Meier, W. A. 1993. "Neue Medien in der Schweiz: ihre Zielsetzungen und Leistungen." In *Medienlandschaft Schweiz im Umbruch. Vom öffentlichen Kulturgut Rundfunk zur elektronischen Kioskware*, edited by W. A. Meier, H. Bonfadelli, and M. Schanne, 203–270. Basel/Frankfurt a. M.: Helbing and Lichtenhahn.

Meier, W. A., H. Bonfadelli, and J. Trappel, eds. 2012. *Gehen in den Leuchttürmen die Lichter aus? Was aus den Schweizer Leitmedien wird*. Münster: Lit.

Melody, W. H. 1990. "Communication Policy in the Global Information Economy: Whither the Public Interest?" In *Public Communication. The New Imperatives. Future Directions for Media Research*, edited by M. Ferguson, 16–39. London/Newbury Park/New Delhi: Sage.

Melody, W. H., and R. Mansell. 1983. "The Debate over Critical vs. Administrative Research: Circularity or Challenge." *Journal of Communication* 33, no. 3: 103–116.

Meyen, M., and M. Löblich, eds. 2007. *"Ich habe dieses Fach erfunden." Wie die Kommunikationswissenschaft an die deutschsprachigen Universitäten kam. 19 biografische Interviews.* Köln: Halem.

Napoli, P. M., and N. Gillis. 2006. "Reassessing the Potential Contribution of Communications Research to Communications Policy: The Case of Media Ownership." *Journal of Broadcasting and Electronic Media* 50, no. 4: 671–691.

Padrutt, C. 1977. *Zur Lage der Schweizer Presse.* Zürich: Publizistisches Seminar der Universität Zurich.

Pool, I. d. S. 1974. "The Rise of Communications Policy Research." *Journal of Communication* 24, no. 2: 31–42.

Puppis, M., and M. Künzler. 2011. "Coping with Change: The Reorganization of the Swiss Public Service Broadcaster SRG SSR." *Studies in Communication Sciences* 11, no. 2: 167–190.

Saxer, U. 1983. "Probleme der Kabelpilotprojekt-Begleitforschung aus der Sicht der Kommunikationswissenschaft." *Media* Perspektiven, no. 12: 825–833.

———. 1989. *Lokalradios in der Schweiz. Schlussbericht über die Ergebnisse der nationalen Begleitforschung zu den lokalen Rundfunkversuchen 1983–1988.* Zürich: Seminar für Publizistikwissenschaft.

———. 1993. "Die Medien-Gesamtkonzeption als Steinbruch? Zur rechtlichen Steuerbarkeit von Mediensystemen." *Zoom Kand*, no. 1: 5–9.

Schade, E. 2005. "Was leistet die Publizistikwissenschaft für die Gesellschaft? Eine Rückschau auf wichtige Forschungsvorhaben zur Ausgestaltung der Medienlandschaft Schweiz." In *Publizistikwissenschaft und öffentliche Kommunikation. Beiträge zur Reflexion der Fachgeschichte*, edited by E. Schade, 13–45. Konstanz: UVK.

———. 2006. "Schweizerische Medienkonzentrationsdebatte in den 1960er bis 1980er Jahren: Ein Rückblick auf zentrale Positionen der Politik und Wissenschaft." In *Medienkonzentration Schweiz: Formen, Folgen, Regulierung*, edited by H. Bonfadelli, W. A. Meier, and J. Trappel, 253–278. Bern: Haupt.

Verhulst, S., and M. Price. 2008. "Comparative Media Law Research and Its Impact on Policy." *International Journal of Communication*, no. 2: 406–420.

CHAPTER TWENTY

"... please grant success to the journey on which I have come"

Media Reform Strategies in Israel

NOAM TIROSH, *Ben Gurion University of the Negev, Israel*
AMIT M. SCHEJTER, *Ben Gurion University of the Negev, Israel*
Penn State University, United States

MEDIA REFORM STRATEGY

Media reform has not been high on the agenda of social reform movements in Israel historically, nor has it emerged as one following the social protests of 2011. All three reform strategies identified by Hackett and Carrol (2006)—internal, alternative media, and structural change—had been tried over the years. It emerges that the only strategy with an impact has been the use of alternative media in its most extreme form: defying the law and launching unlicensed electronic media services. This has been true in particular in the case of radio policy, for which there are a few examples from the 1970s to the 2000s, and cable television.

The transition of Israeli society from a social democracy to a neo-liberal economic regime, and the subsequent privatization of numerous former government services, gave rise to an emergent civil society starting in the 1990s. At the same time, privatization and liberalization processes were most dramatic in the ICT sector, whose structure transformed from being dominated by single government-controlled providers of media and telecommunication services, to one encompassing a competitive landscape of multiple operators and broadcasters of a commercial nature.

The response of civil society to this structural change has been complex and the media reform strategies employed in order to encounter it and promote a more civic, less commercial, and less concentrated media environment cover the whole gamut of civic responses. Single-issue NGOs have sprouted to tackle issues ranging from freedom of information, to different minority groups' representation concerns, to political balance; media-minded NGOs were formed to serve concerns of the whole political spectrum from the far right to the radical left; and multi-issue NGOs, some of which have been around for many years, have also taken issue with ICT-related matters ranging from freedom of expression to media concentration.

The role of media reform movements within civil society is unique. On the one hand, social movements utilize the media to advance their cause and adjust to the emergence of new interactive and collaborative media forms. Nevertheless, not all of these social movements demand reform in the media structure itself, a mandate left to the media reform movement. Israel too has seen this inconsistency, in which social movements that utilize the media to advance their cause ignore the need for media reform, while at the same time an independent media reform movement rises.

Hackett and Carroll (2006) point to three different strategies employed by what at the time of their study was still a fledgling media reform movement. First is an internal reform of the media field in which cultural producers and media workers, including journalists, serve as the prime agents of the change sought, either as unions or as professional associations (ibid., 52). The term *alternative media* can sum up the second strategy, according to which organizations try to bypass the established corporate media by creating "autonomous spaces in civil society" (ibid., 52). And finally, there is the attempt to effect change of state rules, regulations, and policy; forms of media representation; and enhance media literacy education.

This chapter begins with a description of the historical and economic context of the development of media in Israel and the role of the three strategies in reforms that took place along that history. It also focuses on an offshoot of the second strategy—alternative media that succumb to illegal practices—that has had a noticeable effect on the system. We then discuss the limited success of media reformers in Israel and consider the reasons that have made reverting to illegal actions the most influential strategy impacting the system and bringing about change.

CIVIL SOCIETY AND MEDIA REFORM IN ISRAEL

The rise of neo-liberalism and the parallel growth of civil society that characterized the world in the 1980s did not bypass Israel. According to Yishai (1998), these political and cultural changes in Israel also effected the relations between the state and the civil society. The shift changed from a policy of active inclusion of groups the state considered in line with its ideological mission, through a policy of active exclusion and delegitimization of groups it considered challenging to its ethos, to a policy of passive exclusion of the so-called radical groups, decreasing interference with civil society altogether. The media as well chose to marginalize civil society groups that advocated cooperation between Jewish-Israelis and Palestinians and awarded attention to protest groups whose message was more "culturally resonant" with the hegemonic narrative (Lamarche 2009). Ben-Eliezer (1998, 373) goes as far as saying that never in Israel had the "civil society" operated *against* the state apparatus, but rather in line with Gramscian assumptions, it "contributed to the construction of a system of domination which combined coercion and consent and, in fact, limited potential range of knowledge, action or dissent."

Media reform organizations sprouted alongside other civil society organizations in Israel during this same period of neo-liberal transition and economic growth. Their strategies, which mostly correspond to the three strategies identified by Hackett and Carroll (2006), are herein described.

STRATEGY 1: INTERNAL REFORM

With regard to the first strategy, unionization and professional association, one can identify a major role for organizations of creative workers in effecting media policy in Israel. Organizations such as the Israeli Documentary Filmmakers Forum and the Actors' Union were very active in parliamentary debates surrounding quotas and other content policies, which were central during the liberalization of the media sector and the launch of commercial channels prior to the rise of digital platforms in the 2000s.[1] They were also active in demanding changes in the public-service broadcasting arena, discussed further on.

Journalists' unions were founded already during the pre-1948 colonial era, and were prevalent in private newspapers. Since the founding of the state they became a major influence at the Israel Broadcasting Authority (IBA) as well as in the respective media organizations. The rise of neo-liberalism alongside the growing competition from the newly founded electronic media outlets, led both the newspapers and the IBA to offer more differentiated pay arrangements to journalists, further deteriorating solidarity among employees in the industry. The journalists' union still has a role in effecting policies (in particular hiring) at the IBA, however it has lost much of its clout in both the private print and broadcast industries (Nossek 2009).

Following the massive social protest activity in 2011, a new journalists' union, the Journalists' Organization,[2] was founded in 2012 with the support of the central labor union, the *Histadrut*. It had close to two thousand members and was recognized as the official representative of the media workers in fourteen media outlets by the end of August 2013. However, in its own description of its activities it only mentions that it had formed a team of journalists and academics whose goal is to propose reforms in Israeli media. No document addressing these reforms has been published as of yet. One major achievement of the new union was the January 2014 court decision reinstating the head of the union at the daily *Ma'ariv*, who was fired in July 2013 because of his unionizing activities.

In 2013, workers of two of the large private mobile operators—Bezeq's subsidiary *Pelephone* and the largest operator, *Cellcom*—as well as those of *Hot*, the cable operator, unionized, again with support of the *Histadrut*, and much to the discontent and disapproval of their owners. Indeed, the effect of this wave of unionization on the industry, as a reform strategy, is yet to be assessed.

STRATEGY 2: ALTERNATIVE MEDIA

Before the Internet, alternative media were legally possible in Israel, but only in print; however, the colonial Press Ordinance still in place required a license. The descriptor *alternative*, is thus challenging. The most prominent among non–establishment outlets historically was probably the weekly magazine *Ha'olam Hazeh* (This World), which was purchased in the 1950s by muckraker journalist Uri Avneri, who turned it into a platform advancing a radical agenda promoting his career as a Knesset member between 1965–1974 and again between 1979–1981 (Meyers 2007).

Avneri, of course, was not alone. Peri (2004) succinctly describes Israeli media development along the lines of social cleavage, and the flourishing of media within marginalized social groups such as immigrants and religious and ethnic minorities, which he calls "alternative." While these outlets provide an alternative to the mainstream media in that they serve distinct cultural communities and their identity building, they do not necessarily subscribe to a social or media change platform (ibid.). At the same time, Tsfati and Peri (2006) point out that consumption of non-mainstream (sectorial and foreign) media in Israel is congruent with skepticism of mainstream media fare. Nevertheless, while non-mainstream media provide minorities in Israel with a distinct identity and with an alternative viewpoint, they are not known for being focused on reform in the media landscape itself.

What may be a unique media reform strategy that typifies the Israeli landscape is reverting to tactics that are on the verge of, or often beyond, legality. This is especially true when analyzing reforms that have taken place in the electronic media space for both radio and television. The earliest instance of such an activity was the launch of the Voice of Peace (VoP), which, broadcasting from a ship in the Mediterranean Sea from 1973–1993, was the first alternative radio station targeting Israeli audiences. While aiming to contribute to peace in the region, the station broadcast mostly light music. Indeed, we deem this activity as "illegal," since the reason the VoP chose to broadcast from the waters was due to the inexistence at the time of a legal procedure to acquire a broadcast license. In response, the Broadcasting Authority, at the time still a government-controlled monopoly, launched a new station in 1976, *Reshet Gimmel* (Network C), adopting the light character and foreign rock music programming of the rogue competitor (Caspi and Limor 1999, 134).

The next documented event took place in 1981, when the owners of a ship named *Odellia* planned to launch a commercial television channel from the sea, again due to the lack of a legal procedure to acquire a license. The government responded to the threat of the *Odellia* by initiating an amendment to the Wireless Telegraphy Ordinance that prohibited television broadcasting from the seas. The new law made it illegal to assist such broadcasts by selling advertising time in Israel on their behalf. Indeed, the *Odellia* incident did not lead to an immediate change in the media offering in the country; however, it did affect media law, and later, in the 1990s, influenced laws regarding transborder broadcasting (Schejter 2009).

The next instance in which illegal broadcasts contributed to media reform took place with the launch in 1988 of *Arutz Sheva* (Channel Seven), which also started as an offshore radio broadcast from a ship, but whose broadcasts were moved to a studio in a Jewish settlement in the occupied territories. Despite an attempt to legalize it through legislation proposed in 1998, it was shut down by a court order in 2003. *Arutz Sheva*, in addition to its Hebrew musical offering, promoted the West Bank settlers' and the Israeli right wing's agenda (Porat and Rosen-Zvi 2002). Its impact on the broadcasting landscape was also content-based, pushing IBA's *Reshet Gimmel* (Network C) transition from the international flavor it was founded upon in the 1970s to only Israeli music in 1997.[3]

A popular Israeli movie from 1995, *Lovesick on Nana Street*, documents what is perhaps the most blatant example of communication policy change driven by illegal (or at the minimum, unauthorized) alternative media activities, the course that led to the enactment of the cable law in 1986. When the amendment to the Telecommunications Law enacting the cable service was introduced in the Knesset, its initiator said the amendment was designed to eliminate 250 pirate cable television stations around the country. The law was meant to "create order," member of Knesset (MK) Meir Sheetreet said, by providing for supervision of broadcast material and promoting a service of educational, "proper entertainment" and regional programs (Knesset Records, March 27, 1986, 2356). MK Victor Shem-Tov, then head of the opposition socialist United Workers Party (Mapam), described the existing situation as "anarchy where irresponsible people, lacking motivation to educate, enter homes and broadcast movies to children and adults with no educational direction—this the state cannot tolerate" (ibid., 2357).

While the Voice of Peace, *Arutz Sheva*, and the *Odellia* incident, all were launched to broadcast beyond the state's borders, a new stream of alternative media outlets broadcasting from within the state and in clear defiance of the law sprouted in Israel in the 1990s. These were hundreds of so-called pirate radio stations, many of which sought to bring about an alternative voice on the airwaves mostly representing *Mizrahi,* religious and Arab-Israeli voices (Limor and Naveh 2008), all suffering from under- or nonrepresentation in the electronic media landscape. A government committee appointed by the minister of transportation equated these broadcasts with acts of terrorism due to their interference with air control transmissions. The end result of this movement was the establishment of yet another government-sanctioned committee to propose solutions, and the eventual licensing in 2009 of a commercial radio station targeting religious audiences and broadcasting in three regions (unlike all other commercial radio stations, which broadcast only in one region).[4] Illegal or pirate broadcasting thus has had an enduring effect on the Israeli broadcasting landscape.

The introduction of digital media in Israel made the need for "illegal" or "pirate" measures obsolete, and brought about a massive introduction of alternative media sites, offering textual, audio, and audiovisual services. These outlets represent a wide range of political positions and target specific audiences with content focused on distinct tastes, as well as attempt to compete with the traditional media outlets' web presence by providing general interest sites. A few of these outlets that can be mentioned include the generalist News1, left-wing outlets such as +972 in English and *Haokets* (The Sting) in Hebrew and the right-wing outlets such as *Arutz Sheva* (Channel Seven) in Hebrew and its English version.[5] More recent years have seen the introduction of alternative content initiatives, some more profit-oriented such as Saloona, a content and opinion site targeting women, and Holes in the Net, a blogging platform; others more focused on providing alternative information and culture such as the alternative music site Café Gibraltar, or the alternative news site Social TV.[6] However, during this stage of institutionalization of an online media sphere, "the hegemony of old existing power centers is reproduced" (Caspi 2011, 343), as the most accessed news websites and portals—Ynet, Walla!, and Mako—are merely a reflection (as far as ownership and content is concerned) of the same media that dominated the market prior to this time.

ALTERNATIVE MEDIA DURING THE PROTEST MOVEMENT OF 2011

Protests in general and social protest in particular are not strange to the Israeli political culture. However, the protests of summer 2011 were like no other protest event in the history of Israel. The demonstration organizers claimed for the first time to represent a social class—self-described as the middle class—and not an identifiable pressure or interest group of the past. The protest was sparked by a twenty-five-year-old student whose rental lease had run out and who as a result perched a tent in the center of Tel Aviv's Rothschild Boulevard on July 14, 2011, and created a Facebook event inviting her friends to join her—which they did. The incident developed into a massive social movement—called J14 (according to its website)—that protested a variety of economic issues and injustices, and gathered under a common cry, "the people are demanding social justice" (Filc and Ram 2013).

The protest movement, though spontaneous, generated through different avenues a reform agenda. While media reform was not a major issue on the protesters' agenda (Schejter and Tirosh 2015), media reform strategies of the three types made their way to the public and political sphere. Historically situated—the protest followed immediately the Arab Spring and occurred shortly before the Occupy movements sprouted all over the world—J14's agenda lies somewhere in between these two international phenomena: seeking structural change, yet focusing on social gaps and on an economic agenda.

The protest movement launched its own website,[7] which was kept updated until August 2012, and listed all of the initiatives that were derived from the protesters' activities.[8] Indeed, *protest movement* in itself is an elusive term, as due to its nature, it did not have a democratically elected representation nor did it have a formal structure, yet the J14 website was universally identified with the movement and there is no other source that challenges that claim.

Virtually all the Internet activities generated by the social protest were different formats of alternative media. In some cases the websites or Facebook pages served the internal communications of the protest movement, such as the protest movement's "situation room" and blogging platform,[9] which closed down in December 2012. In other cases, they served to raise awareness of the social protest by providing activists and the general public with alternative information: the Civil Press site had multiple u-stream (video) channels operated by activists across the country;[10] the Israel Independent Press was an English language resource about protest activities;[11] and *Radio Beit Ha'am* broadcast weekly alternative audio news and music.[12] J14 listed on its website also the Megafon-news website, which was indeed launched after the July 14 incident, yet describes itself as an independent unaffiliated cooperative of journalists.[13]

The protest movement saw making institutional information available as an important goal. Within this genre, two initiatives, which can also be seen as alternative media outlets, were launched in 2012: Politiwatch, which believes "the media are the watchdog of democracy" and that "there is a moral obligation [on the media] to stick to objective and factual reporting at all times."[14] The website focuses on statements politicians make and checks their truthfulness. Sand, another website, in addition to truthfulness of political talk also rates politicians based on a truthfulness criterion.[15] J14 also mentioned the Open Knesset initiative;[16] however, this initiative existed long before the J14 movement.

The quest for institutional reform of the media has traditionally existed to a certain extent among Israeli civil society organizations, even though the organizations most often seek reform in terms of the representation of ideas and communities and very little activity focused on media rules and regulations or enhancing media literacy. Conspicuous among the representation agenda organizations is the Public's Right to Know, as its name translates from the Hebrew, which identifies itself in English as "Israel's Media Watch" and whose aims according to its website, are:

> strengthening and realizing Israel's democracy by informing the Israeli public about Israel's media, the extent to which they abide by the media codes of ethics, decency and objectivity in reporting; systematic research and surveillance of the media and exposition of political and cultural media bias; and deepening public and institutional involvement in upholding the media codes of ethics and defense of the private citizen against the increasing power of the media against him or her.[17]

Israel's Media Watch is mostly identified with Israel's political right as its present and past presidents are current ministers in the Likud and Israel Beiteinu–led governments or former ambassadors representing Likud-led governments.

At the other end of the political spectrum stands *Keshev*, the Center for the Protection of Democracy in Israel, whose aims also can be seen in line with the third strategy, yet with a somewhat different nuance of "promoting a more moderate media and public discourse through educational activities, by counseling journalists and by publishing research on Israeli media coverage."[18] *Keshev*'s leadership is more politically identified with the Left than manned by party-affiliated individuals as Media Watch is; that, however, can be attributed to the fact that Israel's Far Left, unlike the Likud, has rarely, if ever, taken part in governing the state.

Agenda, the Israeli Center for Strategic Communications, also belonged to this category of organizations. It worked, according to its now defunct website to "reprioritize and reframe social change issues within the Israeli public debate and media" and promoted "the establishment of a new public discourse where other voices are heard and a central place is given to subjects such as equality, minority rights, social, economic and environmental justice, health and human rights."[19] Agenda's goal was "to shape public opinion and lead to social and conceptual change through the media, through full cooperation with the organizations for social change"; thus, although it was supported by the progressive New Israel Fund and the Ford Foundation, it did not carry a partisan affiliation. Agenda folded in the summer of 2012 into an organization that emerged after the 2011 protests, Hasdera, jointly with a third NGO, Uru.[20] Their joint collaboration in the new media sector is discussed further on.

I'lam, the Media Center for Arab Palestinians in Israel, is a nonprofit focusing on media freedoms for Palestinian Israelis.[21] The organization describes its goals as "promoting media rights and empowerment" among Palestinian Israelis and states that it encourages "Israeli media institutions and practitioners to adopt more professional, unbiased standards in their coverage of the Arab citizens of Israel." I'lam, like its counterparts, does not describe its activities as aimed at media regulatory reform, but rather sees its advocacy of media regulatory bodies as aiming to "build a culture for objective, pluralistic and human rights-sensitive media work in the Hebrew media landscape" as well as "capacities of Arab journalists/editors in the active role they will be invited to play in advocacy."

REVOLUTION 101: AN INSTITUTIONAL REFORM STRATEGY THAT (ALMOST) WORKED

What characterized the reform plans of the Israel Broadcasting Authority (IBA) since the early 1990s was that the government initiated them. Furthermore, none of these reforms were ever implemented. However, one significant change in IBA almost occurred between 2001–2010. It was initiated by a media reform movement spearheaded by moviemaker Doron Tsabari, whose efforts are captured in the 2010 documentary *Revolution 101*, on which the following description is mostly based.

Tsabari's reform strategy between 2001 and 2008 developed gradually. While his initial motivation for the reform was ignited by his personal experiences as a freelancer at IBA, he quickly understood the need to have a movement behind him, and an inroad to the political establishment beside him. In that sense, Tsabari's strategy combined the first and third of Hackett and Carroll's (2006) strategies: he used his position as an insider in the industry to rally fellow directors, screenwriters, and other media professionals, but at the same time he mobilized an experienced lobbyist and a seasoned lawyer who specializes in public interest law and with their help was able to enlist a member of Knesset to front the effort and present the legislative changes he sought. Interestingly, all the politicians with whom Tsabari communicates in front of the cameras (as seen in the documentary) insist on having the support of a popular public movement, so that they themselves can have the needed backing to move the wheels of change from within the political establishment. At the beginning of his reform efforts, Tsabari got himself elected as head of the directors' and screenwriters' union, which gave him a movement to represent. With the help of the policy professionals he learned what weaknesses of IBA were those that led it to its current sorry form, and also what tactics he can use to destabilize it and its resistance to change. In the process he was instrumental in the late (all-powerful director general of the IBA) Uri Porat's decision to resign in 2001 (Lavi 2001) and was a major force behind the maneuvering of the government to fire his successor Yosef Bar'el in 2005 after petitioning the High Court of Justice (Balint 2005; Ge'oni 2004).

Tsabari used the court system when he needed to force the government to make decisions (Ge'oni 2004); he organized demonstrations to attract media attention, including a violent storming of IBA meetings and press conferences (Balint 2004); he was interviewed in the media, including a double spread in Israel's most popular weekend supplement "7 Days"; and was even at a certain point appointed as a member of IBA's plenum by the minister in charge who wanted to cooperate with him in his quest for change (Carmel 2006). However, as Tsabari testifies in the film, he learned very early on that only institutional change can bring about a real reform, and while he spent much energy discrediting IBA and its management in order to gain policy-makers' support, his main effort was to enforce passage of a new law. Indeed, in October 2008, the Knesset passed in a first reading an almost revolutionary reform of IBA, based on Tsabari's draft. Not surprisingly, both the remit of IBA and its organizational structure were at the center of the proposed legislation.

Still, the combination of luck and strategy that helped Tsabari initiate the legislation ran out. Before Tsabari's version of the law finished going through the three readings required, elections were held on February 10, 2009, marking the return to power of the conservative Likud over the somewhat more centrist Kadima party, which supported the reform. By October 2010, the Knesset passed the first version of a significantly different reform at IBA, more in line with the Likud's traditional position and in April 2012 the Knesset passed the Israel Broadcasting Authority Law (Amendment No. 27), which outlines a comprehensive reform that completely diverts from the original model making the IBA more (instead of less) dependent on the government, and its remit more (instead of less) nationalistic and aligned with the state's agenda rather than with allowing critical and pluralistic voices to be heard.

The changes that the legal documents went through from first draft to three readings in the Knesset to final legislation demonstrate how reform can be a two-edged sword that leads to nondesirable consequences, at least as far as maintaining the public media's independence both with regards to its cultural remit and to the control over its decision-makers and their appointment. Interestingly, the 2012 reform was never implemented. By July 2014, a new Likud-led government initiated and passed yet another reform, this time involving the liquidation of the IBA and the creation of a new Public Broadcasting Corporation. That reform too is yet to be implemented.

The aforementioned social protest of 2011 brought about not only alternative media but also a plan for structural changes in public administration in general, albeit less so in the governance of media. A group perceived as the leadership of the protest, formed an alternative committee to the government's initiated committee, which was headed by two professors of Ben-Gurion University of the Negev: Avia Spivak, and Yossi Yonah. The Spivak-Yonah report was published as a book in summer 2012. We conducted a textual analysis of the report, identifying every instance in which media reform or any other type of reform that may affect the media and telecommunications industries was mentioned (Schejter and Tirosh 2015). As the following description reveals, the report dealt sparsely with issues relating to the media and with the need for media reform.

The Spivak-Yonah report was in fact an amalgamation of a number of reports, each written by a separate committee of experts, assigned to report and make recommendations on a different topic. There was no specific section devoted to media and telecommunications, and related issues were only mentioned briefly in the report.

Indeed, the Spivak-Yonah report, which was written mostly by economists, cannot be seen as a reform strategy per se. Perhaps the fact that a large, social reform movement was formed—which was seen as breaking traditional social lines and creating a new "subjective collective" (Filc and Ram 2013)—seems like the right moment to propose a comprehensive reform in media and telecommunications that would propose an egalitarian communications policy; however, that was not the case. The report saw computer communications as a means for providing government services more efficiently and recommended initiating "an operative plan for the improvement of service to the citizen by instilling quality methods, client satisfaction polls, decentralized service and information through the Internet" (ibid., 85) in the chapter on public administration. The report also mentioned in the context of education the need for "learning and computer communication spheres" (ibid., 349). In addition, it proposed in the chapter dealing with employment, "raising consciousness for workers' rights" through advertising "in the relevant media for the populations suffering the most from violation of their rights, such as populations for which Hebrew is not their main language" (ibid., 309). The fact that there may be less access to ICTs among these very populations was discussed only tangentially elsewhere and in the chapter on welfare, as the report recommended strengthening the capability of citizens to take part in the development of policy. The report stated in the context of welfare that "for this [cause] the availability and connectivity of people to Internet networks . . . is important—in particular the poor who suffer technological exclusion" (ibid., 280).

One more mention of the media industries appeared in the chapter on lowering monopolistic profit and lowering prices in the report on the economy, in which the authors stated that "out of concern to the freedom of the press and the electronic media, the holdings of large media by owners of large businesses in other sectors of the economy, should be prohibited" (Filc and Ram 2013, 114). Indeed, even for the protesters' alternative agenda, the media and their concentration—albeit a freedom of the press issue—were framed in terms of the economy which serves the status quo perception of media as industries rather than cultural institutions.

CONCLUSION

All three strategies of media reform identified by Hackett and Carroll (2006) have been employed in Israel over its sixty-eight-year history, yet to little if any results. Unions and professional organizations have mostly lost their effectiveness and have remained mainly players in public debates over structural change. Tsabari's reform reveals that eventually structural change in the political system can turn media reform over its head, while using the same reform apparatus in order to achieve different goals. At the same time, alternative media that operated within the boundaries of the law have contributed to the identity of minority groups—religious, ethnic, cultural, and political—yet they did not contribute to a reform of the media landscape at large. The government initiated all institutional changes, and while civil society organizations have been active in the debates surrounding them, their actual effectiveness is yet to be determined empirically.

Paradoxically, the one media reform strategy that had an actual effect on the mainstream media's behavior and on the media setting were the alternative media which acted in defiance of the law, whether for benevolent, ideological, or even criminal motives. That is definitely not a strategy one would normally recommend; however, its effectiveness cannot go unnoticed when a determined constituency is seeking change. In the Israeli case, the "illegal" radio, in particular the "pirate" radios of the 1990s and 2000s, were the driving force behind providing voices to those excluded from the radio dial.

What is it that makes the "illegal" and "pirate" outlets so successful in Israel and could that genre of media reform and of alternative media creation serve as a strategy in other societies? One can only speculate regarding these questions. Indeed, Israeli society is characterized by an absence of formalism in interpersonal relations as well as in citizen-government relations. Schejter and Cohen (2002, 38) describe this as the "basic audacious (*chutzpadic*) temperament of the population," which together with Israelis' infatuation with new technology has led them to quickly adopt electronic gadgets and novelties even before they were fully utilizable. Hence, for example tens of thousands of Israelis had television sets before television broadcasts were initiated and, as mentioned above, tens of thousands of Israelis subscribed to illegal cable services despite their illegality. Clearly very large numbers of Israelis (though no official numbers can attest to that) tuned into the pirate radio stations. The popularity of the "illegal" services was also the reason for their success in impacting policy.

Another reason for individuals reverting to alternative cable and radio operations when the government fails to provide them can be seen as a reaction to the elongated period in which no alternative and nongovernment-affiliated media were available to Israelis, which coupled with the loss of trust in leadership (Katz, Haas, and Gurevitch 1997). Indeed, "the Israeli public has adopted an approach to solving social problems by unilateral initiatives" (Ben-Porat and Mizrahi 2005), a phenomenon also apparent in land and agriculture policies, in which it earned the double-entendre maxim "creating facts on the ground" (Temper 2009).

One "near-success," Tsabari's Revolution 101, should also be assessed in this context. How did Tsabari come so close to affecting reform that he was able to see the passing of a draft bill in which his vision of public broadcasting was much at the center? Unlike the alternative media examples, this is a standalone case in a long history in which only the government has been successful in initiating and seeing through reforms in media; one can only refer to it as an accidental blip on the reform radar. This reading of the reform landscape is yet strengthened by the fact that early in 2014 a government-appointed committee proposed a reform very much along the same lines of Tsabari's and by May of the same year it had already passed a first reading in the Knesset and has a good chance of making its way through the process. Tsabari himself was cited as supporting the new initiative even though he himself did not take a part in forming it.

Are alternative, pirate, and "illegal" media the most viable media reform strategy in Israel for the future as well? Alternative media are probably less of a driving force for reform when it comes to the contemporary media territory, in particular because the plethora of voices over the net can easily drown alternative voices and because affluent and established voices can better promote their presence even over new media. Indeed, the new media era does not promise a stronger and more influential reform movement, nor has it necessarily the potential to bring about a democratization of the conversation on media reform.

NOTES

This study has been supported by a Career Integration Grant awarded by the Marie Curie FP7 program of the European Union and by the I-CORE Program of the Planning and Budgeting Committee and the Israel Science Foundation (grant no. 1716/12). Both authors contributed equally to this study. The title of this chapter is from Genesis 24:42 (New International Version translation).

1. Documentary Filmmakers Forum, www.fdoc.org.il/english; Actors' Union: http://shaham.org.il/.

2. Journalists' Organization, www.itonaim.org.il.

3. That transition can also be attributed to the launch of a network of commercial radio stations at the same time, which offered mostly foreign music, thus pushing the IBA to develop a different niche (Schejter and Elavsky 2009).

4. Commercial radio station targeting religious audiences, www.kol-barama .co.il/.

5. News1, www.news1.co.il/; +972, http://972mag.com/; Haokets, www .haokets.org/; and Arutz Sheva, www.inn.co.il/ and www.israelnationalnews.com/.

6. Saloona, http://saloona.co.il/; Holes in the Net, www.holesinthenet.co.il/ holesinthenet-media; Café Gibraltar, www.cafe-gibraltar.com/; and Social TV, http://tv.social.org.il/.

7. See www.j14.org.il.

8. See http://j14.org.il/polimap/.

9. The Facebook "situation room": https://www.facebook.com/j14live/info; blogging platform (closed): http://k1789.org/?p=2646

10. Civil Press, civilpress.tv/.

11. Israel Independent Press, www.facebook.com/IsraelIndependentPress.

12. Radio Beit Ha'am, http://j14.org.il/j14live/beithaamradio.

13. Megafon-news, http://megafon-news.co.il/.

14. Politiwatch, www.politiwatch.co.il/.

15. Sand, www.sandtalks.co.il.

16. Open Knesset, http://oknesset.org; www.haokets.org.il; and www.tv .social.org.il.

17. Public's Right to Know, www.imw.org.il.

18. Keshev, the Center for the Protection of Democracy in Israel, www.keshev .org.il.

19. Agenda, the Israeli Center for Strategic Communications, http://agenda .org.il/english/.

20. Hasdera, www.hasdera.org.il/what_we_do_eng/.

21. I'lam, the Media Center for Arab Palestinians in Israel, www.ilam-center .org.

REFERENCES

Balint, A. 2004. "Confrontation Between Creators and Channel 1 Seniors." *Ha'aretz*, April 29. www.haaretz.co.il/gallery/1.962354.

———. 2005. "The Government Approves Bar'el's Dismissal with a Large Majority." *Ha'aretz*, May 3. www.haaretz.co.il/misc/1.1007467.

Ben-Eliezer, U. 1998. "State Versus Civil Society? A Non-Binary Model of Domination Through the Example of Israel." *Journal of Historical Sociology*, no. 11: 370–396.

Ben-Porat, G., and S. Mizrahi. 2005. "Political culture, alternative politics and foreign policy: The case of Israel." *Policy Sciences* 38, nos. 2–3: 177–194.

Carmel, A. 2006. "Doron Tsabari and Yigal Galai to Serve in IBA's Plenum." *Ha'aretz*, December 7. http://b.walla.co.il/?w=/3050/1020151.

Caspi, D. 2011. "A Revised Look at Online Journalism in Israel: Entrenching the Old Hegemony." *Israel Affairs* 17: 341–363.

Filc, D., and U. Ram. 2013. "Dafni Leaf's July 14th: The Rise and Fall of the Social Protest." [In Hebrew.] *Theory and Critique*, no. 41: 17–43.

Ge'oni, Y. 2004. "Directors Doron Tsabari and Uri Inbar: We Will Petition the High Court of Justice if Bar'el's Removal Procedure Is Not Completed." *Globes*, November 15. www.globes.co.il/news/article.aspx?did=854293.

Hackett, R. A., and Carroll, W. 2006. *Remaking Media: The Struggle to Democratize Public Communication*. New York: Routledge:

Katz, E., H. Haas, and M. Gurevitch. 1997. "20 Years of Television in Israel: Are There Long-Run Effects on Values, Social Connectedness, and Cultural Practices." *Journal of Communication* 47, no. 2: 3–20.

Lamarche, K. 2009. "Political Activism and Legitimacy in Israel: Four Groups Between Cooperation and Transgression." In *Civil Organizations and Protest Movements in Israel: Mobilization Around the Israeli-Palestinian Conflict*, edited by E. Marteu, 73–90. New York: Palgrave Macmillan.

Lavi, A. 2001. "Porat's Final Battle." *Haaretz*, June 24. www.haaretz.co.il/misc/1.711842.

Limor, Y., and C. Naveh. 2008. *Pirate Radio in Israel*. Haifa: Pardes. [In Hebrew.]

Nossek, H. 2009. "On the Future of Journalism as a Professional Practice and the Case of Journalism in Israel." *Journalism* 10, no. 3: 358–361.

Peri, Y. 2004. *Telepopulism: Media and politics in Israel*. Stanford, CA: Stanford University Press.

Porat, I., and I. Rosen-Zvi. 2002. "Who's Afraid of Channel 7: Ideological Radio and Freedom of Speech in Israel." *Stanford Journal of International Law* 38, no. 1: 79–96.

Schejter, A. M. 2009. *Muting Israeli Democracy: How Media and Cultural Policy Undermine Freedom of Expression*. Urbana: University of Illinois Press.

Schejter, A. M., and A. Cohen. 2002. "Israel: Chutzpah and Chatter in the Holyland." In *Perpetual Contact*, edited by J. Katz & M. Aakhus, 30–42. Cambridge: Cambridge University Press.

Schejter, A. M., and C. M. Elavsky. 2009. "'. . . and the children of Israel sang this song': The Role of Israeli Law and Policy in the Advancement of Israeli Music." *Min-Ad: Israel Studies in Musicology Online* 7, no. 2: 131–153. www.biu.ac.il/hu/mu/min-ad/.

Schejter, A. M., and N. Tirosh. 2015. "'What Is Wrong Cannot Be Made Right'? Why has media reform been sidelined in the debate over 'social justice' in Israel." *Critical Studies in Media Communication* 32, no. 1: 16–32.

Spivak, A., and Y. Yonah. 2012. *To Do Things Different: A Model for a Well-Ordered Society*. Tel Aviv: Hakibutz Hameuchad Publishers. [In Hebrew.]

Temper, L. 2009. "Creating Facts on the Ground: Agriculture in Israel and Palestine 1882–2000." *Historia Agraria* 48: 75–110.

Tsfati, Y., and Y. Peri. 2006. "Mainstream Media Skepticism and Exposure to Sectorial and Extranational News Media: The Case of Israel." *Mass Communication and Society* 9, no. 2: 165–187.

Yishai, Y. 1998. "Civil Society in Transition: Interest Politics in Israel." *Annals of the American Academy of Political and Social Science* 555, no. 1: 147–162.

CHAPTER TWENTY-ONE

Legislating for a More Participatory Media System

Reform Strategies in South America

CHERYL MARTENS, *Bournemouth University, United Kingdom and Universidad de las Américas*
OLIVER REINA, *University Bolivariana de Venezuela and CLACSO, Argentina*
ERNESTO VIVARES, *FLACSO, Ecuador*

MEDIA REFORM STRATEGY

South America has been an important site for media reform over the past decade. The reforms taking place across the continent center on making media more widely accessible to civil society via the promotion of community and alternative media. Examining two South American countries in particular, Argentina and Venezuela, it is possible to see that media reforms have aimed at redistributing media power to a much wider range of civil society actors. The key strategy that brought major support for the reforms in the case of Argentina was the joining of forces of a large number of groups to form the Coalition for Democratic Radio Broadcasting (La Coalición por una Radiodifusión Democrática). Following a wide range of debates, the coalition's twenty-one action points for media reform, presented to the government for the first time in 2004, were incorporated into legislation in 2009. The most significant strategy used in these reforms has been the coalition's call for citizen involvement in media processes and policies. In Venezuela, the strategies have revolved around reviewing and rethinking wider educational, informational and cultural frameworks and their role in redistributing media power and fostering the new approaches to community and alternative media.

Few regions, over the past decade, have as extensively challenged the position of big media companies in the domestic field or made such advances in promoting a more participatory media environment as South America. Media reform debates have been especially significant in countries such as Venezuela, Argentina, Uruguay, Bolivia, Brazil, and Ecuador, but are also gaining momentum through the efforts of civil society groups in countries such as Colombia and Chile.

On the one hand, media reform has been a necessary step for the new democracies of South America, as they seek to distance themselves from the media laws of the dictatorship years. The debates concerning media reform are particularly related to disputes over approaches to development, whereby major media corporations of the region largely legitimate development strategies orientated toward capitalist growth. Corporate media in these so-called pink-tide countries have therefore entered into a power struggle with neo-populist and neo-developmentalist governments (Martens and Vivares 2013).

In this chapter, we are interested in examining reform strategies that center on making media more widely accessible to civil society via the promotion of community and alternative media. We focus on the cases of two South American countries in particular: Argentina and Venezuela. First, we consider the development of audiovisual communication reforms in Argentina—that were aimed at breaking up the Clarin Group's media monopoly and redistributing media power to a much wider range of civil society actors—and the main strategies promoted by the Coalition for Democratic Radio Broadcasting (La Coalición por una Radiodifusión Democrática). Second, we consider the strategies associated with the development of alternative and community communication in line with reforms aimed at the construction of a participatory, democratic project and the diversification of the traditional accumulation of power in Venezuela. The Venezuelan case also moves the discussion beyond media reform to consider the wider components of the social environment, supporting reforms through educational, informational and cultural framework capable of fostering the growth of community and alternative media.

We argue that the current debates and growing polarization concerning media reforms in South America are intimately tied to the struggle across the region to move away from neo-liberal policies to alternative, redistributive forms of development and governance. We do not take for granted the existing order of development and institutions, nor do we consider these to be defined by universal or ahistorical rules. We argue, rather, that these are historically contingent and unfolding in a context of struggles over development with different winners and losers, particularly when it comes to media and democracy. Dynamic and in continuous change, development is a key arena of political struggle defined by the rise, consolidation, and decline of different social forces, political orders, and different paradigms of what development should be (Payne 2005).

The restructuring of media in South America is also approached here as a cultural process in terms of articulators of the communication practices—hegemonic and subaltern—of social movements (Martin-Barbero 1993, 164). Media corporations in South America, for example, often negate neo-populist government approaches, drawing on neoliberal cultural premises concerning market intervention, while governments seek to regulate and discipline these powers. These governments may not necessarily offer adequate political responses. However, what is especially noteworthy is how these media reforms have been based on the momentum of a wide range of social movements opposed to the concentration of media power.

BACKGROUND

Media theorists argue that it is possible to distinguish two main stages in history with regard to the social significance of the mass media in Latin America. First, between 1930 and 1950, the mass media played a decisive role in both conveying and challenging "the appeal of populism, which transformed the mass into the people and the people into the nation" (Martin-Barbero 1993, 164). Second, after 1960, the political function of the media was replaced by an economic one and although the state maintained a rhetoric that the airwaves were public, responsibilities for their management was effectively transferred to the private sector (Martin-Barbero 1993). These stages are now giving way to a third stage, whereby media interests are so tightly intertwined with the private sector (to which they have long belonged) that it is possible to see a promotion of corporate over public interests (Martens and Vivares 2013).

The following sections explore the development and some of the key media reform strategies of Argentina and Venezuela; both countries have contested neoliberal forms of governance through the introduction of a range of media reforms that are aimed at popular media participation.

ARGENTINA: FROM CONCENTRATION TO REFORM

Argentina's media reforms coincide with the projects of government and civil society initiatives to restrict media power and diversify media ownership after decades of monopolistic concentration promoted by the Argentine dictatorship (1976–1983) and weak subsequent democracies. A significant shift in media concentration took place in 1977, when the military dictatorship confiscated the newsprint corporation Papel Prensa from the Graiver family, owners of 75 percent of the holdings, turning the newsprint industry over to three corporations: Clarin, La Nacion, and La Razon, now known as the Clarin Group. This made it possible for the dictatorship to gain control of newspapers, making a monopoly of the major producers of press paper in the country (Papaleo 2009; Sel 2010). The Clarin Group continued to grow in the years subsequent to the dictatorship and throughout the neoliberal years of weak democracy in Argentina.

Pushed by a major debt crisis, the Argentine democracy at the end of 1989 started a new process marked by the opening of the sector to external actors and the privatization of the whole spectrum of telecommunications and audiovisual services, mostly the major public TV channels as well as the National Company of Telecommunications. The growth of the media sector continued into the 2000s but was marked by the transfer of the ownership of the major media companies into the financial hands of international operators (Postolsky 2010). At the end of the process, two main groups dominated the national market: The Clarin Group and Investment Equity Citicorp (IEC). After that the most important events are related to the arrival of the Goldman Sachs Group, who became partners with the Clarin Group, and the dissolution of the IEC.

As media monopolies grew in size and power in Argentina, civil society calls for media reform of the dictatorship laws also became stronger. Following the end of the dictatorship in 1983 there were several proposals for media reform and supportive civil society movements grew in the 1990s and 2000s as telecommunications and audiovisual technologies rapidly developed and gained power. By 2000, a wide range of civil society organizations had joined forces to form the Coalition for Democratic Radio Broadcasting (La Coalición por una Radiodifusión Democrática) seeking to challenge the unfettered power of corporate media. The coalition first took their concerns to the Nestor Kirchner administration in 2004 but it was not until 2009, following a sbattle between the government and the country's main corporate interests, that civil society organized a specific set of media strategies in the promotion of twenty-one points for a more democratic media. The proposed points were distributed and debated throughout a wide range of civil society organizations and later in debates with government officials and members of the public.

ARGENTINA'S LAW NO. 26.522: THE
REDISTRIBUTION OF POWER AND BROAD
PUBLIC PARTICIPATION

With a broad base of public support, drawing together more than three hundred civil society actors, following six months of public debate, on October 10, 2009, the Cristina Kirchner administration passed Law No. 26.522. The law was clearly aimed at moving away from the neoliberal policies of postdictatorship. Breaking up media monopolies also made it possible for the Kirchner government to promote its economic agenda of "accumulation *with* social inclusion." The new regulations fixed the limits of concentration of media monopolies by redistributing audiovisual services equally between corporate, community, and government media, set limits on foreign broadcasters, and established quotas for national content and promotes diversity of media channels at local, national, regional, and international levels (Sel 2010).

The media reform strategies and struggle for the democratization of communication were based on twenty-one points presented to the government by the Coalition for Radio Broadcasting, which had come about following several years of forums and public debates and distribution of information about media democracy by members of the Coalition. The tactics used included the use of the media channels of the participants of the coalition and links with both national and international organizations. In addition to the redistribution of audiovisual services, main strategies integrated into the law include:

Limit to the number of licenses a single company or network can operate. Any operator over this quota must divest within the year.

Seventy percent of radio content and 60 percent of broadcast television content should be produced in Argentina.

Cable television networks should include universities, municipalities and provinces within their coverage area of services. This is designed to promote local creative efforts of painters, musicians, and other artists, as well as to support Argentina's national and local culture.

The creation of a registry for foreign channels.

Limits to the percentage of foreign ownership in local radio and television broadcasting to 30 percent as long as non–Argentina-based ownership does not result in direct or indirect control of the company.

These reforms, based on the notion of the right to communication, stand out for their high level of civil society engagement and ability to generate unity around these particular points despite the differences among actors. With nearly every sector of society involved in the debates of this media law, the twenty-one points presented demonstrating a wider project of popular participation and mobilization. The law also introduced key changes in the administration of the law, setting up the decentralized National Authority for Audiovisual Services (Article 10). It also established a Federal Audiovisual Communication Counsel, consisting of representatives to serve two-year terms from each province and the City of Buenos Aires, three representatives each from, corporations, unions, and nonprofits, one representative each from universities, Indigenous groups, public media, and organizations representing minority rights.

A main oppositional tactic has been to draw on competing discourses concerning freedom of expression. Thirteen media organizations and the International Press Association signed a joint statement in position to the reforms, declaring that: "Far from achieving diversity, these regulations carry with them censorship that will restrict the supply of content to citizens" (Peters 2009). If we look more closely at this statement and the Clarin Group's legal appeals, however, the "supply of content" that is being referred to is corporate media's unrestricted hold on media channels.

Corporate media did not report that the reform's reference to freedom of expression, was not random, but based on a nation-wide consultative process within Argentina. Reformers and policy-makers working on the legislation drew on a wide range of reports and resources which highlight the key role of media diversity in fostering freedom of expression, such as the UN special session on freedom of expression and the African Commission's Human Rights and Peoples Declaration of Freedom of Expression in 2002.

In terms of civil society participation, substantial revisions were made to the law on several occasions through the contributions and lobbying of a wide range of civic groups. Some of the key revisions came from groups interested in promoting indigenous rights and the protection of minors. The legal inclusion of *Pueblos Originarios* or Indigenous peoples, for example became much more integrated into the document due to Indigenous civil society participation. Inclusion, however was not just in reference to communication and identity, but involvement in decision-making processes. These articles were added as a direct result of the concrete proposals of these groups.

The protection of minors offers a second example of civic engagement strategies with regard to drafting the media reform legislation. Several groups representing children's rights became involved in broadening and clarifying Articles 70 and 71, which deal with the protection of children and adolescent rights in relation to discriminatory messages.

This aspect of the law was also an important area for the mobilization of a massive youth movement in Argentina, which beyond the media legislation resulted in a significant number of young people becoming involved in politics through engagement with this specific law.

In total there were over one hundred separate proposals made by a wide range of competing civil society groups, which shaped the final articles of the law (Orlando 2011). This new media law thus demonstrated historic levels of broad civic engagement and participatory democracy within the legislative process.

COMMUNITY MEDIA IN VENEZUELA

Venezuela shares several aspects with other regional cases, including that of Argentina, in terms of its highly concentrated private media ownership. Since the 1920s, two groups have principally dominated the media landscape in Venezuela. Diego Cisneros Organization (ODC-Venevisión), which has over eighty media-related holdings and Grupo 1BC (Radio Caracas Televisión), have monopolized audiovisual services within country. Since the Chavez administration in 1998, there have been attempts at reversing this power through legislative media reform, including the Law on Social Responsibility for Radio, TV, and Electronic Media and the Law on Popular Communication, currently under discussion in the National Assembly.

The development of community communication and subsequent media reform in this area has been related to specific political moments, some of which have promoted the diversification of participants in media outlets both as an initiative of communities themselves and as a consequence of the favorable conditions created by public policies implemented in this field. The communication practices in both cases tend to be different from each other.

In Venezuela, movements for an alternative and community communication in the 1970s and 1980s were initiatives that, in most cases, took place in isolation and usually were the result of particular efforts by small groups. Linked by their intentions to increase the number, quality, and purpose of participation in media communication, these initiatives were most of the time disconnected and did not count on legal or state support. In the midst of adverse conditions, it was common for struggles for community media not to reach a successful conclusion. More recently, these efforts have continued to produce the same results and have moved to marginal spaces within the Venezuelan media system.

The history of alternative communication began with the experiences of local or sectoral print, cine-clubs, and street art. These first steps were joined later by the expansion of broadcast media, which often represented a barrier that prevented regular allocation of frequencies. In 1984, with the adoption of the NTSC standard and the advent of color TV in Venezuela, there was a window of opportunity when many radio and television licenses were first granted. Between 1984 and 1999, more than three hundred radio licenses were granted (including seventy AM licenses) and nearly fifty television stations (VHF and UHF). However, over 80 percent of the licenses were granted to private operators (Rodriguez 2005).

Within this framework, licenses were nevertheless granted to TV and radio stations whose reach was limited and they were labeled as "community media." In the 1990s emerged the pioneering cases of Catia TV and TV Caricuao in Caracas, Teletambores in Aragua state, Tarmas TV in Vargas state and TV Michelena in Táchira state, as well as community radio stations such as La Voz de Guaicaipuro and Radio Perola in Miranda state, Radio Parroquiana in Zulia state, Radio Cool in Bolívar state, Radio Chuspa in Vargas state and Radio Rebelde in Caracas. All these initiatives had something in common: an attempt to open spaces to new discourses and new, more horizontal communication practices to counter information and content broadcast by the traditional media outlets.

The prospects for community media were boosted by the election of Hugo Chávez as president in 1998. Since then, alternative and community communication has been woven into the Venezuelan social fabric through media reforms in line with the changes aimed at the construction of a participatory democratic project and the diversification of the traditional accumulation of power in the country.

Encouraged by the participatory and protagonist democracy of the Bolivarian political project, the grassroots media that operated in the Venezuelan countryside formed a group and created the Venezuelan Community Media Network (RVMC to use its Spanish acronym). In the capital, Caracas, the few media outlets that broadcast illegally formed a group they called *Onda Libre* (Free Wave). These two groups organized two important initiatives for the organization of alternative and community sectors of communication: the Meetings for Free Communication in 2000, and the National Conference of Community Media in 2001. These forums helped to organize popular communication groups in the country, to assess their situation, and to go deeper into their role as organizers of the work of social groups in the communities.

Among the most important activities of the collectives Onda Libre and the Venezuelan Community Media Network was their active participation in the National Constituent Assembly that drafted the new Constitution, aimed at ensuring the legal presence of the state to support community media. Their observations were reflected in the Organic Law on Telecommunications, passed in 2000, largely due to pressure from grassroots organizations with the active participation of the movements Comité Librecomunicación and Comité por una Radiotelevisón de Servicio Público. The passing of both the Organic Law on Telecommunications and the Regulation of Nonprofit Community Radio and TV Broadcasting of Public Service paved the way for a new media landscape and granted community media outlets institutional recognition as legal subjects.

Community media over the past decade, in contrast to its clandestine existence in the 1980s, have thus operated under more favorable conditions, with the support of a legislative framework and the protection of fundamental rights. This represents a qualitative leap in the democratization of the radio electric spectrum, which is defined in Venezuela as a public good. This characteristic brings community media to the same level of commercial or public media: all users of a good whose collective nature involves the fulfillment of social responsibilities defined by law.

Historically, community media have undertaken projects with a deep connection to the communities where they were created and live, where they are linked to the citizens, and whose programs widely and democratically reflect the political and ideological diversity of their geographical area of coverage. These are projects whose objectives were distorted and carried out as the particular heritage of a small group that makes decisions behind the scene, in flagrant violation of the *raison d'etre* of these initiatives. We must add to their different profiles another element of diversity: their attempts to get organized as networks and associations, thus facilitating—and, as we shall see, sometimes hindering—their performance of community media.

The two biggest networks are the National Association of Community, Free and Alternative Media (ANMCKA, Spanish acronym), and the Venezuelan Network of Community Media (RVMC). Sometimes these associations have expressed differences with each other and their affiliate media, thereby hindering the implementation of joint, collective measures in favor of their specific communities and being forced to examine their priorities and objectives within the framework of the law that regulates them and the Venezuelan communication system as a whole.

In order to deepen and promote ties with local, communal, and national identities, further laws were created by the Chavez administration. These laws include the 2002 Regulation and the 2010 Law on Social Responsibility on Radio and TV (also known as the Resorte Law), which were further boosted by the recent entry into force of the six organic laws on the People's Power that make up the Framework for the Organic Law for the People's Power, and the Organic Laws on Communes, the People's Public Planning Law, the Social Comptroller Law, the Law on Communal Economic System and the Law on the Municipal Public Power. These laws are aimed at presenting different vertices of the same fundamental objective: To promote and strengthen the practice of communal self-management and a state-commune co-management, that is conscious and wide-reaching.[1] The objective is to establish specific conditions for organizing communes as participatory and redistributive units of power, which can plan productive projects of each community together with the regional and local executive power, increasing the quality, financial, and material audit of projects implemented and people involved in them.

These measures also aim to build a functional economic system with relative national autonomy, organized from the communes toward the aforementioned self-governance and co-governance. This process highlights the importance of the use of communication that, based on the principles now exposed, is in line with the manner of distribution of the on-going social power. In this regard, community communication needs to be consolidated in order to organize these emerging social processes.

Beyond the undeniable importance of these media reforms, the possibility of these ideas becoming more widely adopted can come about only if these new regulations are also viewed as life projects, as spaces of transformation, and, we emphasize, as possibilities of links that can bring new feelings, and, therefore, new rules, imposed not only by their legal nature but also by the awareness of their importance for consolidating the national project clearly defined in a constitutional framework.

Several years after the adoption of these media reforms, various sectors are now discussing the need to go beyond the legal instruments regulating the exercise of community communication. However, discussions on the table extend beyond reforming legislation and are focused on how to best develop a communication model that strengthens the human, and emotional aspects of society.

These aspects are closely linked to community participation, in accordance with a state policy oriented to a life of solidarity, collective interests, the achievement of social justice, and the rearrangement of a country based on participatory democracy from grassroots sectors and people's organizations.

TOWARD A POPULAR COMMUNICATION

The attempts to transform the communication field we have described arise in a historical moment in which the Venezuelan and Argentine political processes seek to consolidate the structural differences in society and take a heterodox approach to development. In this respect, the achievements in areas such as health, education, public welfare, income distribution and social justice have been recognized nationally and internationally, highlighting the favorable comments of the United Nations regarding the early implementation of the commitments made when subscribing to the Millennium Development Goals.

This is significant in the case of Argentina and other countries in South America since powerful media and other corporate actors, under heterodox models of development, have, in many regards, taken on the role of opposition players entering into a dialectic of conflict with neopopulist and neodevelopmentalist governments (Martens and Vivares 2013). The main evidence that supports this case is the fact that similar struggles concerning corporate versus state media power are *not* evident to the same extent in countries whereby neoliberal models dominate, such as in Chile and Colombia.

While some nations (including the United States and some in Europe) continue to apply structural adjustment plans with major cuts to social spending, both Argentina and Venezuela have increased investment in the social sphere. Both states prioritize education, health, and housing, that is, areas that have not been fully developed given the constant sabotage and disruptions, which have been part of media reform struggles. In the case of Venezuela this was particularly the case in the 2002–2003 period—the most controversial period of its contemporary history—through actions perpetrated by both internal factors and external allies.[2]

Regarding the transformations taking place following the implementation of media reforms, communication itself has become an important area of social reflection and is now more widely viewed as integral to social relations, transcending each dimension of community life. Communication, in its deployment of symbolic forms and representations facilitates public and articulated social and linguistic processes and practices that open spaces to emancipation, new social imaginaries, new discourses, and perceptions as well as constructions of the social reality.

A popular communication approach demands the reworking of how we conceive of information flows, communication, education, and culture, and how we might consider them as an integral unit, taking into account the specific features of each of these dimensions and understanding them as edges of an organic unit. Communication cannot take place without informing; education cannot develop without the support of culture; and culture cannot be constructed without communication. Likewise, communication must be viewed from a critical approach and solidly reflect on the experiences achieved. This necessarily implies transcending media as the main subject of debate, in order to engage with communication as a dynamic social and cultural process, as a producer of symbolic forms and social meanings.

CONCLUSION

Media reforms introduced in Argentina and Venezuela have been based around strategies that allow for the incorporation of greater citizen participation in media, beginning with citizen participation in the creation of media reforms. Despite the challenges inherent in confronting corporate media power, most marginalized voices via processes of media reform movements have been given unprecedented media space and new possibilities for participation, offering new possibilities for the democratization of the media and the public sphere more generally.

Argentina's 2009 antimonopolistic media legislation demonstrates how civic groups can collaborate with governments to lessen monopoly power, and restructure and regulate neoliberal power. Powerful media such as the Clarin Group continue to seek to delegitimize governmental actions as populist, antidemocratic policies, personalizing their attacks on the figure of executive power. However, while the battles with media corporations have polarized opinions, they have also been important in consolidating a wide range of views as to media and citizen participation.

As the trajectory of media reforms in Venezuela demonstrates, however, in order to successfully implement reforms and make media accessible to the wider population, strategies for media reform need to also include an examination of the social frameworks that support the interaction of information, communication, education, and culture.

Strategies for media reform therefore, have the potential to foster new kinds of communication, which implies the assertion of politics, endowing media channels with more ethical content. Communication defines the kind of society we live in as well as the ways in which communities participate in that society. Structural change through media reform and supporting social frameworks, have the potential to foster the conditions for the growth of full human potential through solidarity and humanistic media practices. In sum, it is not possible to separate communication from politics. For media to be accessible we must work from a wide range of angles to create media reform strategies based on the social redistribution of power.

The process of building democratic and widely accessible communication, beyond market-led models, is open and unfinished, with more questions than answers.. However, as the cases of Argentina and Venezuela demonstrate, media reform which, on the one hand, limits media monopolies and, on the other, promotes the development of community media—offers unique opportunities for allowing a wider range of actors (many of whom, until now, have had little or no media visibility) and permitting them to become subjects rather than objects in the creation of their media and wider social contexts.

NOTES

1. The Organic Law of People's Power defines self-management as "the set of actions through which organized communities directly assume the management of projects, execution of works, and services to improve the quality of life in its geographical area." It also notes that co-management is "the process through which organized communities coordinate with the Public Power, at any level and authority, the joint management for the execution of works and services needed to improve the quality of life in its geographical area (*Official Gazette*, December 21, 2010).

2. An example of these destabilizing actions against the Venezuelan government are described in the book *Documentos del Golpe* [The Coup Documents] (Caracas: El Perro y la Rana, 2009), http://archivopopular.org/sites/default/files/DocumentosGolpe/LosDocumentos5taEd.pdf.

REFERENCES

Martens, C., and E. Vivares. 2013. "Media Reform in South America." In *Media/Democracy*, edited by A. Charles, 125–135. Cambridge: Cambridge Scholars.
Martin-Barbero, J. 1993. *Communication, Culture and Hegemony: From the Media to Mediations*. London: Sage.
Orlando, R. 2011. *Medios privados y nuevos gobiernos en Ecuador y Argentina*. Quito: FLACSO.
Papaleo, O. 2009. "Clarín compro Papel Prensa con la familia Graiver secustrada." *Miradas del Sur*, September 26.
Payne, A. 2005. *The Global Politics of Unequal Development*. London: Palgrave.
Peters, C. 2009. "Argentine Government Fast-Tracking Controversial Media Law." International Press Institute, September 16. www.freemedia.at/

newssview/article/argentine-government-fast-tracking-controversial-media
-law.html.

Postolsky, G. 2010. "Continuidades, Desplazamientos y transformaciones en las politicas de Comunicación en Argentina." In *Politicas de Comunicación en el Capitualismo Contemporaneo*, edited by S. Sel, 135–154. Buenos Aires: CLACSO.

Rodriguez, E. 2005. *Psicoterrorismo mediático*. Caracas: Publicaciones del Ministerio de Comunicación e Información.

Sel, S. 2010. "Actores Sociales y Espacio Publico." In *Politicas de Comunicación en el Capitualismo Contemporaneo*, edited by S. Sel, 183–210. Buenos Aires: CLACSO.

CHAPTER TWENTY-TWO

Public Service Broadcasting in Egypt

Strategies for Media Reform

RASHA ABDULLA, *The American University in Cairo, Egypt*

MEDIA REFORM STRATEGY

The Egyptian Radio and Television Union (ERTU) functions as a state broadcaster. To transform ERTU into a public broadcaster, Egypt should carry out major media reforms. An independent regulatory council of media experts and professionals should be established to devise policies and regulate the media according to universal standards. ERTU's Board of Trustees would need to be restructured to include reputed professionals with a vision to transform ERTU into a public broadcaster. The board should abolish the current ERTU charter and set up a new one that guarantees ERTU independence and focuses on noncommercial political, social, and cultural content that caters to all sectors of society. The new charter should abolish the designation of ERTU as the sole broadcaster in the country, and allow for the establishment of private terrestrial channels. A new code of ethics should be formulated based on a commitment to professionalism, objectivity, inclusiveness, and diversity. ERTU should become financially independent from the state, with the public providing funding through subscriptions or taxes. The number of channels operated by ERTU should be decreased to one or two, and the staff provided with training and an efficient media strategy to follow. Research should be constantly utilized to study audiences and viewership habits. These changes require political will or political change.

Egypt to date does not have a public service broadcasting system. The closest thing it has to public service broadcasting is the state-funded and state-controlled

Egyptian Radio and Television Union (ERTU). As a country that has historically been known as a leader in media content and production in the Arab world, and a country trying to make its rough transition into democracy in the wake of several waves of a revolution, this situation is not acceptable today. This chapter suggests ways in which the state broadcaster in Egypt could be turned into a proper public broadcaster. In doing so, it first examines the current media structure in Egypt, focusing on ERTU, and the legal framework in which the media operates. It points out the strengths and weaknesses in the system, and makes concrete suggestions as to how to reform the media sector and establish a proper public broadcaster that functions with the best interest of the public, rather than the regime, in mind.

IS EGYPT STILL A MEDIA LEADER?

Historically, Egypt has developed a reputation for being a media leader in the Arab world. Egyptian content, particularly drama, is a major export to the region, which is the reason why the Egyptian dialect of Arabic is the most widely understood all over the Arab world. Egypt was the first country to have radio broadcasting, with amateur service starting in the 1920s and formal service in the early 1950s. The country also was also a pioneer in television broadcasting, with transmission starting in 1961. Cairo is widely called the "Hollywood of the Middle East"; and actors and singers do not consider themselves popular before they have made their mark on the Egyptian market (Abdulla 2006, 2013c; Boyd 1999).

The political regimes that have controlled the country since its independence from the United Kingdom in 1952 realized early on that the media was a way to control the masses as well as to bring about development. The media at that time were nationalized and placed under the control of the government. Gamal Abdel Nasser used it to position the country and himself personally as leader of the Arab world. In doing so, he invested in media development to make sure the infrastructure and the quality of content would be unmatched in the region. Special attention was given to entertainment content, in which Nasser embedded his political messages. Radio stations like Voice of the Arabs and Middle East Broadcasting became extremely popular, and television drama transcended political boundaries with the aid of high-quality content and powerful transmitters. Television was used strategically to hook people to the screen, and the mere presence of a television set became a status symbol, sometimes even before transmission was introduced to the area. Today, over 98 percent of urban homes and 90 percent of rural ones own one or more of the eighteen million television sets in Egypt (ITU 2011).

Egypt was also the first country in the Arab world to possess a satellite channel in 1990. In 2001, it started allowing private channels onto the market. However, obtaining a license was restricted to businessmen with close ties to the regime and continues to have no clear criteria to this day. A lack of diversity of content and threats to media freedom are still major problems, and public broadcasting in the sense of a broadcaster that serves the public before the ruler is yet to be found.

There are now also several countries with programming that competes with Egyptian media. In recent years, historical drama productions out of Syria (before the political unrest started), as well as romantic Turkish soap operas dubbed into Arabic became very popular in the Arab world, and with the stagnant standard in Egyptian production, began to rock the "media leader" status of Egypt. The credibility of news programs on Egyptian television amidst the current political upheaval is another threat facing the country's own media outlets.

ERTU'S INDEPENDENCE AND LEGAL FRAMEWORK

Generally speaking, Arab broadcast media are still mostly government-owned, and so are Egypt's media. The government-controlled channels of ERTU rely on the state for their financial survival, and it is not unusual therefore to receive direct or indirect directives or guidelines on what to say or what to cover. Since these channels are the closest Egypt has to a public service broadcaster, the outcome is an environment that to this day cannot be called free or independent; it is a media apparatus that is primarily at the service of the regime or the government rather than the people (Abdulla 2010).

ERTU is governed by its initiating charter, Law 13 of 1979, last modified by Law 223 of 1989. The entity started operating in 1971 and is the sole authority responsible for broadcasting in Egypt both in terms of radio and television. ERTU has different sectors: radio, television, specialized channels, local channels, news, sound engineering, production, economic and financial affairs, general Secretariat, and a *Radio and Television* magazine. Other sectors may be established according to the needs of the business, based on a decision of the ERTU board of trustees. However, the charter is outdated and problematic on many fronts and modifications made in 1989 had the sole purpose of giving absolute power over the media to the Minister of Information.

Historically, Egyptian constitutions have, in theory, guaranteed freedom of expression. However, the government as well as security forces regularly interfere in media freedom. For many years, this has taken place through emergency laws as well as articles from the penal code. Despite marginal freedom, which has been earned by the private TV channels, the state maintained a way of monitoring and controlling content through unspoken, yet well-known red lines for broadcasters.

As per its charter, ERTU is the sole entity allowed to own a radio or television broadcasting station in Egypt. However, in the early 2000s, the government decided to allow a few private satellite and radio broadcasters into the market. In order to circumvent that legal obstacle, Egypt set up a Media Production City (MPC) on the outskirts of Cairo, and turned it into a "free zone." What that meant was that private broadcasters could operate from within this free zone without having to adhere to the laws making ERTU the sole broadcaster. However, what this also meant was that the entity in charge of approving broadcasters would be the General Authority for Investments and Free Zones, an entity with no media expertise or functions whatsoever. There are no known criteria as to how broadcasting entities are approved for licensing, although interviews with ERTU officials indicated that a good deal of the process is political in nature and requires the approval of state security (Abdulla 2013b, 2013c).

By virtue of the ERTU charter, the mission of ERTU is spelled out as follows:

> The Union aims to fulfill the mission of broadcast media, both audio and visual, in terms of policy, planning, and execution within the framework of public policy of the society and its media requirements. In doing this, it will utilize state of the art technology and the latest applications and developments in the fields of using audio-visual media to serve the community and the attainment of its objectives (ERTU 1979, Article 2).

In order to fulfill this broad mission, and in the spirit of a public broadcaster, the charter lists thirteen objectives that ERTU aims to achieve, including the following:

> (1) Providing an efficient service of radio and television broadcasting and ensuring that the service is directed at serving the people and the national interest within the framework of values and traditions inherent to the Egyptian people and in accordance with the general principles of the Constitution.
>
> . . .
>
> (3) Disseminating culture, and including aspects of educational, cultural, and humanitarian nature within programs in accordance with the high standards of the Egyptian, Arab and international vision, to serve all sectors of society, and to devote special programs for children, youth, women, workers and peasants, a contribution to the Building of the human culturally and pursuant to the cohesion of the family.
>
> . . .
>
> (5) Contributing to expressing the demands and the daily problems of the people and raising issues of public concern while giving the opportunity to express the different viewpoints, including those of political parties, and presenting the efforts to objectively deal with topics of concern.
>
> (6) Publicizing the discussions of the Parliament and local councils and committing to broadcast whatever official communication the government asks

to broadcast, as well as all that relates to the state's public policies, principles and national interest.

(7) Committing to allocate radio and television broadcasting time for political parties on election days to explain their programs to the people as well as allocating time on a regular basis to reflect the main focal points of public opinion. (ERTU 1979, Article 2)

As these articles demonstrate, on paper, the ERTU charter clearly spells out many of the rights and responsibilities that public service broadcasters enjoy in other parts of the world. In fact, the charter was modeled after the BBC, and therefore does so quite eloquently. It emphasizes important duties that include serving the people, disseminating culture, serving all sectors of society, providing educational, cultural, and humanitarian content, providing content for children, youth, women, workers and peasants, raising issues of public concern, and expressing diversity of opinions. Unfortunately, the reality of implementation is quite a different matter.

These seemingly well-written public broadcasting articles aside, state control and centralization of the media dominates the ERTU charter, starting with Article 1, which states that ERTU is the sole responsible entity for radio and television broadcasting in Egypt, and is located in the capital, Cairo. The article also specifies that ERTU is the only entity that has the right to establish radio and television stations, and that the Union has the right to monitor and censor its productions. Article 2 states that ERTU "is committed to broadcasting whatever the government officially asks the Union to broadcast" which legally skews the Union toward the government, and is a significant indicator of state control over the media apparatus.

To complement the charter, ERTU has a code of ethics (ERTU 1989). However, this still does not safeguard its independence. Furthermore, there is no independent body to monitor media performance and compliance with such codes. The sole entity for communication regulation in Egypt, the National Telecommunication Regulatory Authority (NTRA), is only concerned with telecommunications, and does not regulate broadcasting. The Ministry of Information was the only body to deal with issues of concern in the national media, which made it unlikely that the ministry would have allowed any content that would be deemed offensive to the regime. While there is no written regulation to this effect, there is a culture of self-censorship that has developed over the years and has deep roots within the media sector.

Unlike the usual concerns covered in a code of ethics, the ERTU code of ethics is not concerned with issues of professionalism, objectivity, equal time, and other universal standards of broadcasting. Rather, the code limits freedom of speech in an unprecedented manner. Listing thirty-three vaguely worded prohibitions, the code bans criticism of state officials or the "state national system" and outlaws material that "creates social confusion" or contains any "confidential information."

The ERTU charter was amended once in 1989 with the aim of giving absolute powers to the minister of information to control the media. According to the charter, the minister supervised ERTU and made sure its objectives are implemented and its services to the public properly offered. The charter also gave much power to the minister over the board of trustees of ERTU, which does not have much autonomy. Basically, no decision by the board could come into effect before it was approved by the minister. Moreover, the president of the ERTU board of trustees is appointed by the president of Egypt upon the recommendation of the prime minister. The president of the Republic also determines the president of the board's salary and terms of compensation, as well as the duration of the board presidency. Interestingly, even though the Ministry of Information was abolished in June 2014, the charter has not yet been changed.

The charter also details the criteria of the membership of the ERTU board of trustees. The board is comprised of active public figures, including intellectuals, religious leaders, artists, scientists, journalists, economists, as well as engineering, financial, and legal experts. Their appointment, salary, compensation, and the term of membership are determined by the prime minister. The heads of the different ERTU sectors, who are appointed by the minister of information, are ex-officio members of the board, together with the head of the State Information Service.

The charter details the specific responsibilities of the board of trustees. Theoretically, the board is responsible for developing the general policies of ERTU, adopting the principal strategies and plans needed to implement these policies, as well as monitoring and evaluating the functions of ERTU. The board is also responsible for approving the administrative and financial regulations and the annual budget planning and preparation; approving the general scheme of content programming, including foreign content; bringing in foreign broadcasting experts for training or consultations; adoption of the rules that govern the commercial activities of ERTU; expressing opinions on broadcasting legislation; and approving the establishment of music and theatrical troupes that provide content to ERTU. In reality though, interviews with members of the board showed that it does very little. The meetings are a mere formality, and no serious discussions take place. The main function of the board was to serve as a body that approved the ministry's policies and actions, so that it would seem that these policies come from a board rather than one person, the minister of information (Abdulla 2013b, 2013c). So there has always been a potential symbiotic relationship of pleasing the government to stay on the board.

Funding remains a major problem. ERTU is currently about $2.9 billion in debt. ERTU is therefore dependent on the government for funding, which is a major obstacle to its editorial independence. The ERTU charter states that ERTU shall have an independent annual budget, issued by a decree from the president of Egypt. The charter stipulates that ERTU's income consists of "Proceeds from fees prescribed by law for the benefit of radio and television broadcasting; Proceeds from the commercial activities of ERTU sectors and its services; Funds allocated by the State for ERTU; Subsidies and grants; Loans within the limits and rules established by the prime minister and leaving a surplus income for each financial year to the next year; Union's share of surplus/profits for the companies it owns." The charter also stipulates that "the government shall decide annually upon the subsidy amount that it grants the Union and deposit it in the Union's account at the Central Bank of Egypt" (ERTU 1979).

In other words, the Egyptian government is in full control of ERTU finances on several fronts. The government decides ERTU's annual funding, as well as on the salaries and compensations of all those with any authority over the Union, namely the board of trustees, the chair of the board, and, until recently, the minister of information. Financial autonomy is therefore nonexistent. Moreover, the compensation of those in charge is directly related to the government's satisfaction with the performance of ERTU. The Union thus acts as a tool to serve the government, rather than the people.

The 2014 Egyptian constitution states that an independent regulatory body is to take charge of the media in Egypt. However, such a body has not yet been established. Under the current political circumstances in the country, there are no signs that such a body would be truly independent. Most likely, it will be a replacement for the Ministry of Information, with all the rights and responsibilities that the ministry had carried out for decades.

ERTU'S EDITORIAL POLICIES

In terms of editorial policies, the charter does not guarantee the basic components that make for an efficient public broadcaster. There is a mention of ERTU responsibility toward political parties in Article 2 of the charter, which states that ERTU commits to "giving the opportunity to express the different viewpoints, including those of political parties"; and "Committing to allocate radio and television broadcasting time for political parties on election days to explain their programs to the people as well as allocating time on a regular basis to reflect the main focal points of public opinion" (ERTU 1979, Article 2). However, these articles were never strictly observed. There is also no mention of inclusiveness or cultural diversity in terms of content or of the agents featured on ERTU. There is no commitment on behalf of ERTU to represent the different sectors of the Egyptian society, or to represent the different groups of minorities, including religious, socioeconomic, age, ethnic, special needs, cultural, or ideological minorities.

The lack of commitment to inclusiveness also applies to the representation of gender on Egyptian television. The ERTU charter does not make reference to what percentage of programming should be directed to women (or to any other group). It only refers to having "special programs" for children, youth, women, workers, and peasants. A recent study of diversity on different Egyptian newspapers and television channels examined the content of the main talk show on ERTU. The lack of diversity was staggering. Out of eighteen episodes of the evening talk show that were analyzed featuring 120 individuals, only 8 were females (6.7 percent), while 112 were male (93.3 percent). This means that for every woman that appears on the main talk show of Egypt's public television, 14 men appear. Religious minorities were also severely under-represented, as were all other minority groups (Abdulla 2013a).

ERTU'S CODE OF ETHICS

ERTU does have a code of ethics (ERTU 1989). The document is comprised of thirty-one statements, all of which begin with the phrase "It is prohibited to broadcast . . . ." In contrast to most codes of ethics, which generally put forth some moral obligations which are not legally binding, ERTU's code presents a set of prohibitions that is not very practical. While a few statements do protect human rights—such as "It is prohibited to broadcast any program that encourages discrimination on the basis of color, race, religion, or social status," and "It is prohibited to broadcast any program that may hurt the feelings of the handicapped or the developmentally challenged"—the code mainly presents a series of what not to do rather than presenting guidelines on what is morally correct or encouraging the inclusion of minorities. If ERTU was to actually abide by all the prohibitions on the code of ethics, it would literally broadcast nothing at all. Some of the prohibitions of the code represent a clear violation of freedom of speech and expression. These include vague statements such as:

It is prohibited to broadcast any program that criticizes the state national system.

It is prohibited to broadcast any program that criticizes national heroism.

It is prohibited to broadcast any program that criticizes Arab nationalism and its struggle, values, and national traditions.

It is prohibited to broadcast any program that criticizes officers of the courts, military officers, or security officers as well as religious leaders.

It is prohibited to broadcast any program that criticizes state officials because of their performance.

It is prohibited to broadcast any program that favors divorce as a means to solve family problems.

It is prohibited to broadcast any program that contains materials that may cause depression or spread the spirit of defeat among individuals or communities.

It is prohibited to broadcast any program that creates social confusion or criticizes the principles and traditions of Arab society.

It is prohibited to broadcast any pictures of horror.

It is prohibited to broadcast any confidential information.

It is prohibited to broadcast any scientific, technical, professional, or religious advice or commentary unless it has been reviewed or is broadcast by experts in the field.

It is prohibited to broadcast any program that criticizes other broadcast programs. (ERTU 1989)

TRANSITIONING TO DIGITAL TERRESTRIAL TELEVISION?

Is ERTU Ready for the Switch to Digital Broadcasting? The short answer is: No!

The recent *Mapping Digital Media* report for Egypt indicates that ERTU is far from being ready for the transition to digital terrestrial television (DTT) broadcasting. The transition to digital broadcasting requires advanced equipment, and much training of human resources in the technicalities of DTT, steps that ERTU does not seem to be very concerned about. As the report states, "It appears that no serious steps have been taken along the road to make this transition (to digital broadcasting) happen" (Abdulla 2013b).

The one step that Egypt has taken in 2009 was to commission the German Fraunhofer Institute to do a feasibility study to design a "road map" for the transition to digital terrestrial broadcasting in the country. The study indicates that the transition is a process that requires long-term planning, and advises Egypt to carry out the following steps to be able to make the transition in 2014:

Set up a joint coordination group Digital Transition

Decide on a regional strategy

Decide on a business model for network build-up

Decide on set-top boxes

Decide on programs and channel line-up

Decide on spectrum use, including the Digital Dividend. (Fraunhofer 2009)

Egypt has not committed to meet the 2015 deadline set internationally by the International Telecommunication Union (ITU) for the transition to DTT, although there have been press statements in 2008 and 2009 whereby the executive president of the National Telecommnunications Regulatory Authority indicated Egypt would meet the deadline. Amr Badawi emphasized that the success of the project requires cooperation between the parties concerned, including ERTU, the Ministry of Information, Telecom Egypt, and content producers, as well as telecom operators, equipment manufacturers, and distributors of television sets (Al Hayat 2008, 2009). However, to date, and particularly with the political instability in the country, there has not been any public discussion of the transition and there is no evidence that it is a priority for the country.

MAIN CHALLENGES FACING ERTU AND MEDIA REFORM

Currently, the main reason for the absence of media reform in Egypt, is the lack of political will to do so. The political instability in the country, which brought about four regimes in four years, makes it easier for each incoming regime to try to control ERTU and use it to its own advantage. The media scene is currently extremely monotone, and ERTU is playing its usual part of serving as the mouthpiece of the regime. It is not in the best interest of the regime to change this picture, and so any change will have to come from the few dissident voices inside ERTU and be supported by the huge masses outside ERTU.

However, the main problem with ERTU itself is the mentality of being there to serve the regime rather than the public. ERTU employees and media personnel think of themselves as working for the state media, and as such, their first priority is the state which employs them. Many of them actually believe that there is nothing wrong with ERTU being the mouthpiece of the state because it constitutes the state media. The problem, therefore, is one of approach and a way of thinking on behalf of not only the regime but the entire ERTU as well. This is not going to be rectified overnight, but it is one of the main challenges facing ERTU, and one that needs to be changed as soon as possible.

The sheer numbers of those who work at ERTU is another major challenge. There are currently between forty-two thousand and forty-six thousand individuals working at ERTU, many of whom constitute what is known as disguised unemployment. ERTU could easily and efficiently function with fewer than ten thousand professionals. Many of these currently employed by ERTU are friends and/or relatives of two main ministers of information who belonged to the Mubarak regime (Abdulla 2013c).

In terms of the legal framework governing ERTU, the initiating charter and code of ethics are severely problematic. There are no specifications as to the exact duties of ERTU as a public service broadcaster. Both documents were clearly written with the aim of serving the regime rather than the public, and therefore they make it legally binding for ERTU personnel to abide by the directives of the state. On the other hand, the documents lack guidelines on media inclusiveness, objectivity, or diversity. There is also no requirement for ERTU to cover the issues that commercial media may not cover, such as culture, science, and technology, women's and children's programs, and so on, nor does the charter stipulate that content should be dedicated to minorities such as people with disabilities, the educationally unprivileged, religious minorities, and others. As explained earlier, the code of ethics is but a long list of what not to broadcast, and does not refer to any universal standards of ethical broadcasting.

Funding is another major problem. As long as ERTU is directly dependent on the state for its financial well-being, as well as for the compensation packages of its main staff and board of trustees, the Union cannot achieve editorial independence.

SUGGESTED STRATEGIES FOR ERTU REFORM

There are several recommendations that need to be carried out for media reform in Egypt and for ERTU to start functioning as a proper public service broadcaster. Most importantly, these strategies need political change or the political will to bring them about. In the absence of political will or change, it will be extremely difficult to turn ERTU into a proper public broadcaster, since the main problem now is that it is used to serve the regime rather than the public.

Abolishing the Ministry of Information is a good step. This entity only exists in repressive countries, with the purpose of keeping the media under state control. All over the Arab world, the ministers of information are the main players who control the media and make sure the whole media apparatus is loyal to the regime. Ministries of Information believe that it is their prerogative that government messages be delivered to the public through state-controlled media. In Egypt, the ministry should now be replaced by a truly independent regulatory council of media experts and professionals, which could devise media policies as well as monitor and regulate the media. Such monitoring should take place in accordance with universal ethical and professional standards, adapted to Egyptian culture. The standards would ensure there are enough guarantees in place to safeguard diversity of content and diversity of agents featured. However, if the new regulatory body is independent only on paper, not much change will occur.

It is of utmost importance to make sure that the role of the regulatory council would be to monitor the media with the aim of ensuring professionalism rather than controlling content, as is usually the case in the Arab world. In addition, the council would be responsible for devising criteria for satellite (and eventually terrestrial) channel ownership in Egypt, as well as for receiving and approving applications, and for assigning spectrum frequency, which is currently handled by the National Telecommunications Regulatory Authority.

There have been attempts to establish such a council in Egypt following the revolution of January 25, 2011. I was involved in one such attempt by the Egyptian Initiative for Media Development, in which thirty prominent media experts and professionals participated. However, the problem was agreeing on a structure for the formation of the board that would enable the council to be truly independent. I believe the formation of the council is what could make or break the effort. It is therefore unacceptable to have a large percentage of the council board members selected by the president or the prime minister. The selection is equally problematic if made by parliament members if the parliament does not reflect a true representation of Egyptian society. The majority of the council members should be comprised of independent media experts and reputed media figures with significant experience in the field and a reputation for objectivity and impartiality. It would be ideal for Egypt to establish such a council only after taking some steps toward the transition to democracy, so that the council could be set up under a democratically elected president and parliament. Otherwise, the council risks becoming just a seemingly more independent body that carries out the responsibilities of the Ministry of Information.

The first task for such a regulatory council should be to restructure ERTU's board of trustees so that its members have the interest and experience needed to reform ERTU. Although the board has been changed post January 25, 2011, most of the old faces remain. The changes need to be significant, and the new faces need to have professional credentials, a scientific mentality to utilize research, and a vision to transform public service broadcasting in Egypt.

The board's most important and most primary task should be to abolish the ERTU initiating charter and code of ethics, and establish a new law to govern the structure, content, and activities of ERTU, along with a new code of ethics. This should happen with the broad objective of transforming ERTU into a real public service broadcaster. This will entail a paradigm shift in the way ERTU functions so that it operates primarily with the interests of the people in mind rather than with the interests of the regime or the government. As such, the new charter should designate ERTU as the official public service broadcaster of Egypt. It should cancel the stipulation that currently exists by which ERTU is obliged to broadcast any material provided to it by the government since this represents an official and automatic affiliation of ERTU to the government, rather than to the people. The new charter should also formulate a clear idea of the functions and responsibilities of the public service broadcaster. As such, it should stipulate that the main content provided by ERTU be focused on the noncommercial political, social, and cultural buildup of the nation and its people. ERTU should become a voice for all strata of the society rather than a voice for the government or the regime. This should take place according to set criteria and minimum percentages of media coverage that match the demographic, cultural, and political structure of Egypt. All these criteria are currently nonexistent. The new charter should abolish the designation of ERTU as the sole broadcaster in the country, and instead should allow for the establishment of private terrestrial channels, a change which will be inevitable in the age of digital broadcasting.

A new code of ethics should be formulated, largely by those working at ERTU, to develop a sense of ownership in relation to the new document, thus encouraging them to strive to follow its guidelines. The new code should omit the inhibitions on broadcasting present in the current code, and instead focus on universal ethical guidelines, including issues of professionalism and objectivity as well as inclusiveness and diversity.

The financial aspects of ERTU need to be settled in order for the Union to achieve financial independence from the state. As a public service broadcaster, the basic funding of ERTU should come from the public either through subscriptions or through using some of the taxes already collected by the government, which would be dedicated to the ERTU budget, including ERTU board of trustees salaries. Examining other models of financing of public service broadcasters worldwide would help determine if a particular model would be best suited for Egypt. As part of dealing with the current financial problems associated with the ERTU budget, a new compensation system needs to be worked out, whereby the huge gap in income between ERTU staff would be decreased. While there will always be media professionals who are more expensive than others, there should be clearly established criteria for compensation that are publicly disclosed in a transparent manner, and that are based on merit rather than nepotism and loyalty to the regime.

One of the main reform tasks will be to deal with the forty-two thousand to forty-six thousand individuals currently working at ERTU. These should be filtered by committees of media experts, with the aim of retaining those who are efficient professionals in their fields. The remaining work force, as well as anyone appointed through nepotism, should be provided with a fair, early retirement plan and asked to submit their resignations. This will not happen overnight, but rather over several years. The number of channels operated by ERTU should be decreased to one or two, and those constituting the new work force should be provided with sufficient state-of-the-art training and an efficient media strategy to follow. Training should be conducted by public service broadcasting experts, covering all aspects of the broadcasting process, be it managerial, editorial, or technical in nature.

Finally, throughout the process, the use of research cannot be ignored or underestimated. Egypt in general lacks proper, independent, scientific ratings research. ERTU should either establish an in-house independent research center or enlist the services of one that uses the latest technologies to conduct audience and viewership studies.

CONCLUSION

The current political leadership in Egypt seems to lack the political will to carry out much media reform in the country. The current media system allows the regime to control key media institutions and use them to its own advantage. The media scene lacks diversity and ERTU continues to play its usual part of serving as the mouthpiece of the regime. Under the circumstances, it seems that any real media reforms might have to come from the few dissident voices inside ERTU and be supported by the masses outside ERTU.

REFERENCES

Abdulla, R. 2006. "An Overview of Media Developments in Egypt: Does the Internet Make a Difference?" *Global Media Journal—Mediterranean Edition* 1, no. 1: 88–100.

———. 2010. "Arab Media Over the Past Twenty Years: Opportunities and Challenges." In *The Changing Middle East: A New Look at Regional Dynamics*, edited by B. Korany, 59–84. Cairo: American University in Cairo Press.

———. 2013a. *A General Look at Media Diversity in Selected Egyptian Newspapers (March–April 2013)*. London: Media Diversity Institute. www.media-diversity.org/en/additional-files/documents/MDI%20Final%20Report%20on%20Media%20in%20Egypt.pdf.

———. 2013b. *Mapping Digital Media: Egypt Report*. London: Open Society Foundations.

———. 2013c. *Public Service Broadcasting in the MENA Region: Egypt Report*. Paris: Panos Paris Institute and the Mediterranean Observatory of Communication.

Al Hayat. 2008. "Promises (that Digital Broadcasting) Reaches Egyptians by 2015." [In Arabic.] http://international.daralhayat.com/archivearticle/200758.

———. 2009. "Arab TV Terrestrial Broadcasting to Become Digital; Egypt Engaged." [In Arabic.] http://international.daralhayat.com/international article/8809.

Boyd, D. 1999. *Broadcasting in the Arab World: A Survey of the Electronic Media in the Middle East*. Ames: Iowa State University Press.

Egyptian Radio and Television Union (ERTU). 1979. Law no. 13 of 1979 modified by Law no. 223 of 1989. www.ujcenter.net.

———. 1989. Code of Ethics. The General Communication Plan 1989–1990. http://tbsjournal.arabmediasociety.com/Archives/Spring05/printerfriendly/ERTUPF.

Fraunhofer. 2009. "TK Egypt: Strategies for the Transition Including a Roadmap." Fraunhofer Institute for Systems and Innovation Research ISI. www.isi .fraunhofer.de/isi-de/t/projekte/archiv/bb_tk_egypt.php

International Telecommunications Union (ITU). 2011. *World Telecommunication/ ICT Indicators Database 2010*. Geneva: ITU.

CHAPTER TWENTY-THREE

Impunity, Inclusion, and Implementation

Media Reform Challenges in Thailand, Myanmar, and the Philippines

LISA BROOTEN, *Southern Illinois University, Carbondale, United States*

MEDIA REFORM STRATEGY

In Thailand, significant obstacles to reforming the country's mixed system of government-controlled broadcast media and the private print sector include the extreme polarization of Thai society and coups d'état in 2006 and 2014, increased harassment of journalists, and the problematic use of lèse-majesté provisions and other laws to censor political dissent. In Myanmar, rapid changes have provoked a reshuffling and realignment of the internal and returning elements of the exile media as plans are developing for public, commercial, and community media sectors. In addition to protecting press freedoms, reform efforts include changing the long-held culture of secrecy and attitudes toward information, encouraging government officials to provide news and accept critique, and including grassroots perspectives and especially ethnic minority voices in the media. The Philippines, although considered one of the freest media environments in the region, remains one of the most dangerous places in the world to practice journalism; reformers in this context have focused on countering the culture of impunity, especially in the killings of journalists in rural areas of the country, promoting transparency and ensuring the proper implementation of existing laws.

This chapter compares media reform efforts in Thailand, until recently considered among the freest media environments in Southeast Asia; Myanmar (formerly Burma), once a repressive pariah state but now cited as potentially a leading force in contemporary regional democratization efforts; and the Philippines, whose media system looks great on paper but has proven deadly on the ground. The chapter analyzes the political underpinnings of reform efforts in each country, which involves in each case a struggle for cultural change away from militarized ways of approaching problems and the control of information, toward more transparency and freedom. This is in part because each of these countries is facing violence in which media have in one way or other become targets or face difficulties requiring protections often taken for granted elsewhere. The specific reform strategies used in these three countries include working to end impunity in cases of violence against journalists; protect press freedom; increase transparency in information and media policymaking and enforcement; gain the right to access information about government actions; and increase the diversity of viewpoints through promoting and protecting various forms of grassroots access to media so as to broaden people's "right to communicate," or to be *heard* as well as to receive information. The exploration of these efforts offers insights into media reform obstacles and opportunities in conflict-affected contexts in which media remain targets of violence and repression.

MEDIA REFORM EFFORTS

The media system in Thailand, in Myanmar, and in the Philippines, as in any country, is a product of its unique history. The Philippines and Burma were colonies; the Philippines was colonized by the Spanish, then the Japanese, and then by nearly fifty years of US occupation, and the British ruled the former Burma from 1824 until 1948. Thailand, while never colonized, has a history of absolute monarchy followed since 1932 by a constitutional monarchy. Each country's media system has developed in close relationship with its military. All of these factors affect the media landscapes of these three countries, heavily influencing their development and regulation, and in turn, their reform strategies.

In the Philippines, advocates are calling for an end to the culture of impunity both in response to the endemic violence against critical journalists but also to corruption within the profession itself. Despite recent setbacks in Myanmar, reformers are urging a continued opening and uprooting of old militarized ways of thinking about information and control, and inclusion of ethnic minority voices in both content and media policy-making. In Thailand, the 2014 coup installed a military caretaker government, increasing repression and abuses of civil rights and obstructing efforts to loosen government control over broadcast media and counter the overuse of problematic laws to quell dissent. Each of these involves different strategies for reform.

THAILAND

The Thai government has historically been a paternalistic institution, affected by the widely accepted palace constructions of the country's monarchs as the benign fathers of the land. In 1932 the country's absolute monarchy was replaced by a constitutional monarchy, whose reigning kings have since overseen the country as it has experienced eighteen coup attempts, eleven of them successful (the most recent were the overthrow of Prime Minister Thaksin Shinawatra in 2006 and his sister, Yingluck Shinawatra, in 2014).[1] Each coup involves some form of media censorship or intimidation, but the 2014 coup introduced military-style interrogations not seen in the most recent coups in 1991 and 2006. While the polarization of media since the 2006 coup reflected, reinforced, and in some instances inflamed the ongoing conflict, events since the 2014 coup have significantly reduced media freedoms. While Thailand has traditionally had a lively print press largely free from direct censorship, the broadcast media have been controlled by the military or by state-affiliated civilian agencies that provide concessions to private operators. This has only begun to be challenged.

Reform efforts since the early 1990s have been resisted by each ruling administration, but have been especially difficult since the 2006 and 2014 coups. The country's polarization has essentially pitted the "yellow shirts"—made up of military generals, royalists, and many educated middle-class opponents of twice-elected Thaksin Shinawatra—against the "red shirts" (many of them Thaksin supporters), who are varied but include many from rural areas of the northeast. The ill health of the much revered eighty-seven-year-old King Bhumibol Adulyadej has exacerbated tensions, especially given anxieties regarding the impending royal succession. This cultural anxiety is evidenced by the use and abuse of lèse-majesté provisions, which criminalize criticism of the monarchy through the Constitution and Article 112 of the Criminal Code. This is arguably the most pressing freedom of speech issue facing Thailand today, a problem exacerbated by the recent coups.

After the 2006 coup, the military rescinded the 1997 Constitution, dissolved parliament and appointed a 250-seat National Legislative Assembly (NLA) in its place. They imposed martial law—restricting freedoms of speech, the press, and assembly—and occupied television newsrooms, in a few cases for weeks or even months. They also closed down more than three hundred community radio stations. After the coup, "the 'mainstream' terrestrial, free-access Bangkok-based television remained pro-government throughout the conflict, switching sides according to whether Thaksin or anti-Thaksin forces were in power" (Thompson 2013, 14), although some noted a "yellow" bias among the mainstream private press (Askew 2012). In August 2007, a new constitution was approved after a contentious drafting process and national referendum. Subsequently, one elected and one appointed pro-Thaksin prime minister were both removed from office on legal technicalities, before the 2008 election of a coalition government under the Democrat Party's Abhisit Vejjajiva, and then the 2011 election of Yingluck Shinawatra, Thaksin's sister, as prime minister.

During these years, rhetorical battles raged between rival media outlets, and media found themselves the target of increasing physical attacks. In the mid-2000s, the mainstream Thai media largely split along yellow and red lines. Both factions established cable stations, and radio became an important political medium but also a forum for extreme views and hate speech. Attacks in 2010 included a fire at the national TV station Channel 3, a grenade attack on the state-run TV station NBT, and deaths and injuries among journalists covering the clashes. Crackdowns in 2010 included many red shirt–affiliated community radio stations, which were "met with silence, if not approval, by the majority of Thai mainstream media and among many of the educated middle class" (Pravit and Jiranan 2010, 165; see also AHRC 2011). Facebook and Twitter also became sites for polarized discourse.

Media closures, censorship, harassment of and attacks on journalists, and repression of activists increased following the 2014 coup, which unseated Yingluck Shinawatra. This led to the disbanding of the 2007 Constitution, the appointment of the military-installed National Council for Peace and Order (NCPO), headed by General Prayuth Chan-ocha, and arbitrary detentions and "attitude adjustment" sessions for hundreds of dissidents. Self-censorship increased as the regime targeted newer, more politicized media channels that had recently challenged the compliant mainstream print media and the state or army controlled broadcast stations. These media were eventually allowed back on air, but were provided with only temporary, one-year licenses and prohibited from discussing politics. All independent radio stations were temporarily closed, affecting some 6,000 community radio stations still waiting on trial licenses from the National Broadcasting and Telecommunications Commission (NBTC), but not the 525 government-owned stations. Many media remain highly polarized, contributing to the country's divisions, and red shirt–affiliated community stations have remained a primary target of government shutdowns. The junta also blocked foreign network transmissions, banned journalists from interviewing its critics, pressured online service providers to suspend services for anyone hampering the work of military officials, and has reportedly blocked at least 1,500 websites. The NCPO has employed arbitrary arrests and detentions, enforced disappearances, and harassment to silence hundreds of dissidents.

Frequent coups in Thailand mean frequent new constitutions and therefore inconsistencies between constitutional provisions and the organic laws intended to operationalize them. For example, a controversial new Broadcasting Operation Act that came into effect in March 2008 under the coup-appointed NLA provided for three categories of broadcasters—public, private, and community broadcasters—without specific provisions for how they would operate. The law left these details to the regulatory body that was at the time not yet established (Brooten and Klangnarong 2009). Prior to the NBTC commissioners taking office in October 2011, community radio stations were operating without licenses and therefore were vulnerable to arbitrary closure. Even after the establishment of the NBTC, radio licensing has not been a priority. After the 2014 coup, all community radio stations were required to sign a memorandum of understanding in order to obtain temporary licenses, pending a process of review for more permanent status, In addition, while the constitution provides for freedom of expression, the Computer Crime Act places multiple restrictions on this right, including the very vague lèse-majesté provisions, indicating the lack of transparency in Thailand's Internet policy. Community radio groups and Internet-based independent media need to negotiate this uncertain policy environment and remain constantly vigilant to promote and defend their communication rights (Brooten and Klangnarong 2009).

The country has seen an increasing number of charges and arrests for lèse-majesté under Article 112 of the country's penal code and the Computer Crimes Act (CCA), and in several cases, for content the alleged perpetrators hosted or published online but did not themselves author. This climate has truncated public discussion about the royal family, has significantly stifled free speech, and has provided a means for rivals to demonize each other. In 2009, for example, the Abhisit administration established a highly controversial Internet monitoring project, leading to a sharp increase in lèse-majesté cases. The government's online surveillance capabilities were considerably strengthened under Prime Minister Yingluck Shinawatra, who approved a directive giving the national police the ability to collect evidence without a court order (Crispin 2013). The NCPO has attempted to implement a single gateway for Internet traffic to better control local and foreign websites it deems problematic. The Ministry of Information and Communication Technology has blocked tens of thousands of Facebook pages for posting materials seen as critical of the monarchy. Even Internet users who "like" online content critical of the royal family can be charged under the Internal Security Act or the Computer Crime Act (Crispin 2013).

The polarization of Thai society and media that reinforce this, the regulatory uncertainty around many of the media laws, the chilling effects of the lèse-majesté provisions and other laws, and recent increases in harassment and attacks on journalists—all work to silence dissent and promote self-censorship and therefore are targeted for reform.

Reform Efforts

For four days in May 1992, massive street demonstrations erupted in Thailand after Army Commander Suchinda Kraprayoon was appointed as prime minister, despite announcing in February during the coup he had led that he did not intend to hold power. Violence broke out between angry protestors and the police and army, resulting in 52 officially acknowledged deaths, over 3,500 arrested, 500 injured, and 200 disappeared. The mainstream press largely supported opposition protestors, many journalists resisted censorship, and 33 were injured covering the protests. The violence stopped when King Bhumiphol called General Suchinda and opposition leader Chamlong Srimuang to a televised audience, urging them to resolve the crisis. This royal intervention is often cited as a key moment in Thailand's democratic development, and the 1992 Black May events as jumpstarting Thailand's media reform movement.

The Thai public raised severe criticism of the five television stations that kowtowed to the military demands for censorship, and began to call for privately owned broadcast news organizations (Ramasoota 2007) and for more grassroots, people's media. This led to the establishment of iTV, Thailand's first nongovernmental television station, which eventually became Thai PBS, the Southeast Asian region's first nongovernmental, noncommercial public service broadcaster. The Black May events also ushered in the country's most progressive constitution, dubbed the 1997 People's Constitution, which guaranteed the right to freedom of expression and the public's right to *access* the airwaves in addition to media professionals' right to conduct their work free from state interference. This led to the establishment of community radio as an element of the Thai media landscape, especially as a 2000 law defining broadcast frequency allocations required that 20 percent of the airwaves be set aside for community radio and television.

Thaksin Shinawatra had a contested relationship with the media and media reform efforts. He was well liked among rural villagers for his populist appeal and was elected with an impressive victory for his Thai Rak Thai (TRT) Party in the 2001 and 2005 general elections, but he disliked critique and reacted strongly to it. He worked to silence dissent through his cronies' manipulation of large advertising budgets; through government manipulation of the advertising budgets of state enterprises, benefitting supportive media; and through the use of multiple high-stakes libel and defamation suits to intimidate journalists. He threatened foreign journalists with expulsion and banned an issue of the *Economist* for its critical coverage. Under his administration, the Anti-Money Laundering Office (AMLO) began financial investigations of 247 prominent journalists and civil society activists (Thitinan 2003). As Thaksin's attacks on mainstream media became legend, civil society reformers began looking to community radio and the Internet as options for improved communication.

Thaksin also diluted the community radio movement in 2003 by passing a cabinet resolution allowing these stations to air six minutes of advertising per hour. This made community radio suddenly appealing to politicians and business owners who had previously rented airtime from government-owned stations, and confused the Thai public over the meaning and purpose of community radio. The number of community stations jumped from about five hundred to more than two thousand within three months (Brooten and Klangnarong 2009). Since then, community radio proponents have worked to more clearly define community radio and develop their networks of local, nonprofit, publicly accessible stations. They are working to influence the NBTC in the development of rules and procedures for community radio licensing.

Activist groups have campaigned against the 2007 Computer Crimes Act's draconian provisions, prompting a government-appointed review of the law's use in lèse-majesté cases by the Human Rights Commission, criticized for its lack of transparency and independence. In addition, journalist and advocacy organizations opposed the closed-door redrafting of the CCA in 2013—the new draft law would eliminate judicial oversight for blocking websites, for example—and called for a more open, consultative process in redrafting the law. In late 2015, the NCPO is reconsidering its plan to establish a single Internet gateway after activists hacked government websites in protest and business organizations objected. Many journalists and reform advocates have called for the decriminalization of libel. The Article 112 Awareness Campaign had its kickoff event in March 2011, but lost steam as both the Abhisit and Yingluck administrations vowed publicly not to address the lèse-majesté issue and to do what they can to protect the monarchy, a trend reinforced by the 2006 and 2014 coups.

The main thrust of media reform efforts has shifted from the early 1990s strategies of promoting independent and people's media alongside regulatory change; regulatory change remains key, but much attention is now focused on the polarization of media and violence against journalists, and the cultural shift it will require to change this. Media reform in Thailand is significantly obstructed for the foreseeable future, as libel is a criminal offense, and the overuse of lèse-majesté provisions (and resultant self censorship) cannot be seriously addressed until politicians and lawmakers address the issue openly. This is, however, a matter far too touchy to be raised by most politicians wanting to stay in office. In addition, the regulatory process has yet to catch up with practices on the ground, and this means that media in Thailand that can provide people with access in fulfillment of their right to communicate, primarily community radio and the Internet, must continue to operate in legal limbo. The few NBTC commissioners charged with representing the public interest concur with critics worried that the regulatory body is stacked with those representing military and business interests. While the NCPO has organized a National Reform Council whose mandate includes media reform, most council members represent established players and focus on institutional, legal and regulatory change; very few represent new media sectors that provide citizen access. Unfortunately, until stable laws and regulations are established, along with a way to safely discuss the future of the monarchy in Thai society as well as the needs of rural or otherwise marginalized sectors of society, Thai people will need to struggle to protect their right to be heard.

MYANMAR

The British established direct rule of (formerly) Burma beginning in 1824, ousting the monarchy and separating religion and the state, all under a pretense of colonial benevolence. Yet these moves squashed what had been a very lively press under the Burmese King Mindon, an advocate of press freedom, and the growing use of media to protest imminent colonial rule. The country's first indigenous press law protected freedom of the press and promoted a critical and active press prior to colonization by the British, in contrast to the rest of Southeast Asia. After independence from the British in 1948, Burma had one of the freest press systems in Southeast Asia, with constitutional protections for press freedom in the 1947 constitution.

This lasted until the takeover by the military in a 1962 coup, when the media quickly became a government monopoly—especially the broadcast and daily newspaper sectors—with all media becoming the victim of direct and severe censorship laws. During massive protests against military rule in August and September 1988, there was a brief "democracy spring," in which the media landscape opened and flourished with critical assessments of the political situation. But the military clamped down on September 18, taking control again of the media landscape with its censorship body, the Press Scrutiny Board, later renamed the Press Scrutiny Registration Division (PSRD). Nevertheless, while many journalists were forced to practice self-censorship, there have always been Burmese journalists risking their safety to challenge the authorities.

Prior to recent changes beginning in 2010, the Burmese had learned to live in an environment of surveillance and fear, in which to speak was often dangerous and important information therefore most often was circulated in hushed voices, rumors, and jokes. A private sector of weekly and monthly journals developed that grew significantly during the late 1990s and early 2000s, but these publications had to be cleared by the PSRD. Until very recently, less than 1 percent of the total population has had access to the Internet, and those who did took risks if they used it to access political information. There was virtually no space for independent Burmese journalism. Even the types of critical commentary found in traditional Burmese entertainment were strictly curtailed, and there was no government tolerance for media devoted to social change. Independent media were only able to develop and grow outside of the country.

Independent media were established in the border areas after the massive uprisings against military rule in 1988, when dissident students met up with ethnic minority rebel groups also struggling against the brutal military regime. Some of these dissidents established media in exile, and their work over the last twenty-five years has helped pave the way for the reforms of today (Brooten 2008, 2011, 2013). The exile Burmese media can be conceptualized as having consisted, for the most part, of two groups: a prominent, better funded and developed, ethnically "unmarked" media; and a sector of smaller, less developed and less well-funded, ethnic-focused media. There also existed an important sector of international broadcasters with Burmese services, such as BBC, Voice of America, and Radio Free Asia. By the 2007 uprisings, dubbed the Saffron Revolution, and the 2008 Cyclone Nargis, the internal and external media had developed significant networks that could get news from inside the country, and then broadcast it to the world and back into the country.

In 2010, military control was loosened and a rapid and largely unexpected series of changes began. The government released political prisoners, including journalists and bloggers; closed the prepublication censorship board; stopped blocking international and Burmese exile news websites; decreased its surveillance and harassment of journalists; and allowed freedom for the publication of many kinds of publications and stories that would not have passed the censors in the past. Also, the government permitted the establishment of independent professional journalists' associations, established an interim Myanmar Press Council charged with improving conditions for journalists and helping to draft new media laws, and began licensing private daily newspapers, thus breaking the government monopoly on the dailies. It also began issuing licenses for ethnic language media, after decades of preventing the use of ethnic languages in schools and private media. The government also extended invitations to the exile media groups, many of which have now opened offices in Yangon or elsewhere in addition to or replacing their offices in exile. Since 2014, however, the country's rapid rate of reform has declined sharply, in some cases reducing the space that had recently been gained. Concerns include the increasing harassment and attacks on journalists and, in the case of journalist Par Gyi, suspicious death at the hands of the military; arrests, detention, and imprisonment of journalists; use of laws against journalists, especially defamation and the penal code; surveillance and intimidation of journalists from nationalist groups; and a worrying increase in hate speech on social media. Media practitioners have learned that emerging from decades of military rule and strict censorship takes time and requires challenging militarized approaches to communication and media.

Reform Efforts

The country's current media reforms reflect both the significance of recent changes and the continuing struggle against common attitudes toward information control prevalent since military rule began in 1962. Until recently, journalists from inside were rarely able to travel to neighboring countries for training or workshops, attending only those few held inside the country clandestinely. Foreign funding for media development has increased significantly since 2010, bring with it a greater number of conferences, workshops, and training sessions devoted to media inside Myanmar. In addition, expectations of a rapid increase in mobile phone and Internet access means the influence of online media will increase, worrying those concerned with the recent increase in online hate speech.

Media reformers are largely focused on the revision of the legal and regulatory framework for media; the safety of journalists; establishing sustainable financing; and challenging old notions of control by pushing for greater government transparency and access to government information as well as active citizen involvement in media and decision-making. These efforts have involved ongoing negotiations between government officials, the interim Myanmar Press Council (MPC), other groups of journalists, and the public. In March 2014, the media and the printing and publishing registration laws were both ratified by parliament and signed into law after a tumultuous start in which the MPC and local media associations protested the government's draft laws for ignoring MPC amendments. Critics charge that the Printers and Publishers Registration Law, while removing the legal threat of imprisonment for journalists, also gives the government the right to withhold media licenses and bans reporting harmful to "national security" or that "insults religion" or violates the constitution. In October 2014, a broadcast bill passed with little debate despite some concerns over the makeup of its mandated broadcasting council. Other draft laws such as the Public Service Media Bill and a telecommunication law have undergone much debate but as of late 2015 remain unresolved. The public service draft law proposes that the state-run media become public media, including state-run daily newspapers, worrying critics who believe the result will be a publicly funded, pro-government media conglomerate. In addition to new laws, the high cost of entry to the media sector limits opportunities for a wide diversity of media owners. This is evidenced by the number of newly licensed daily papers that folded within a year or so of struggling with the saturated market and the lack of substantial ad revenue against the stronger, often government or military-affiliated players.

This difficult economic environment also affects the integration of internal and formerly exile media, and the inclusion of ethnic minority voices in the media landscape. Current reforms are influenced by the newly returned and returning exile media, including the ethnic media, which developed over the last two and a half decades in exile. One particularly active ethnic media network is Burma News International (BNI), established in exile as a consortium of smaller ethnic-focused media. For over a decade, BNI members have produced online news and print publications, adding weekly radio programs in multiple languages along the way. BNI has now set up inside the country and is helping member groups apply for publishing licenses in ethnic languages. Member groups have the capacity to provide news from difficult-to-access rural areas and move beyond the perspectives of the local ethnic leaders, often the primary sources in coverage by urban-based media with little experience and few contacts in the area. BNI has also lobbied for better media laws, insisting that ethnic perspectives be taken into consideration in the drafting process and ethnic representatives be included in the press council. Nevertheless, some ethnic issues remain extremely sensitive, especially the situation in Rakhine State, where violence between Buddhists and Rohingya Muslims has continued sporadically since it broke out in June 2012. Inclusion of ethnic and minority voices seems especially daunting given the increasingly commercial nature of the country's media and the trend toward ownership consolidation, often with conglomerates owning significant nonmedia interests. It remains to be seen if the mandate to cover ethnic concerns becomes a priority for the government, and remains a commitment of the formerly exile media as they are weaned off of foundation support. It also depends on whether the ethnic media themselves can make a financial go of it.

Reform efforts in Myanmar focus on cultural change, as the country struggles to rethink its relationship with information and media. This includes attempts to move away from militarized approaches to information as a means to maintain power, such as harassment, coercion, threats and attacks on media as means of silencing dissent. It also includes opposing the use of libel and defamation suits as a strategy for silencing critics, and countering the difficulties journalists face in getting information from government officials hesitant to speak unless they have permission from supervisors. The process of establishing the press council and drafting media laws has already strained the relationship between the community of media practitioners and the former generals now at the helm in civilian clothes. Press freedom remains a strong demand. These are all significant challenges for reformers. If current efforts are any indication, however, there is much motivation and creativity to draw from among the Burmese peoples.

PHILIPPINES

The Philippine media have been significantly influenced by the various periods of occupation. The Spanish influence is evident in the Philippines' feudal patron-client relationships, and nearly fifty years of US occupation, from 1898 to 1946, have left their mark on the Philippine constitution and laws. All four Philippine constitutions have ensured press freedoms and the right to freedom of expression, and the 1987 Constitution guarantees freedom of information. Yet on the ground, these constitutional provisions and the country's liberal press freedom tradition have not guaranteed free expression and have provided little protection for Philippine journalists and activists, whose frequent killings challenge the widely made claim that the Philippines has one of the freest media systems in Southeast Asia.

The media in the Philippines have historically been controlled by a small elite, especially family clans who maintain their immense power through patronage and intimidation and their relationships with the presidential palace (Coronel et al. 2007; Florentino-Hofileña 2001). The state has often targeted journalists, dissidents, and activists, and even when not a direct threat, the state has participated in the violence by allowing extrajudicial killings to go unpunished. Elements of the police and military have carried out human rights violations in the countryside without distinguishing between armed groups and legitimate, unarmed political parties, according to the governmental Commission on Human Rights

(Article 19 and CMFR 2005). The government also frequently uses antiterrorism measures to suppress dissent, such as withholding permits for public demonstrations or making threats against legal left-wing groups. This can be dangerous for journalists, especially when reporting on violations by military, police, or other armed forces (Arguillas 2001).

There is a widespread culture of impunity, with many killings and few convictions. According to statistics from the press freedom advocacy group, the Center for Media Freedom and Responsibility (CMFR), two days before the commemoration of the International Day to End Impunity on November 2, 2015, gunmen shot the 150th journalist killed in work-related violence since martial law ended in 1986. While gunmen have been convicted in 15 cases, no masterminds have yet been prosecuted, with the exception of alleged masterminds in the ongoing trial involving the brutal 2009 Amputuan massacre (CMFR 2015).[2] In this case, 58 people, including 32 journalists and their staff, were murdered in local election-related violence in what watchdog groups have called the world's worst case of violence against journalists. In 2011, a global campaign marked the anniversary of this massacre by declaring it the International Day to End Impunity. Six years after the massacre, the defense in the Amputuan trial has successfully used delaying tactics and the trial stretches on. In 2014, the Philippines ranked as the third worst country worldwide for the fifth consecutive year in the Committee to Protect Journalism's (CPJ 2013) Impunity Index, which identifies countries where journalists are murdered regularly yet their killers go free. In 2015 the Philippines moved to fourth place, alongside Iraq (second), Syria (third) and South Sudan (fifth).

The culture of impunity extends as well to corruption within the ranks. The Philippine media scene is highly competitive, resulting in salaries that often do not provide a living wage and rendering journalists vulnerable to corruption. Common practices include accepting envelopes from politicians at press conferences, called "envelopment journalism," a pun on "development journalism"; a more contemporary version of this practice called "ATM journalism"; and AC/DC or "attack-collect, defend-collect" journalism, in which one political rival is played against another as a means of financial gain (Florentino-Hofileña 2004). These practices are especially prevalent in the provinces where most of the media killings have occurred.

Reformers argue that the culture of impunity, along with the commercialism and competitive nature of the media, promote sensationalism and undermine civic responsibility and ethics, foster an "us versus them" framing and a focus on personalities rather than issues, and make coverage of issues facing the rural poor an enormous challenge for journalists. News reports are often sensational in order to boost circulation. The prevalence in radio broadcasting of block timing—the practice of buying blocks of airtime, often for political or partisan purposes—contributes to sensational reporting and violence against its practitioners. In addition to physical threats, reporters expected to cover human rights violations or complex stories such as the Muslim insurgency in the southern island of Mindanao have difficulty understanding the issues; they are often under time and budget constraints that make it difficult to travel and meet with those whose situation they are covering. They rely overwhelmingly on government sources, and often leave out the perspectives of those most affected (Arguillas 2000; CMFR 2015; Article 19).

Reform Efforts

The primary issue for media reformers in the Philippines is their challenge to the culture of impunity that continues to rob Philippine journalists of their lives, and in particular, the lack of progress in the Amputuan massacre trial. Reform advocates also continue to struggle to decriminalize libel, to pass a Freedom of Information Bill, and to prevent the passage of the Cybercrime Prevention Bill enacted on October 3, 2012, without any input from journalists or journalist organizations; after public demonstrations against the bill, the Supreme Court suspended the law's implementation for 120 days. The CMFR called the law "the worst assault on free expression since Ferdinand Marcos declared martial law 40 years ago." In February 2013, the Supreme Court extended until further notice the temporary restraining order it had issued in October 2012, and as of late 2015, unresolved attempts had been made to repeal or amend the law. Civil society advocates continue to lobby for a freedom of information bill after various drafts have met with resistance and indifference. Several bills have been introduced as advocates challenge the country's penal code, which makes libel a criminal offense punishable by a prison term and, in some cases, large fines. In October 2011, the United Nations Human Rights Commission declared criminal sanction for libel in the Philippines to be excessive and in violation of the International Covenant on Civil and Political Rights, to which the Philippines is a signatory.

Efforts to involve communities in improving local coverage or contributing to more inclusive media productions have been challenging due to the media's commercial focus. Although local commercial media do exist, there is no regulatory or legislative provision for nongovernmental, noncommercial community broadcasting. The current licensing system requires a congressional franchise, available only to those with financial means. The few attempts at developing independent community media have been either unsustainable or the target of attack. There is a history of development-oriented community media in the Philippines, but this is no longer as vibrant a sector as it once was and is tainted by its history under Ferdinand Marcos, who used development as a means to justify his rule (Lex Librero, personal communication, June 25, 2008). Many development-oriented community radio stations are difficult to sustain once foreign funding is withdrawn, and while some are still functioning, most have by necessity affiliated with local government units, universities, or churches for support, and for this reason are often top down in their approach, and studiedly apolitical.

Reform advocates have worked to encourage media to be more inclusive of minority and nonelite perspectives or to develop alternative means of sharing information. The block-timing practice is one way that journalists can access the airwaves to share alternative or critical viewpoints. A push for civic or public journalism also promotes change by focusing attention on community problems while maintaining a commitment to journalistic objectivity and detachment (Santos 2007, 10). Civic journalism emphasizes what people are doing and can do to solve problems, demonstrating that local people "are not powerless . . . and that their voices matter" (Batario 2004, 14). Social movements and civil society organizations produce a myriad of small-scale media to publicize their work and causes. There also exist several alternative, independent media production groups such as Bulatlat, Kodao Productions, and Southern Tagalog Exposure, that provide alternative viewpoints. And the Philippines is a top user of Facebook and Twitter, both of which play a significant role in the contemporary media landscape in the Philippines.

CONCLUSION

Media reform advocates in each of these countries are constructing contextually specific visions for cultural transformation that will provide space for a broadened set of communication rights. Since each of these countries faces significant conflict, media reform efforts understandably prioritize the protection of media workers (and society generally) from physical harm, but they also focus on challenging impunity, corruption, coercive policies, and a lack of transparency in media and information policies. There is a demand to strengthen the legal and regulatory frameworks to address these issues more effectively. And in all three countries, reform advocates are working to be more inclusive of multiple sources and perspectives in mainstream media, and promoting alternative media and other diverse sources of information. A significant component of reform efforts in each of these countries is the push for a broadened set of communication rights beyond press freedom, including people's "right to communicate," or to have their voices heard in public debates. This is important because it is the collective communication efforts of marginalized groups that tend to illuminate the structural issues most pressing in efforts for social justice. These are the voices policy-makers need to hear most. The effort to increase transparency and expand the space for diverse voices includes the push to decriminalize libel, promote freedom of information, and nurture diverse forms of small-scale media accessible to a wide variety of people. It is only through efforts like this that genuine change will be built, protest by protest, program by program, policy by policy.

1. Establishing a definitive number of coups and attempted coups have proven difficult. These figures are based on a tally made by the community of scholars at New Mandala, http://asiapacific.anu.edu.au/newmandala/2011/03/08/counting-thailands-coups/.

2. The number of killings depends on who is counting, and how they determine that a journalist was killed in the line of duty. CMFR is more conservative in this count; as of early November 2015, the National Union of the Journalists of the Philippines counts 169 journalists killed in the line of duty since martial law ended in 1986.

REFERENCES

Arguillas, C. 2001. "Human Rights Reporting on the Philippines' Rural Poor: Focus on Mindinao." In *Media and Human Rights in Asia: An AMIC Compilation*, edited by AMIC, 31–44. Singapore: AMIC.

Article 19 and Center for Media Freedom and Responsibility (CMFR. 2005. *Freedom of Expression and the Media in the Philippines*. London and Manila: Article 19 and CMFR.

Askew, M. 2012. "The Ineffable Rightness of Conspiracy: Thailand's Democrat-ministered State and the Negation of Red Shirt Politics." In *Bangkok May 2010: Perspectives on a Divided Thailand*, edited by M. Montesano, P. Chachavalpongpun, and A. Chongvilaivan, 72–86. Singapore/Bangkok: ISEAS and Silkworm Books.

Batario, R. 2004. *Breaking the Norms*. Quezon City: Center for Community Journalism and Development.

Brooten, L. 2008. "'Media as Our Mirror': Indigenous Media in Burma." In *Global Indigenous Media: Cultures, Practices and Politics*, edited by P. Wilson and M. Stewart, 111–127. Durham, NC: Duke University Press.

———. 2011. "Media, Militarization and Human Rights: Comparing Media Reform in the Philippines and Burma." *Communication, Culture and Critique* 4, no. 3: 312–332.

———. 2013. "Beyond State-Centric Frameworks: Transversal Media and the Stateless in the Burmese Borderlands." In *New Agendas in Global Communication*, edited by K. Wilkins, J. Straubhaar and S. Kumar, 142–162. New York: Routledge.

Brooten, L., and S. Klangnarong. 2009. "People's Media and Reform Efforts in Thailand." *International Journal of Media and Cultural Politics* 5, nos. 1–2: 103–117.

Center for Media Freedom and Responsibility (CMFR). 2013. "Impunity Reigns as Killings Continue." November 25. Manila: CMFR. http://cmfr-phil.org/endimpunityinph/2013/11/impunity-reigns-as-killings-continue/.

———. 2015. "Alleged masterminds in Palawan broadcaster slay arrested in Thailand." September 21. Manila: CMFR. http://www.cmfr-phil.org/2015/09/21/alleged-masterminds-in-palawan-broadcaster-slay-arrested-in-thailand/.

Committee to Protect Journalists (CPJ). 2014. "Getting Away with Murder." https://cpj.org/reports/2014/04/impunity-index-getting-away-with -murder.php.

Coronel, S., Y. Chua, L. Rimban, and B. Cruz. 2007. *The Rulemakers: How the Wealthy and Well-Born Dominate Congress*. Philippine Center for Investigative Journalism. Pasig City: Anvil Publishing.

Crispin, S. W. 2013. "In Asia, Three Nations Clip Once-Budding Online Freedom." Committee to Protect Journalists. http://cpj.org/2013/02/attacks-on -the-press-internet-opening-is-shrinking.php#more.

Florentino-Hofileña, C. 2001. "Travails of the Community Press." In *Investigating Local Governments: A Manual for Reporters*, edited by C. Balgos, 35–77. Metro Manila: Philippine Center for Investigative Reporting.

——. 2004. *News for Sale: The Corruption and Commercialization of the Philippine Media*. Manila: Philippine Center for Investigative Journalism.

Ramasoota, P. 2007. "Freedom of Expression in Thailand During the Thaksin Administration." In *Communication and Human Rights*, 128–161. Nakornpathom, TH: Office of Human Rights Studies and Social Development, Mahidol University.

Santos, V. 2007. *Civic Journalism: A Handbook for Community Practice*. Manila: Philippine Press Institute.

Thitinan, P. 2003. "Thailand: Democratic Authoritarianism." *Southeast Asian Affairs 2003*, 277–290. Singapore: Institute of Southeast Asian Studies.

Thompson, M. R. 2013. "Does the Watchdog need Watching? Transitional Media Systems in Southeast Asia." Working Paper Series no. 145. Hong Kong: Southeast Asia Research Centre, City University of Hong Kong.

CHAPTER TWENTY-FOUR

Media Reform through Capacity Building

PETER TOWNSON, *Doha Center for Media Freedom, Qatar*

MEDIA REFORM STRATEGY

The rise of online media and citizen journalism, as well as regional coverage of the recent revolutions in the Arab world, reaffirms the importance of upholding international standards of journalism. The Doha Centre for Media Freedom (DCMF) has introduced numerous capacity-building initiatives to improve the media and information literacy among media consumers in Qatar and in the region. In addition, the DCMF organizes training programs across the region, aimed at developing the skills of journalists and defending media freedom in the Arab world. The center also conducts workshops across the wider Arab region, with a particular focus on Syrian journalists covering the ongoing conflict. While the jury is still out as to the success of both programs, some of the effects of promoting the DCMF's Media and Information Literacy (MIL) program throughout Qatar are already visible as more students opt to engage in media production. In a region where the right to information and free expression are too often denied, educating a young generation of media consumers who recognizes the importance of defending these rights is an essential aspect of DCMF's mission.

The Doha Centre for Media Freedom (DCMF) is a nonprofit press freedom organization, which works to support quality, responsible journalism and seeks to guarantee the safety of journalists worldwide, with a specific focus on the Arab region.

Since its re-establishment in 2011, the center has been working in a particularly fragile and restless environment. The Arab Spring, which some hoped would blossom into democratic governance, has adversely brought significant political instability. Unfortunately, one of the most prevalent consequences of this instability has been the increased number of violations committed against journalists and the impunity enjoyed by the perpetrators of attacks against media workers. Indeed, the three countries with the highest numbers of journalists killed in 2013 are all located in this region: Syria, Iraq, and Egypt.

In parallel to these developments, the Arab region has witnessed an exponential growth in social media usage, enabling citizens to use mobile and online platforms as journalistic tools. Yet, growing concerns have been raised about the state of the media in the region: the professionalism of Arab journalists—whether traditional or citizen journalists—working in highly polarized contexts, is increasingly questioned, and so is the quality of the message they deliver, which has in most cases become impossible to separate from its political agenda.

DCMF has responded to these recent concerns through four main programs: training, research and monitoring, emergency assistance, and MIL education. Through its emergency assistance program, the center offers support to journalists who are being persecuted because of their work. Over the past year, the center has been able to assist more than sixty journalists in distress across the globe through relocation, medical, start-up, or legal support. The center is committed to promoting quality responsible journalism and to enabling journalists to continue to carry out their work despite the difficulties they face, and therefore assists journalists in a sustainable manner, with a focus on allowing media workers to continue with or return to their work.

DCMF also engages in advocacy work for journalists in distress, helping to raise awareness of their cases with government officials and joining international campaigns calling on media freedom to be respected. Another essential aspect of DCMF's work revolves around monitoring press freedom violations in the region and covering news stories from around the world on a daily basis on its website.

DCMF's capacity-building efforts have resulted in the training of some eight hundred journalists, and the center has conducted MIL workshops for more than six hundred students in ninety schools in Qatar, Bahrain, Jordan, and Egypt. This program is aimed at producing significant reform within the media sector in Qatar and the region, through developing the capacity of news consumers and creators alike.

MEDIA AND INFORMATION LITERACY TRAINING

The starting point for the center's MIL program was its home country, having noted the need for the development of critical thinking throughout the Qatari youth. Qatar is an incredibly wealthy nation, and much of its wealth has been invested in education with a National Vision for 2030, which places emphasis on developing human potential (GSDP 2008). DCMF firmly believes that entrenching the concept and value of media freedom is essential to developing the nation's human capacity. Like any society, educating a young generation of people who appreciate the importance of information and fully comprehend that information can be manipulated by media producers, is an integral aspect of human development. The distinct lack of critical thinking or MIL being taught in schools in the region has been highlighted by researchers (Abu Fadil 2007) as well as the global realisation that MIL is an increasingly important subject for governments to factor into educational programs (Torrent 2011). According to the United Nations Educational, Scientific, and Cultural Organisation (UNESCO), MIL can be defined as "the ability to access, analyse, evaluate and create media in a variety of forms."

According to initial research carried out at the center, 80 percent of schools in Qatar did not teach MIL prior to 2011, but over 90 percent of students were exposed to various forms of media on a daily basis. As a result, DCMF identified a significant need for MIL education to be introduced to students in the country, and it did so under the slogan: "Reading and writing doesn't do it anymore—you need to be media literate!"

The multicultural nature of Qatar's population necessitates improved understanding of the media and how news and information is produced and presented. With locals making up only around 15 percent of the country's population, Qatar (and especially Qatar's schools) brings together people from all over the world who interact on a daily basis. Providing news which can be consumed by all is an almost impossible task, but through its MIL programs, DCMF hopes that future generations will be able to consume their news critically, regardless of their backgrounds.

The program was initially launched in 2011, incorporating some 35 students from four schools and the initiative has gone from strength to strength. By the end of February 2014, the program will have been run in eighty-nine schools, helping to train some 130 teachers and coordinators and educating nearly 600 students.

DCMF's MIL program does not simply focus on the theoretical aspect of media freedom or consumption, but also brings young people to the newsroom, teaching them the technical aspects of journalism. By working on DCMF's team of junior reporters, these youngsters have the opportunity to learn how news is produced, experiencing firsthand the importance of editorial decisions in terms of the value and reliability of a piece of news.

In line with the program's aims to develop cross-cultural understanding and promote dialogue, the majority of events covered by the Junior Reporters have been related to human rights and multiculturalism in one form or another. For instance, DCMF's team of junior reporters have covered the Fourth Forum of

the Alliance of Civilisations held in Qatar in 2011, the Tenth Interfaith Dialogue Conference, held in Qatar in 2012, and the UN Global Forum on Media and Gender, hosted in Bangkok in 2013, among others. The experience of attending and covering these events will prove valuable for the students in later life as they tackle questions of media ownership and the intentions behind a piece of reporting.

DCMF has been working alongside the Supreme Education Council of Qatar to introduce the subject to schools across the country, but at the moment, it is an extracurricular subject and not a part of the mainstream curriculum. However, the center aims to see the adoption of a curriculum which includes a dedicated subject on MIL by 2015.

As part of DCMF's strategy for developing MIL in the region, the center has conducted a number of training workshops in other Arab countries, namely Bahrain, Egypt, and Jordan. Regional collaboration and development of MIL initiatives remains an integral aspect of the center's strategy, and with increased experience and success from its work in Qatar, the center is looking forward to exporting its program to more countries in the region in the future.

Developing a network of MIL trainers in the region is another important aspect of DCMF's program, and in May 2013 the center organized the first MIL Train the Trainer program for teachers in the Arab region, certifying twelve new trainers, six of whom were Qatari nationals.

In June 2013, the center hosted an experts' meeting on MIL, where the Doha Declaration on Supporting Media and Information Literacy Education in the Middle East was adopted (Townson 2013). The declaration includes a number of recommendations for successfully introducing MIL programs based on case studies discussed during the meeting, such as the importance of developing more Arabic-language training materials for teachers, introducing a system for monitoring and mapping MIL education across the region and forming a steering committee of experts to share expertise and represent the region at international events, among other recommendations.

TRAINING JOURNALISTS

Equally important to developing media freedom in Qatar and the rest of the region is fostering the creation of a media landscape that entrenches standards of professionalism and ethics. Values such as objectivity and neutrality are often disregarded by journalists in the Arab world, with many believing that they are somehow inadequate tools for covering the situations facing people in the region. As a result, DCMF has focused some of its training efforts on promoting international standards of quality journalism in programs it has conducted in Qatar and the wider region.

As well as improving general standards and promoting professionalism throughout the industry, DCMF has also identified the significance of conducting safety training programs, and in 2012, the center launched its Ali Hassan Al Jaber safety training for journalists program (Townson 2012), with a focus on journalists in conflict and danger zones.

The program has been conducted in a number of countries, in partnership with organizations and individuals and tailored to meet the specific needs of journalists in the area. The Ali Hassan Al Jaber Safety Training program for journalists is one of the center's flagship initiatives. By conducting eighteen safety training workshops for journalists from across the region in countries such as Yemen, Egypt, and Turkey as well as further afield in the rural areas of Pakistan, DCMF has managed to provide key skills to journalists who find themselves in danger on a regular basis in the course of their work.

However, the program does not ignore the need to develop professional skills, which are essential for protecting media freedom and media workers in the long term. To this end, the center has also carried out training workshops in Iraq, Jordan, Libya, Palestine, and Qatar. But the reach of the center's capacity building program extends beyond the borders of host countries, and often journalists from a particular country are transported elsewhere due to safety concerns (for instance, workshops in Jordan and Turkey were specifically for Syrian journalists and citizen journalists).

In total, DCMF has trained around eight hundred journalists since its relaunch in 2011, and there are already a number of training workshops planned for the coming months, some of which will focus on areas such as cybersecurity for citizen journalists, to help combat the increased online targeting of journalists by governments.

CONCLUSION

Emphasizing the role of MIL education in a region where journalism has taken various meanings and forms is one of the DCMF's core strategies, as the center sees the discipline as a central tool in promoting quality journalism and guaranteeing access to impartial information. Educating citizens about the role of the media while teaching youth how to evaluate media content with critical thinking is key in shaping well-informed and well-equipped media consumers and producers.

Qatar's recent surge in development and the rapid pace at which the country is growing mean that developing human capacity is absolutely essential. The nation's increasingly significant role in the realm of global politics means that understanding what is being written about Qatar, its foreign policy, and its international involvement is similarly important. Combining these factors with the sharp increase in the prominence of social and online media, it is clear that developing citizens able to discern between fact and fiction, news and views, is no longer a theoretical, but a practical necessity.

DCMF has been pioneering efforts to achieve this goal in Qatar, and has already begun to share its expertise in the field with other organizations in the region. While MIL in general is a relatively young concept, and teaching materials, research, and general information on the subject is somewhat lacking, it is becoming more widely recognized as a pressing concern with the realization that media will continue to play an integral part in the daily lives of future generations. DCMF will continue in its mission to develop culturally aware, media-literate critical thinkers who are able to effectively interpret the vast array of information with which they are presented, and remains strict in its belief that this is a media reform strategy which is essential to defending media freedom and promoting responsible, quality journalism across the globe.

However, it is not the only way of achieving this end and DCMF recognizes the importance of developing the capacities of those behind producing the news. Improving standards of professionalism and upholding journalistic ethics will contribute to the overall protection of media freedom as well as helping journalists themselves at a time when impunity abounds and media workers are targeted on an increasingly regular basis. Improving the professionalism of journalists will also assist consumers and make their jobs of being more media literate a less complicated task, thus contributing toward the defense of media freedom as a whole.

DCMF has always adopted a holistic approach to dealing with issues facing journalists and journalism in the Arab world, and one of the ways it has developed its long-term strategy is by identifying the importance of capacity building. In line with other international plans, such as the UN Plan of Action on the Safety of Journalists and the Issue of Impunity, DCMF has identified the importance of adopting a multistakeholder approach and developing partnerships to achieve its aims. Working alongside organizations with similar aims and sharing expertise is the best way to implement effective and sustainable change.

Training journalists to produce responsible news and training their readers, listeners, and viewers to be able to identify this, has never been more important than it is now.

REFERENCES

Abu Fadil, M. 2007. "Media Literacy: A Tool to Combat Stereotypes and Promote Intercultural Understanding." Research paper prepared for the UNESCO Regional Conferences in Support of Global Literacy.

General Secretariat for Development Planning (GSDP). 2008. *Qatar National Vision 2030*. www.gsdp.gov.qa/portal/page/portal/gsdp_en/qatar_national _vision/qnv_2030_document/QNV2030_English_v2.pdf.

Torrent, J. 2011. "Media Literacy, Congratulations! Now, the Next Step." *Journal of Media Literacy Education* 3, no. 1: 23–24.

Townson, P. 2012. *DCMF Outlines Ali Hassan Al Jaber Safety Training for Journalists Programme*. Doha Centre for Media Freedom. www.dc4mf.org/ en/content/dcmf-outlines-ali-hassan-al-jaber-safety-training-journalists -programme.

————. 2013. *DCMF Media and Information Literacy Meeting Adopts Doha Declaration*. Doha Centre for Media Freedom. www.dc4mf.org/en/content/ dcmf-media-and-information-literacy-meeting-adopts-doha-declaration.

CHAPTER TWENTY-FIVE

Media Reform in Guatemala

MARK CAMP, *Cultural Survival, Guatemala*

MEDIA REFORM STRATEGY

The use of community radio by Indigenous Peoples in Guatemala is a vital source of self-expression. In the light of the continuing refusal of Guatemalan governments to provide Indigenous communities with legal access to community radio stations, the use of legal challenges together with the emergence of citizen-based low-power FM stations mounts pressure for reform. These activities have been complemented by intensive lobbying by local residents for constitutional change and the presentation of a petition that is currently being considered by the Inter-American Commission for Human Rights (IACHR) of the Organization of American States. The legalization of community radio is a crucial battle for freedom of expression for Indigenous groups and an increasingly salient issue for Guatemalan citizens.

Cultural Survival envisions a future where all Indigenous Peoples live by their inherent rights deeply and richly interwoven in their aboriginal lands, native languages, spiritual traditions, and dynamic cultures; and where Indigenous rights are honored through self-determination. Cultural Survival advocates for the rights of Indigenous Peoples and supports Indigenous communities in their self-determining work for cultural and political survival.

The impulse for the founding of Cultural Survival arose during the 1960s with the opening up of the Amazonian regions of South America and other remote regions elsewhere. As governments all over the world sought to extract resources from areas that had never before been developed, the drastic effects this trend had on the regions' Indigenous Peoples underscored the urgent need to defend the human rights of these victims of progress.

Today, the organization promotes the rights of Indigenous communities around the globe. All of the work is predicated on the United Nations Declaration on the Rights of Indigenous Peoples. In 2007, after twenty-five years of negotiation, the UN General Assembly adopted the Declaration on the Rights of Indigenous Peoples. This document, which was created by Indigenous representatives working with government representatives, is the fundamental document spelling out the distinctive rights of Indigenous Peoples. These include the right to live on and use their traditional territories; the right to self-determination; the right to free, prior, and Informed consent before any outside project is undertaken on their land; the right to keep their languages, cultural practices, and sacred places; the right to full government services; and, perhaps most significant, the right to be recognized and treated as peoples.

Cultural Survival partners with Indigenous communities to defend their lands, languages, and cultures. We assist them in obtaining the knowledge, advocacy tools, and strategic partnerships they need to protect their rights. When their governments don't respond, we partner with them to bring their cases to international commissions and courts, and we involve the public and policy-makers in advocating for their rights. In addition, Cultural Survival offers the most comprehensive source of information on Indigenous Peoples on the planet. Our award-winning magazine, the *Cultural Survival Quarterly*, has been published for almost forty years.

A (SOMEWHAT) SUCCESSFUL MEDIA REFORM STRATEGY

The United Nations celebrated the International Day of the World's Indigenous Peoples in 2013 by shining "a spotlight on indigenous media—television, radio, film and social media—and their role in helping *preserve indigenous cultures, challenge stereotypes and influence the social and political agenda*."[1] The role that media play in protecting indigenous peoples' rights to freedom of expression—which includes the right to seek, receive, and impart information, culture, and participation—is recognized not only by the larger international community but also by the inter-American human rights system. Community radio, as a form of media, is a valuable tool to educate children, linked to the preservation of language and culture, and promotes democracy by advancing participation and nondiscrimination.

Despite this recognition of the value of indigenous community radio and the obligation on states to protect indigenous peoples' rights to freedom of expression and culture, the Republic of Guatemala continues to deny access to its indigenous peoples of this form of media. In the seventeen years since the

Agreement on Identity and Rights of Indigenous Peoples (AIDPI) in which Guatemala agreed to provide Indigenous Peoples with access to radio as a medium of expression, numerous attempts have been made to amend Guatemala's current law which does not provide for community radio of any kind. These efforts have been unsuccessful and today, instead of being closer to that goal, Guatemala's Congress is considering legislation to further prevent its indigenous population from exercising its right to freedom of expression by criminalizing community radio stations.

Despite the commitment to and interest in indigenous community radio, current Guatemalan law prohibits such use of the radio frequencies. The current law stands in stark contrast to the commitment Guatemala made in the negotiated Peace Accords ending the decades-long civil war. In the AIDPI, signed in 1995, Guatemala agreed to "make frequencies available for indigenous projects" and "promote . . . the abolition of any provision in the national legislation which is an obstacle to the right of indigenous peoples to have at their disposal communications media for the development of their identity."[2] The CRONOGRAMA Agreement, which established a timetable for the implementation of the Peace Accords, provided that the AIDPI be complied with during the years 1998, 1999, and 2000.[3] After the Government's failure to reform Guatemala's telecommunications law, MINUGUA, the United Nations Verification Mission in Guatemala, rescheduled this process for 2001 and 2002. Eighteen years later, these provisions of the AIDPI remain unfulfilled.

USING THE LEGISLATIVE PROCESS

Guatemala's current law, in effect, only provides for commercial and government radio stations. Under Article 62 of the General Telecommunications Law, radio frequencies are awarded through a public auction. The Superintendent of Telecommunications, the government agency charged with administering the assignment of frequencies, shall always award the frequencies to whomever offers the highest bid at public auction.[4] The highest bid at a public auction is entitled to the radio frequency. The law does not provide for community radio or any other form of nonprofit radio.[5]

More formal attempts to amend the telecommunications law began soon after the law went into effect. In 2002, the first bill (Initiative 2621) was introduced to Congress, but failed to get out of committee. In response to concerns that the entire spectrum of frequencies would be auctioned to commercial radio stations and as a result of pressure by the European Union, the auction of frequencies was suspended on March 24, 2002. A second bill (Initiative 3142) was introduced in 2004, and a third in 2005 (Initiative 3151). Both met the same fate. Meanwhile, the Superintendent of Telecommunications released a study that found that twenty-nine frequencies could be available if a community radio bill passed.

At the urging of the Special Rapporteur on Freedom of Expression of the Inter-American Commission, a National Round Table for Dialog was established to resolve the problem of "illegal radio stations." The Guatemalan government charged COPREDEH (Comisión Presidencial para los Derechos Humanos), the Presidential Commission for Human Rights, with hosting the round table. The round table resulted in a fourth bill (Initiative 4087), which was introduced in Congress on August 3, 2009.

All the while, in dozens of Guatemalan villages, local residents banded together to start small, low-power FM radio stations to serve as the voice of their communities. Whether authorized or not, many of these stations are de facto nonprofit community radio stations. To complicate matters, other unauthorized stations exist for the illicit profit of a private owner. An even larger number of unauthorized radio stations have taken to the airwaves chiefly to promote a particular religious view. Even a casual listener can quickly tell the difference.

Over the course of the next few months, dozens of volunteers who run community radio stations in their villages participated in personally lobbying their representatives in Congress to pass Initiativa 4087. This is especially significant when the history of Guatemala is considered. Guatemala cannot boast a proud record when it comes to responding to citizen action. Only a few decades ago, elections where little more than shams and killings and massacres were the government's response to citizens pressing for reform. The very idea that government officials might respond to pressure from citizens seemed far-fetched. A well-coordinated campaign focused on the eleven members of the Congress's Indigenous Peoples Committee included scores of meetings in Guatemala City and with local constituents in their home districts. Average citizens, shopkeepers, and farmers spent eighteen hours on a bus to meet their congressmen in the capitol. Housewives and teachers knocked on the doors of their congressmen's homes to encourage them to support the bill. The bill was revised by the Congress's Indigenous Peoples Committee and then sent back to the full Congress with a favorable recommendation in January 2010. This was celebrated as a victory because every previous bill had died in committee. But here the tale of legislative "victory" ends; or, more accurately (and optimistically), stalls.

Despite intensive lobbying by indigenous communities and community radio organizations, congressional leadership failed to place the bill on the agenda for debate and a vote. In 2010, the Guatemalan Government body CENAP, the National Advisory Group on the Implementation of the Peace Accords, helped pay to print two thousand booklets exhorting the Guatemalan Congress to approve Initiative 4087.

USING THE JUDICIAL PROCESS

With the legislative process stalled, the community radio movement made use of the courts. On November 17, 2011, Asociación Sobrevivencia Cultural (Cultural Survival's Guatemalan sister organization) presented a case to the Constitutional Court, arguing that the current telecommunication law is unconstitutional on the grounds that it discriminates against Indigenous Peoples by denying them access to radio frequencies. In its March 2012 decision, the Court disagreed with the claims presented by Asociación Sobrevivencia Cultural and held that the General Teleommunications Law does not violate the Constitution of Guatemala. However, the Court "urged" Congress to consider and pass a law granting and regulating radio frequencies to Indigenous communities to promote Indigenous languages, traditions, spirituality, and other cultural expression but this was not based on any constitutional requirement to do so. The power of this "urging" or, more accurately, "exhorting" (exhorto) continues to be debated. Lawyers for the community radio stations interpret this as tantamount to a court order to the Congress to consider and vote on Initiativa 4087.

Having exhausted the legal remedies in Guatemala after receiving an uncertain ruling from Guatemala's highest court, Asociación Sobrevivencia Cultural, partner organization Mujb ab'l Yol, and Cultural Survival with legal representation from the Suffolk University Law School's Indian Law and Indigenous Peoples Clinic, presented a petition to the Inter-American Commission for Human Rights (IACHR) of the organization of American states. As of the date of writing the petition is still being studied by the commission. The IACHR Special Rapporteur for the Freedom of Expression and the UN's Special Rapporteur for the Freedom of Expression have both been vociferously engaged in pressuring the Guatemalan government to legalize community radio.

CONCLUSION

While it is not possible to claim victory for media reform in Guatemala at this time, some incremental progress has been made. A favorable recommendation from a congressional committee is a step toward reform. The exhortation of the Constitutional Court can also be viewed as a positive step as is the engagement of the United Nations and the Organization of American States. A significant advance is the willingness of Guatemalan citizens to lobby their elected representatives. It is worth noting that in 2015, citizens, in response to corruption scandals, staged massive protests that forced Vice President Roxana Baldetti and eventually President Otto Perez Molina to resign. The most crucial, fundamentally important and hopeful fact, however, is that, despite the legal grey area, hundreds of volunteers in dozens of villages continue to show up and go on the air all day, every day to make community radio a reality for tens of thousands of listeners.

1. UN Press Release, HR/5102, August 6, 2012; emphasis added. Retrieved from www.un.org/News/Press/docs//2012/hr5102.doc.htm.

2. The Agreement on Identity and Rights of Indigenous Peoples, done at Mexico City, March 31, 1995, UN Doc. A/49/882S/1995/256, April 10, 1995; sec. III. (H)(2)(b). Retrieved from www.undemocracy.com/A-49-882.pdf.

3. Acuerdo Sobre Cronograma para la Implementación,Cumplimiento y Verificación de los Acuerdos de Paz, sec. 4. Retrieved from www.guatemalaun .org/bin/documents/Acuerdo%20cronograma%20implementaci%C3%B3n,%20 cumplimiento%20y%20verificaci%C3%B3n.pdf.

4. The Guatemalan General Law of Telecommunications, art, 62

5. Although amateur radio enthusiasts are permitted to use frequencies without a license, these frequencies are not broadcast frequencies that are useful for reaching listeners.

CHAPTER TWENTY-SIX

Media Reform in Mexico

MARIUS DRAGOMIR, *Open Society Foundations, Program on Independent Journalism, United Kingdom*

MEDIA REFORM STRATEGY

The Program on Independent Journalism (PIJ; formerly Open Society Media Program) focuses on supporting journalistic initiatives led by individuals or collectives that strive to improve their journalism under difficult circumstances, such as autocracy, violence, repression, or poverty, or in moments of great opportunity, such as first democratic elections, peace agreements, or massive social mobilizations. Prior to 2014, a major part of the program's work was research and advocacy. The program's main goal in this area was improved media policy and legislation to create an environment enabling journalists to carry out their work independently. The program's strategic approach to fulfil this goal was a combination of local and global expertise, knowledge, and engagement. In projects run in-house, the program hired local experts from a variety of fields and professions to write policy reports, usually based on a common methodology that includes templates for policy recommendations. It then supported local civil society groups with experience in the policy-making process to foster dialog with parliaments, governments, regulators and other decision-making bodies and convince them to adopt the policy recommendations spawned by the research. Overall, it was all about chasing the right opening and acting on the right channel at the right time. It was all a game of opportunities.

Mexico was one of fifty-six countries that the Open Society Foundations' Program on Independent Journalism (PIJ) covered in the Mapping Digital Media

(MDM) research and advocacy project between 2009 and 2014. The project was launched by the PIJ at a time of major changes in the journalism industry triggered by the digitization of media operations. The big goal of the project was to understand and help the program's partners, grantees, media outlets and policy makers, understand what digital technology meant for journalism. To do that, we extended our analysis to cover a broad array of issues and sectors.

DO YOUR RESEARCH

We examined the changes in media consumption patterns and mapped out the challenges and opportunities that digitization brought to public service broadcasters. We also canvassed in this project the impact of the Internet on coverage of sensitive issues and analyzed what impact the changes in the frequency allocation process had on media outlets. Analysis of the shifts in the media's financial management and review of how legislation and regulation influenced media and journalism was also part of the research.

The outcome of this project was a set of fifty-six country reports, produced by local experts following the same methodology and guidance. The idea behind the project was to use the findings from this in-depth, methodologically scrupulous study to generate judicious proposals for policies and regulation that would enable a better environment for journalists to work independently.

Beyond this, the project served other functions as well. In Central Asia and the Caucasus, it was heralded as a pioneering effort that helped to build scarce or totally absent research and advocacy capacity.

The project was run by an editorial team in the PIJ's offices in London who worked with a squad of six regional editors in charge of liaising and editing the work of national researchers. The research generated a rich menu of trends experienced by the media, both positive and negative: on the one hand, increased access to news content, improved audience interaction and participation, and a higher number of platforms for civil society groups; on the other hand, extension of dominant positions and dilution of public-interest journalism available freely to the broad public.

Latin America was a focal region in the project, with a team of thirty researchers working in nine countries from Argentina to Brazil to Nicaragua. The most populous Spanish-speaking country in the world, Mexico was included in the project. The country epitomizes some of the most common risks and opportunities facing the media industry: concentration of ownership, close links with political power, and lack of transparency in regulation versus growing Internet reach and a proliferation of fresh voices online.

The Mexican report was written by a team with diverse experience and expertise, which was needed given the broad scope of this study. They included Rodrigo Gomez, a university academic, Gabriel Sosa Plata, a researcher with experience as an academic and a media practitioner, Primavera Tellez Giron, a journalist with a good grasp of the changing Mexican newsrooms, and Jorge Bravo, a journalist with a deep understanding of the Mexican media industry.

The year 2013 became a landmark in Mexico's modern history of competition law as Congress adopted legal amendments that drastically changed the rules of the game in many economic sectors, making it more difficult for powerful businesses to further consolidate their market dominance. It also constrains Mexico's largest broadcasting groups, which were seen as the president's backers. This breakthrough was a success for a coalition of civil society groups led by one of Mexico's best-known nongovernmental media organizations. Here is how they did it.

ENLIST THE RIGHT ADVOCATES

As the report was nearing completion, discussions started with potential partner organizations to take the report's findings to policy-makers. The PIJ had the advantage of significant prior experience in Mexico, where it had supported throughout the 2000s a range of groups with legal expertise, and a well-grounded understanding of the policy process and the main actors involved in it.

The lead in this effort was taken by Aleida Calleja, the former vice-president of the World Association of Community Broadcasters (AMARC). With her, the report's writers and other local experts the PIJ's staff brainstormed in November 2012 about how best to influence the Mexican policy-making process to improve media policy. The key question was: Can we really break the dominance of powerful media companies in the policy-making? Can we really convince a Congress that for years favored heavyweight broadcast groups such as Televisa to open the space to more players?

As elections were expected in Mexico the following year, nobody in power was likely to touch the media and the laws that governed the media industry. The most that could be done was to use this period of six or seven months to lure politicians—both those in power hoping to carry on and those in opposition hoping to replace them—to take the report's recommendations and make them part of their political plank. There was excitement among people in the room, but deep inside many of us were skeptical that politicians would open their ears to such proposals in a country where two media groups, Televisa and Azteca, together held 94 percent of all the broadcast frequencies in the market.

Why would any politician pledge in his or her electoral program to push through legal provisions that would ensure equitable and fair access to broadcast licenses, as the Mapping Digital Media report recommended? Why would anyone in the country's political establishment want to alienate financially mighty media groups that have historically maintained a firm grip on the Mexican audiences?

Ms. Calleja, who had at that time just been appointed to take the helm of the Mexican Association of the Right to Information (AMEDI), asked the PIJ for funds to advocate for the recommendations in the Mapping Digital Media report. By January 2012, the work was underway.

AMEDI channeled its efforts into two areas. On the one hand, they had to reach out to the public. Talking about media and how they function had never happened outside Mexico City. With the report's authors and other experts, AMEDI travelled to eight Mexican states and told students, professors, local government officials, journalists and others why it was important to have diversity in the media and how that can be achieved. Writers of the Mapping Digital Media reports in Argentina and Colombia were flown in for some of these events to share with the audience what they faced in their countries and how they worked on these issues.

The campaign yielded fast responses. The Federal Telecommunications Commission (Cofetel) uploaded the full report on its website. Members of the Commission of Economic Competition, Mexico's antitrust regulator, took part in some of these debates. Lawmakers in the chamber of deputies and the Senate, although busy boosting their chances for a new term in power, joined some of these events as well. The Notimex news agency ran a story about the planned reform, which boosted the coverage of the campaign.

MAKE POLITICIANS COMMIT

The second line of AMEDI's work was interacting with political forces. The organization targeted political parties, the presidential candidates and the Congress. AMEDI involved in this effort the Citizens Coalition for Democracy and Media, comprising over 140 civil society organizations and 200 academics, artists, writers, filmmakers, actors, journalists, lawyers, and opinion leaders.

At the end of May 2012, the coalition asked presidential candidates to answer an open-ended questionnaire on the Right to Information and Freedom of Expression Agenda. The seven questions covered issues such as attacks on journalists, media reform, universal access to broadband Internet, and independent production. All presidential candidates answered the questionnaire almost immediately except for Josefina Vazquez Mota.

Enrique Peña Nieto won the presidency of the country. In autumn 2012, AMEDI and the Citizens Coalition resumed advocacy of the report's proposals. The time had now come for real action: making the law. As the number of congressmen allegedly supported by large media conglomerates in the new congress decreased and a number of AMEDI's allies in the media reform campaign were elected to Congress, the chances of pushing the report's recommendations into law were much improved. In the preelection questionnaire, Enrique Peña Nieto had agreed with three major recommendations from the Mapping Digital Media report, namely: extending universal broadband access, amending the law to ensure transparent allocation of government advertising, and increasing support for community radio. However, counting on Nieto's support was problematic as he was allegedly supported by Televisa group, which has a major interest in staving off attempts to let new players into the market.

Following some four months of intense advocacy with allies in Congress, AMEDI was thrilled to see in June 2013 that most of their requests made it into the law. That month, the Federal Congress approved constitutional reforms affecting telecommunications, broadcasting, economic competition, and information rights areas. The reform was made possible by agreement between the three main political parties in Mexico (PAN, PRI, and PRD) as part of what became known as the Pact for Mexico.

CONCLUSION

Although AMEDI had no direct access to the Pact for Mexico debates, its proposals were defended by its political allies. A total of nineteen of the twenty-one recommendations put forward by AMEDI were incorporated in the new legislation. They included provisions on public service media, support for community media, Internet access, spectrum allocation, and must-carry rules. "Several [of the adopted provisions] exceeded our expectations," Ms. Calleja told us in an e-mail at the time.

The recommendations on dismantling the dominant positions in broadcasting were the only proposals that were not adopted—confirming that large media groups retain a firm grip on parts of the congress. But AMEDI's work in this area did not stop. At the time of writing, it is running a campaign called *No MÁS Poder* (No MORE power) which aims to push through provisions on limiting ownership concentration and make sure that regulators uses their authority granted through the 2013 law to impose such limits.

The advocacy work in Mexico has become a model for the Latin American world. The multiplier effect gained by linking with existing campaigns, by judicious timing, and by efficient mobilization of networks and local organizations all contributed to the outcome. What truly unlocked the dialog with decision-makers, however, was the intelligent engagement with politicians, both before and after the elections.

INDEX

Abhisit Vejjajiva, 298, 300, 302
academic institutions: importance to media reform initiatives, 178
AC/DC journalism, 307
ACTA protests, 82
Actors' Union (Israel), 254
Agenda (Israeli Center for Strategic Communications), 259–260
Agreement on Identity and Rights of Indigenous Peoples (Guatemala), 321
Ali Hassan Al Jaber Safety Training program, 316
Allied Media Projects (AMP), 107, 108–113
alternative media: in Israel, 253, 255–258, 263–264; in Taiwan, 162–163; theory and practice as media reform, 7–8; in Venezuela, 269, 274–277
AMARC. *See* World Association of Community Radio Broadcasters
Amazon, 104
AMEDI. *See* Mexican Association for the Right to Information
American Censorship Day, 96
American Reinvestment and Recovery Act (U.S.), 108, 109, 111
AMP. *See* Allied Media Projects
Amputuan massacre, 307, 308
Anderson, Steve, 171, 178
Annenberg Schools of Communication, 28

Anonymous (hacker group): media reform and, 59–60; online video opposing Bill C-30, 51
Anti-Counterfeiting Trade Agreement (ACTA) protests, 82
Anti-Money Laundering Office (Thailand), 301
antimonopoly media reform: in Argentina, 271–272, 278; in Taiwan, 154, 159–161
Anti-Terrorism Act (Canada), 55n15
Appeal to Reason, The (newspaper), 214
Arab Spring, 313
Argentina: Law No. 26.522, 230, 271–273; media concentration in, 270–271; media reform in, 230–231, 268, 269, 270–273, 277–279
Argentinian Coalition for Democratic Broadcasting, 230
Arutz Sheva, 256
Asociación Sobrevivencia Cultural (Guatemala), 322–323
Assange, Julian, 59, 64, 65, 66, 67, 68
Association of Taiwan Journalists, 161
ATM journalism, 307
AT&T, 101, 102–103
attack/collect, defend/collect journalism, 307
Audio-Visual Communication Services (Argentina), 230

Chaos Communication Congress, 231
Chaos Computer Club, 228
Chartist press, 8
Chávez, Hugo, 274, 275, 276
Cheng, Nylon, 156
Chicago Independent Radio Project, 195–196
children's rights: media reform in Argentina and, 273
China Network Systems, 159
China Times, 159–161
Chomsky, Noam, 7
Chynoweth, Danielle, 186–187
citizen media: FM community radio and Indigenous Peoples in Guatemala, 319, 320–323; OpenMedia and, 173–174; in Taiwan, 162–163; in Venezuela, 268, 269, 273–279. *See also* community radio; low-power FM radio
Citizen Observatory for Gender Equality in the Media (Mexico), 131
Citizen's Bill for a Convergent Law on Broadcasting and Telecommunication (Mexico), 133–134
Citizens Coalition for Democracy and Media (Mexico), 328
civic hacking, 227
civic journalism: Philippines and, 309
civilmedia.tw, 162–163
Civil Press website, 258
civil society advocacy: consensus mobilization dynamics and, 225–226; limitations of classic advocacy, 226–227; media reform and, 223–224; Mexican Association for the Right to Information and, 130–131
civil society–based policy reform initiatives: characteristics and implications, 232–234; examples of, 228–231; goals of, 223; media reform in Argentina and, 273
Clarín Group, 269, 270–271, 273, 278
Class D FM radio stations, 196
Clear Channel, 189n1, 191
Clement, Andrew, 50
clientelism, 124–125
CMFR. *See* Center for Media Freedom and Responsibility
Coalition for Democratic Radio Broadcasting (Argentina), 268, 269, 271, 272
Coalition for Establishing Public Media (Taiwan), 158
coalitions: importance to media reform initiatives, 177–178

code of ethics: Egyptian Radio and Television Union, 285–286, 287–288, 293
COFECOM, 230
collaboration: importance to media reform initiatives, 177–178
Combating Online Infringements and Counterfeits Act (U.S.), 94–95
Comcast: Internet freedom issues and, 101, 102–103, 105; relationship with the FCC, 102
Comité Libercomuniación (Venezuela), 275
Comité por una Radiotelevisón de Servicio Público (Venezuela), 275
Committee to Protect Journalists, 307
communication policy. *See* media policy
communication studies: impact on media policy in Switzerland, 239–249
communicative democracy, 170
community media: in Venezuela, 268, 269, 273–279. *See also* citizen media
community radio: Indigenous Peoples in Guatemala and, 319, 320–323; in the Philippines, 308–309; in Thailand, 299–300, 301–302; U.S. context for, 191–192. *See also* low-power FM radio
COMPASS. *See* Consortium on Media Policy Studies
Computer Crime Act (Thailand), 300, 302
computer hacking, 227
Congress of Industrial Organizations, 217
conjunctural moments, 212–213
connective action, 226
consensus mobilization dynamics: limitations of, 226–227; in media policy-making, 225–226
Consortium on Media Policy Studies (COMPASS), 20, 27–31
constitutive moments, 209. *See also* critical junctures
COPREDEH. *See* Presidential Commission for Human Rights
copyright issues: opposition to the Combating Online Infringements and Counterfeits Act, 94–95; overview of, 93–84; SOPA protests, 92–97; U.S. government seizure of domain names for copyright infringement, 95
Corral, Javier, 130
Council of Europe: IRPC Charter and, 80
Crawford, Susan, 104
Creative Commons license, 227

FCC. *See* Federal Communication Commission

Federal Audiovisual Communication Council (Argentina), 272

Federal Communication Commission (FCC): contemporary media activism and, 214; Free Press and the campaign for net neutrality, 100–102, 103–105; low-power FM radio and, 183–184, 190–191, 193, 194–196; postwar media reform and, 216–217; response to pirate FM radio, 183, 192

Federal Institute of Telecommunications (Mexico), 124, 132

Fifth Estate: cancellation of Canadian Bill C-30 and, 39–40; digitally-mediated, 41–42; estate model and, 40–41; strategies to opposed Bill C-30, 42–53

Fifth Estate (strategies for activating): building and online community of networked individuals, 44–46; developing targeted content to be shared and distributed, 50–53; overview, 39, 40, 43, 53–54; shaping preexisting digital platforms, 46–50

Fight for the Future, 95–96

FM radio: Class D stations, 196. *See also* low-power FM radio

Ford Foundation, 27, 259

Fort Hood Support Network, 196

Foucault, Michel, 60, 146

Fourth Estate: press as watchdog, 41; Wikileaks and, 68–69. *See also* journalism; press

Fraunhofer Institute, 289

Freedman, Des, 140

Freedom of Information Act (Iceland), 229

freedom of the press: in Myanmar, 303, 304–306; in Taiwan, 153–154, 155

Freedom Rings Partnership, 111, 112

Free Press: founding of, 214; Huff on the dangers of working through the system, 9–10; net neutrality campaign, 100–102, 103–105

free radio movement, 192

free speech: IRPC Charter and, 80; media reform in Taiwan and, 155–157; SOPA protests and, 92–97

Free Wave, 275

Gambia, 202

General Authority for Investments and Free Zones (Egypt), 284

General Telecommunications Law (Guatemala), 321, 323

Gen Why Media, 174

Germany: Hamburg Transparency Law Initiative, 228; regional coalitions on net neutrality, 231

Ghana, 202

Gilligan, Andrew, 143

Giron, Primavera Tellez, 327

Go Daddy, 96

Goldman Sachs Group, 271

Gomez, Gustavo, 230

Gomez, Rodrigo, 326

Google: blackout to protest SOPA, 96; EU Hackathon, 232; Open Internet Coalition, 104

Graiver family, 270

Gramsci, Antonio, 146

Gramscian theory: on power and history, 211–214

Greenwald, Glenn, 58–59

Grupo 1BC (Radio Caracas Televisión), 273–274

Guardian, 64, 66

Guatemala, 319, 320–323

Guide on Human Rights for Internet Users (Council of Europe), 80

Guinea, Republic of, 202, 204

Gullah People's Movement, 195

Habte, Rahwa, 195

Hack4YourRights, 232

Hacked Off, 140, 142, 144, 148

Hackett, Robert, 10

hacking: meanings of, 227–228. *See also* phone hacking; policy hacking

Hack the Government, 232

Hall, Stuart, 212

Hamburg (Germany), 228

Hamburg Transparency Law Initiative, 228

Ha'olam Hazeh (magazine), 255

Harman, Harriet, 144

Hasdera, 260

"Healthy Digital Ecology, A," 110

hegemony, 211–212

Herman, Ed, 7

Histadrut (mobile phone operator), 255

historical institutionalism, 213

history: Gramscian approach to, 211–214

Holes in the Net website, 258

Hot (mobile phone operator), 255

Huff, Mickey, 9–10

human rights. *See* IRPC Charter of Human Rights and Principles for the Internet

Human Rights Radio, 194

Hutchins Commission on Freedom of the Press, 216

IACHR Special Rapporteur on Freedom of Expression, 322, 323
IBA. *See* Israel Broadcasting Authority
Iberoamerican University, 127
Ibrahim, Zane, 191
ICCPR. *See* International Covenant on Civil and Political Rights
Icelandic Modern Media Initiative, 229–230
IGF. *See* Internet Governance Forum
Independent Media Center website, 44
Independent Press Standards Organisation, 144
indigenous media: community radio in Guatemala, 320–323
Indigenous Peoples: Cultural Survival and, 319–320; media reform in Argentina and, 273; United Nations Declaration on the Rights of, 320
Indymedia website, 44
Institute for Public Representation, 195
institutional reform organizations: in Israel, 259–261
Institutional Revolutionary Party (PRI; Mexico), 125, 128–129, 134
Inter-American Commission for Human Rights, 319, 322, 323
interest group approach: to media policy-making, 145–146
internal reform: in Israel, 254–255
International Covenant on Civil and Political Rights (ICCPR), 77, 308
International Day of the World's Indigenous Peoples, 320
International Day to End Impunity, 307
International Federation of Journalists, 160
International Journal of Communications, 112
International Press Association, 272–273
International Telecommunication Union, 82–83
Internet: blackouts to protest SOPA and PIPA, 96–97; Computer Crime Act in Thailand, 300, 302; digitally-mediated Fifth Estate and, 41–42 (*see also* Fifth Estate); Free Press and the campaign for net neutrality, 100–102, 103–105; IRPC Charter and (*see* IRPC Charter of Human Rights and Principles for the Internet); J14 protest activities, 258; OpenMedia and, 115–117, 118–119,

172, 173–174; opposing forces in the history of, 102–103; usage-based billing and Stop the Meter campaign, 172
Internet Corporation of Assigned Names and Numbers, 83
Internet Governance Forum (IGF): Dynamic Coalition on Net Neutrality, 231; IRPC Charter and, 73, 74, 77, 78, 81, 86, 87; multistakeholder process and, 225
Internet Rights and Principles Coalition (IRPC), 73
Investigating and Preventing Criminal Electronics Communications Act (Canada), 42
Investment Equity Citicorp, 271
IRPC. *See* Internet Rights and Principles Coalition
IRPC Charter Booklet, 74, *76*
IRPC Charter of Human Rights and Principles for the Internet (IRPC Charter): achievements and sustainability of, 86–87; aims of, 73–74; Article 13, *78*; collaborative development, 74; issue of distinguishing between rights and principles, 85–86; lessons learnt, 84–86; media reform and, 72–75; origins of, 76–77; outlook for, 84, 87–88; techno-historical and geopolitical context in the development of, 82–84; "Ten Punchy Principles," 74, 80, *81*; timeline of, 77–81
Israel: alternative media, 253, 255–258, 263–264; media reform in, 252–64; social protest of 2011, 257–258, 262–263
Israel Broadcasting Authority (IBA), 254, 260–261
Israel Broadcasting Authority Law, 261
Israeli Center for Strategic Communications (Agenda), 259–260
Israeli Documentary Filmmakers Forum, 254
Israel Independent Press, 258
Israel's Media Watch, 259
isvictoewswatchingme.com, 51
iTV (Thailand), 301

J14 protest, 257–258, 262–263
Jefferson, Thomas, 53
Jensen, Robert, 47–48
Jordan, 313, 315
journalism: culture of impunity in the Philippines, 296, 297, 306–309; journalist training by the Doha Centre for Media Freedom, 312, 315–316, 317;

Media Mobilizing Project (MMP), 107, 108–113

media monopolies: formation in Argentina, 270–271; media reform in Argentina and, 271–272, 278; media reform in Taiwan and, 154, 159–161

media observatories: in Latin America, 126; Mexican Association for the Right to Information, 129–130, 131–132

media policy literacy: challenges to advancing media policy pedagogy, 21–27 (see also media policy pedagogy); definition and components of, 20–21; media reform and, 19–20, 31–32

media policy-making: balancing activism and engagement with policy making, 138, 139, 149–150; civil society advocates and consensus mobilization dynamics, 225–226; complex environment of, 225; cycle model, 145; "interest group approach to," 145–146; limitations of classic advocacy, 226–227; multistakeholder processes, 225; as "order of discourse," 146–147; policy hacking, 223–235; in Switzerland, impact of communication studies on, 239–249

media policy pedagogy: challenges to advancing, 21–27; COMPASS program, 27–31

media policy research: impact on media policy-making in Switzerland, 239–249

media privatization: in Taiwan, 154, 155–157

Media Production City (Egypt), 284

media reform: balancing activism and engagement with policy making, 138, 139, 149–150; challenges of, 3; civil society-based advocacy and, 223–224; critical junctures and, 138–139, 209, 213–214; as a critique of mainstream media's content and structure, 6–7; definition and dimensions of, 4–12; democratic media activism, 8–12; digital justice coalitions and the Broadband Technology Opportunities Program, 107–113; Fifth Estate and (see Fifth Estate); Free Press and the net neutrality campaign, 100–102, 103–105; Hackett and Carroll's five action frames for, 156–157; IRPC Charter and, 72–75; in Israel, 252–264; in Latin America, 268–279; media policy literacy and, 19–20, 31–32; OpenMedia and, 115–119 (see also OpenMedia); organizational strategies for, 167,

176–179; policy hacking, 223–235; social justice and, 147–149; SOPA protests and, 92–93; strategy-implementation gap, 3–4; theory and practice of alternative media, 7–8; through capacity building in Qatar, 312–317; in the U.S., history and future of, 209–211, 214–220; Wikileaks and, 58–69

Media Reform Coalition (MRC): founding principles of, 139–141; Leveson inquiry and media reform, 140–145; media reform and social justice, 148–149

media reform initiatives: in Canada, 167–179 (see also Canada); organizing principles and guidelines, 167, 176–179; in West Africa, 199–206

Mediation and Protest Movements (Cammaerts et al.), 9

Media Watch Foundation (Taiwan), 153, 154, 157, 162, 164

Meetings for Free Communication (Venezuela), 275

Megafon-news website, 258

Mehr Demokratie, 228

memes, 51

mergers. See media monopolies

Mexican Association for the Right to Information (AMEDI): advocacy for reforms recommend by the Mapping Digital Media project, 328–329; as a media observatory, 129–130, 131–132; successful strategies of, 130–134; telecommunications reform and media democratization, 123, 124, 129–130

Mexico: media reform and the Program on Independent Journalism, 325–329; sociopolitical context behind media reform, 126–129; telecommunications structure and reform, 123–134

Michigan State University, 111

Middle East Broadcasting, 282

MIL. See Media and Information Literacy program

Milberry, Kate, 50

Ministry of Information (Egypt), 285, 290, 291

MINUGUA. See United Nations Verification Mission in Guatemala

Mitchell, Greg, 58–59

MMP. See Media Mobilizing Project

mobile phone industry. See wireless industry

Molina, Otto Perez, 323

Moore, James, 118
Mota, Josefina Vazquez, 328
"Moving Toward a Surveillance Society"
    report, 53
Mozilla, 232
Mujb al'l Yol, 323
Multilateral Agreement on Investment
    (MAI), 44
multiple system operators (MSOs): in
    Taiwan, 156
"Muzzling the Scientists" panel discussion, 170
Myanmar, 296–297, 302–306
Myanmar Press Council, 304, 305

Namecheap, 96
Napoli, Philip, 19
Nasser, Gamal Abdel, 282
Nation (magazine), 220
National Advisory Group on the Implementation of the Peace Accords (Guatemala), 322
National Association of Broadcasters
    (U.S.), 186, 187
National Association of Community,
    Free and Alternative Media (Venezuela), 276
National Authority for Audiovisual
    Services (Argentina), 272
National Broadcasting and Telecommunications Commission (Thailand),
    299, 300, 302
National Cable and Telecommunication
    Association (U.S.), 102
National Cheng-Chi University (Taiwan),
    158
National Communications Commission
    (Taiwan), 159–161
National Communications Commission
    Organization Act (Taiwan), 159
National Company of Telecommunications (Argentina), 271
National Conference for Media Reform
    (U.S.), 58–59
National Conference of Community
    Media (Venezuela), 275
National Council for Peace and Order
    (Thailand), 299, 300, 302
National Lawyer's Guild (U.S.), 195
National Reform Council (Thailand),
    302
National Security Agency surveillance
    (U.S.), 83
National Taiwan University, 160, 161

National Telecommunication Regulatory
    Authority (Egypt), 285, 290
National Telecommunications and Information Administration (.U.S), 108–
    109, 110, 112
National Union of Journalists (UK), 142
NBC, 216, 217
NBT TV (Thailand), 299
Netflix, 104
Netherlands, 231
net neutrality: Free Press and the campaign for, 100–102, 103–105; media
    activism and, 214; OpenMedia and,
    172; regional coalitions in Europe, 231
net rights. *See* IRPC Charter of Human
    Rights and Principles for the Internet
Network C (Israel), 256
New Israel Fund, 259
Newman, Russell, 28
Newspaper Guild, 214
New Transparency: Surveillance and
    Social Sorting project, 50
New World Information and Communication Order, 214
New York Times: Wikileaks and, 61, 64,
    66, 67
Nicholson, Rob, 40, 43, 47
Nieto, Enrique Peña, 328, 329
"No China Times Movement," 161
No Internet Lockdown campaign, 173
No MORE Power campaign (Mexico),
    329

Obama, Barack, 183, 187
Occupy movement, 139
ODC-Venevisión, 273–274
Odellia (ship), 256
OFCOM (Switzerland), 244–245, 247,
    248, 249
Oliver, John, 104
Onda Libre, 275
OneAmerica, 194–195
online community of networked individuals, 44–46
online petitions, 47
online videos: in the opposition to Bill
    C-30, 50–51
Open Internet Coalition, 104
OpenMedia: assessment of, 176; coalition
    building and, 177; growth and success
    of, 173–174; leadership in opposition
    to Bill C-30, 40; media reform in Canada and, 168, 171–174; mission and
    purpose of, 115–117, 118–119; origins of,

public service broadcasting: lack of in Egypt, 281–282; suggested strategies for transforming ERTU into a public broadcaster, 281, 291–294; in Thailand, 301

Public's Right to Know, 259

Public Television Station (Taiwan), 157–159

Qatar, 312–317

Quadrature du Net, La, 231

Quadruple A model: applied to Wikileaks, 63–69; overview, 63

Quality News Development Association, 163

radio: alternative and community media in Venezuela, 274–277; alternative media in Israel, 255–257; block timing in the Philippines, 307–308, 309; civil society–based policy reform initiatives in South America, 230–231; fight for low-power FM in the U.S., 182–188 (see also low-power FM radio); media reform challenges and strategies in Thailand, 299–300, 301–302; ownership in the U.S., 188–189n1, 191–192; pirate FM stations in the U.S., 183; postwar media activism in the U.S., 216–217; reform in the Republic of Ghana, 204; regulation in Guatemala, 321–322; structure and content in Egypt, 282–283 (see also Egyptian Radio and Television Union); structure and reform in Mexico, 123–134. See also broadcast media

Radio and Television Act (Switzerland), 240, 244

Radio Beit Ha'am, 258

Radio Broadcast Preservation Act (U.S.), 193

Radio Caracas Televisión, 273–274

Radio Chuspa, 274

Radio Cool, 274

Radio Free Asia, 303

Radio Handbook (CIO), 217

Radio Mutiny, 192

Radio Parroquiana, 274

Radio Perola, 274

Radio Rebelde, 274

Random Hacks of Kindness, 232

Reddit, 96

Regulation of Nonprofit Community Radio and TV Broadcasting of Public Service (Venezuela), 275

Reimagine CBC, 168, 174–176, 177, 179

Reshet Gimmel, 256

Resorte Law (Venezuela), 276

Revolution 101 (documentary film), 260–261, 264

rights advocacy, 156

Right to Information and Freedom of Expression Agenda questionnaire (Mexico), 328

Rivera, Angélica, 128

Rogers Communications, 117–118

Rural Utility Service (USDA), 109

Saloona website, 258

Sand website, 258

satellite television: in Egypt, 282–283

Sheetreet, Meir, 256

Shem-Tov, Victor, 256

Siepmann, Charles, 217

Sifry, Micah, 59

Simon Fraser University, 170, 171, 178

Smith, Gordon, 186

Snowden, Edward, 69

social justice: low-power FM radio and, 193–194; media reform and, 147–149

social media: using to advance activism objectives, 46–50. See also Facebook; Twitter

social movement theory: alternative media theory and, 8; democratic media activism and, 9–10

Social TV website, 258

Solis, Beatriz, 130

SOPA. See Stop Online Piracy Act

SOPA protests: advocacy and initiatives, 95–96; blackout of major websites, 96–97; context of copyright fights, 93–94; IRPC Charter and, 82; lessons learned, 97; media reform and, 92–93

Southern Tagalog Exposure, 309

Spivak, Avia, 262–263

Spivak-Yonah report, 262–263

Stop Online Piracy Act (SOPA), 94, 98n3

Stop Online Spying campaign, 44–46, 47

Stop the Meter campaign, 172, 173, 177

Stop the Squeeze campaign, 117–118

Student Group for Observing the Public Television (Taiwan), 158

student movements/protests: J14 protest in Israel, 257–258; #YoSoy132 in Mexico, 126–128

subaltern radio, 194

Suchinda Kraprayoon, 300, 301

Suffolk University Law School, 323

University of British Columbia, 171
University of Michigan, 28
University of Windsor, 171
(Un)Lawful Access (online video), 50
unlicensed FM radio, 183, 192. *See also*
    pirate radio
Uru, 260
Uruguay, 230–231
usage-based Internet billing, 172

Valentines Day card, 48–49
Vancouver (Canada): Media Democracy
    Day, 169–171
Venezuela: media reform in, 268, 269,
    273–279; Organic Law of People's
    Power, 276, 279n1
Venezuelan Community Media Network,
    275
Venezuelan Network of Community
    Media, 276
Verizon, 101, 102–103
Vikileaks, 52
Voice of America, 303
Voice of Peace, 255–256
Voice of the Arabs, 282
Voice of the Listener and Viewer, 142

Wales, Jimmy, 96
Want Want China Times Media Group,
    160–161
Want Want Corporation, 159–161
Washington Post, 61
wereport.org, 162, 163
West Africa, 199–206
West Philadelphia Pirate Radio (WPPR),
    192
Wheeler, Tom, 101, 104

Wikileaks: journalistic and legal implica-
    tions, 60–62; media reform and,
    58–69; phases of, 63–69
Wikipedia: blackout to protest SOPA, 96
Williams, Raymond, 196
wireless industry: OpenMedia and
    reform in Canada, 117–118; unioniza-
    tion in Israel, 255
Wireless Telegraphy Ordinance (Israel),
    256
WLBT radio case, 214
World Association of Community Radio
    Broadcasters (AMARC), 230
World Conference on International Tele-
    communications, 82–83
World Environmental Education Con-
    gress, 24
World Health Organization: Wikileaks
    and, 63–64
World Summit on the Information Soci-
    ety (WSIS), 73, 214, 225, 226
World Trade Organization, 44
World Wide Web: open protocol for,
    102; websites in the opposition to Bill
    C-30, 51
WPPR (West Philadelphia Pirate Radio),
    192
WQRZ-LP radio, 186
WSIS. *See* World Summit on the Infor-
    mation Society
WTRA radio, 193–194
Wu, Tim, 104
Wyden, Ron, 95

Yingluck Shinawatra, 298, 299, 300, 302
Yonah, Yossi, 262–263
#YoSoy132 hashtag, 126–128